Science for Engineering

To Yvonne

Science for Engineering

J. O. Bird, BSc(Hons), CMath, FIMA, CEng, MIEE, FCollP, FIEIE

Newnes
An imprint of Butterworth-Heinemann Ltd
Linacre House, Jordan Hill, Oxford OX2 8DP

℞ A member of the Reed Elsevier plc group

OXFORD   LONDON   BOSTON
MUNICH   NEW DELHI   SINGAPORE   SYDNEY
TOKYO   TORONTO   WELLINGTON

First published 1995

British Library Cataloguing in Publication Data
Bird, J. O.
    Science for Engineering
    I. Title
    500

ISBN 0 7506 2150 8

Library of Congress Cataloging in Publication Data
Bird, J. O.
    Science for engineering/J. O. Bird.
    p. c.m.
    Includes index
    ISBN 0 7506 2150 8
    1. Engineering. 2. Science. I. Title.
    TA145.B53   1994
    500.2'02462–dc20   94–31866
                                        CIP

Composition by Scribe Design, Gillingham, Kent
Printed and bound in Great Britain

Contents

Preface

Science for Engineering aims to develop in the reader an understanding of fundamental science concepts and to give a basic mechanical, thermal and electrical engineering systems background for student engineers. More specifically, the aims are to describe engineering systems in terms of basic scientific laws and principles, to investigate the behaviour of simple linear systems in engineering, to calculate the response of engineering systems to changes in variables, and to determine the response of such engineering systems to changes in parameters.

The text covers the GNVQ mandatory units *Science and Mathematics for Engineering (Intermediate)*, i.e. GNVQ 2, and *Science for Engineering (Advanced)*, i.e. GNVQ 3. However, it can be regarded as a basic textbook in engineering science for a much wider range of courses; for example, it will be useful for Technical and Further Education Departments in Australia, or East and West Africa Examining Councils and for similar technical examining authorities in English-speaking countries worldwide.

Mathematics is a tool used by scientists and engineers. Solving simple equations and simultaneous equations, evaluating and transposing formulae, plotting and analysing graphs, using trigonometry, and determining irregular areas are all involved in the 'Science for Engineering' studies contained in this text. The first part, Mathematics for engineering, which contains typical practical examples in these areas, is intended as preparation for the Science for mechanical, thermal and electrical engineering parts which follow it.

Each topic considered in the text is presented in a way that assumes in the reader little previous knowledge of that topic.

Theory is introduced in each chapter by a reasonably brief outline of essential definitions, formulae, laws, procedures, etc. The theory is kept to a minimum, for problem-solving is extensively used to establish and exemplify the theory. It is intended that readers will gain real understanding through seeing problems solved and then through solving similar problems themselves.

Science for Engineering contains some 500 worked problems, together with 325 multi-choice questions, the latter being in the form expected of students for National testing of their 'Science for Engineering' units. Also included are some 400 short answer questions, the answers for which can be determined from the preceding material in that particular chapter, and some 700 further questions, all with answers in brackets immediately following each question. Nearly 400 line diagrams further enhance the understanding of the theory. All of the problems – multi-choice, short answer and further questions – mirror practical situations found in science and engineering.

I would like to express my appreciation for the friendly co-operation and helpful advice given by the publishers, to colleague Huw Fox for his valuable checking of the manuscript, and to Mrs Elaine Woolley for the excellent typing of the manuscript.

'Learning by Example' is at the heart of *Science for Engineering*.

John O. Bird
Highbury College
Portsmouth

Part 1

Mathematics for Engineering

1

The solution of simple equations

At the end of this chapter you should be able to:

- solve simple equations

- apply the solution of simple equations to practical science and engineering applications

1.1 Introduction

An equation is simply a statement that 2 quantities are equal. For example, 1 m = 1000 mm or $y = mx + c$.

To 'solve an equation' means 'to find the value of the unknown quantity'. For example, if $2x = 6$, then $x = 3$ is the solution of the equation.

There are numerous examples in science and engineering where simple equations need to be solved. In fact, you will find examples of such practical equations in every chapter of this book.

In section 1.2 we learn the rules by manipulating simple algebraic equations. In section 1.3 we put the rules to practical use.

1.2 Worked problems on the solution of simple equations

Problem 1. Solve the equation $4x = 20$

Dividing each side of the equation by 4 gives:

$$\frac{4x}{4} = \frac{20}{4}$$

(Note that the same operation has been applied to both the left-hand side (LHS) and the right-hand side (RHS) of the equation so the equality has been maintained.)

Cancelling gives $x = 5$, which is the solution to the equation. Solutions to simple equations should always be checked and this is accomplished by substituting the solution into the origi-

nal equation. In this case, LHS = 4(5) = 20 = RHS.

Problem 2. Solve $\frac{2x}{5} = 6$

The LHS is a fraction and this can be removed by multiplying both sides of the equation by 5. Hence

$$5\left(\frac{2x}{5}\right) = 5(6)$$

Cancelling gives $2x = 30$.
Dividing both sides of the equation by 2 gives:

$$\frac{2x}{2} = \frac{30}{2}$$

i.e. $x = 15$

Problem 3. Solve $a - 5 = 8$

Adding 5 to both sides of the equation gives:

$$a - 5 + 5 = 8 + 5$$

i.e. $a = 13$

The result of the above procedure is to move the '−5' from the LHS of the original equation, across the equals sign, to the RHS, but the sign is changed to +.

Problem 4. Solve $x + 3 = 7$

Subtracting 3 from both sides of the equation gives:

$$x + 3 - 3 = 7 - 3$$

i.e. $\qquad x = 4$

The result of the above procedure is to move the '+3' from the LHS of the original equation, across the equals sign, to the RHS, but the sign is changed to –. Thus any term can be moved from one side of an equation to the other as long as a change in sign is made for that term.

Problem 5. Solve $6x + 1 = 2x + 9$

In such equations the terms containing x are grouped on one side of the equation and the remaining terms grouped on the other side of the equation. As in problems 3 and 4, changing from one side of an equation to the other must be accompanied by a change of sign.

Thus since $\qquad 6x + 1 = 2x + 9$
then $\qquad 6x - 2x = 9 - 1$
$$4x = 8$$
$$\frac{4x}{4} = \frac{8}{4}$$

i.e. $\qquad x = 2$

Check: LHS of original equation = $6(2) + 1 = 13$
RHS of original equation = $2(2) + 9 = 13$
Hence the solution $x = 2$ is correct.

Problem 6. Solve $4 - 3p = 2p - 11$

In order to keep the p term positive the terms in p are moved to the RHS and the constant terms to the LHS. Hence

$$4 + 11 = 2p + 3p$$
$$15 = 5p$$
$$\frac{15}{5} = \frac{5p}{5}$$

Hence $\qquad p = 3$

Check: LHS = $4 - 3(3) = 4 - 9 = -5$
RHS = $2(3) - 11 = 6 - 11 = -5$
Hence the solution $p = 3$ is correct.

If, in this example, the unknown quantities had been grouped initially on the LHS instead of the RHS then:

$$-3p - 2p = -11 - 4$$

i.e. $\qquad -5p = -15$
$$\frac{-5p}{-5} = \frac{-15}{-5}$$

and $\qquad p = 3$, as before

It is often easier, however, to work with positive values where possible.

Problem 7. Solve $3(x - 2) = 9$

Removing the bracket gives:
$$3x - 6 = 9$$
Rearranging gives: $\qquad 3x = 9 + 6$
$$3x = 15$$
$$\frac{3x}{3} = \frac{15}{3}$$

i.e. $\qquad x = 5$

Check: LHS = $3(5 - 2) = 3(3) = 9 = $ RHS
Hence the solution $x = 5$ is correct.

**Problem 8. Solve
$4(2r - 3) - 2(r - 4) = 3(r - 3) - 1$**

Removing brackets gives:
$$8r - 12 - 2r + 8 = 3r - 9 - 1$$

Rearranging gives:
$$8r - 2r - 3r = -9 - 1 + 12 - 8$$

i.e. $\qquad 3r = -6$
$$r = \frac{-6}{3} = -2$$

Check: LHS = $4(-4 - 3) - 2(-2 - 4) = $
$-28 + 12 = -16$
RHS = $3(-2 - 3) - 1 = -15 - 1 = -16$
Hence the solution $r = -2$ is correct.

Problem 9. Solve $\dfrac{3}{x} = \dfrac{4}{5}$

The lowest common multiple (LCM) of the denominators, i.e. the lowest algebraic expression that both x and 5 will divide into, is $5x$.

Multiplying both sides by $5x$ gives:

$$5x\left(\frac{3}{x}\right) = 5x\left(\frac{4}{5}\right)$$

Cancelling gives:

$$15 = 4x \qquad (1.1)$$

$$\frac{15}{4} = \frac{4x}{4}$$

i.e. $\dfrac{\mathbf{15}}{\mathbf{4}}$ or $x = 3\dfrac{3}{4}$

Check:

$$\text{LHS} = \frac{3}{15/4} = 3\left(\frac{4}{15}\right) = \frac{12}{15} = \frac{4}{5} = \text{RHS}$$

(Note that when there is only one fraction on each side of an equation, 'cross-multiplication' can be applied. In this example if $\dfrac{3}{x} = \dfrac{4}{5}$ then $(3)(5) = 4x$, which is a quicker way of arriving at equation (1) above.)

Problem 10. Solve $\dfrac{2y}{5} + \dfrac{3}{4} + 5 = \dfrac{1}{20} - \dfrac{3y}{2}$

The LCM of the denominators is 20.
Multiplying each term by 20 gives:

$$20\left(\frac{2y}{5}\right) + 20\left(\frac{3}{4}\right) + 20(5) = 20\left(\frac{1}{20}\right) - 20\left(\frac{3y}{2}\right)$$

Cancelling gives:

$$4(2y) + 5(3) + 100 = 1 - 10(3y)$$

i.e. $8y + 15 + 100 = 1 - 30y$

Rearranging gives:

$$8y + 30y = 1 - 5 - 100$$
$$38y = -114$$
$$y = \frac{-114}{38} = -3$$

Check:

$$\text{LHS} = \frac{2(-3)}{5} + \frac{3}{4} + 5 = \frac{-6}{5} + \frac{3}{4} + 5$$
$$= \frac{-9}{20} + 5 = 4\frac{11}{20}$$
$$\text{RHS} = \frac{1}{20} - \frac{3(-3)}{2} = \frac{1}{20} + \frac{9}{2} = 4\frac{11}{20}$$

Hence the solution $y = -3$ is correct.

Problem 11. Solve $\sqrt{x} = 2$

Wherever square root signs are involved with the unknown quantity, both sides of the equation must be squared. Hence

$$(\sqrt{x})^2 = (2)^2$$

i.e. $x = \mathbf{4}$

Problem 12. Solve $2\sqrt{d} = 8$

To avoid possible errors it is usually best to arrange the term containing the square root on its own. Thus

$$\frac{2\sqrt{d}}{2} = \frac{8}{2}$$

i.e. $\sqrt{d} = 4$

Squaring both sides gives: $d = \mathbf{16}$, which may be checked in the original equation.

Problem 13. Solve $x^2 = 25$

This problem (and problem 14) involves a square term and thus are not simple equations (they are, in fact, quadratic equations since their highest power of x is 2). However the solutions of such equations are often required and are therefore included for completeness.
Whenever a square of the unknown is involved, the square root of both sides of the equation is taken. Hence

$$\sqrt{x^2} = \sqrt{25}$$

i.e. $x = 5$

However, $x = -5$ is also a solution of the equation because

$$(-5) \times (-5) = +25$$

Therefore, whenever the square root of a number is required there are always two answers, one positive, the other negative. The solution of $x^2 = 25$ is thus written as $x = \pm 5$ (which reads as 'plus or minus 5').

Problem 14. Solve $\dfrac{15}{4t^2} = \dfrac{2}{3}$

'Cross-multiplying' gives:

$$15(3) = 2(4t^2)$$
$$45 = 8t^2$$
$$\frac{45}{8} = t^2$$

i.e. $t^2 = 5\dfrac{5}{8} = 5.625$

Hence $t = \sqrt{(5.625)} = \pm\textbf{2.372}$ correct to 4 significant figures.

1.3   Practical problems involving simple equations

Problem 15. A copper wire has a length l of 1.5 km, a resistance R of 5 Ω and a resistivity ρ of 17.2×10^{-6} Ω mm. Find the cross-sectional area, a, of the wire, given that $R = \rho l/a$

Since
$$R = \frac{\rho l}{a}$$
then
$$5\,\Omega = \frac{(17.2 \times 10^{-6}\ \Omega\ \text{mm})(1500 \times 10^{3}\ \text{mm})}{a}$$

From the units given, a is measured in mm². Thus

$$5a = 17.2 \times 10^{-6} \times 1500 \times 10^{3}$$

$$a = \frac{17.2 \times 10^{-6} \times 1500 \times 10^{3}}{5}$$

$$= \frac{17.2 \times 1500 \times 10^{3}}{10^{6} \times 5}$$

$$= \frac{17.2 \times \cancel{15}^{3}}{10 \times \cancel{5}_{1}} = 5.16$$

Hence the cross-sectional area of the wire is 5.16 mm².

Problem 16. The temperature coefficient of resistance α may be calculated from the formula $R_t = R_0(1 + \alpha t)$. Find α given $R_t = 0.928$, $R_0 = 0.8$ and $t = 40$

Since $R_t = R_0(1 + \alpha t)$ then
$$0.928 = 0.8[1 + \alpha(40)]$$
$$0.928 = 0.8 + (0.8)(\alpha)(40)$$
$$0.928 - 0.8 = 32\alpha$$
$$0.128 = 32\alpha$$

Hence $\quad \alpha = \dfrac{0.128}{32} = \textbf{0.004}$ or 4×10^{-3} in standard form

Problem 17. The distance s metres travelled in time t seconds is given by the formula $s = ut + (1/2)at^2$, where u is the initial velocity in m/s and a is the acceleration in m/s². Find the acceleration of the body if it travels 168 m in 6 s, with an initial velocity of 10 m/s.

$s = ut + (1/2)at^2$, and $s = 168$, $u = 10$ and $t = 6$

Hence $\quad 168 = (10)(6) + \dfrac{1}{2}a(6)^2$

$$168 = 60 + 18a$$
$$168 - 60 = 18a$$
$$108 = 18a$$
$$a = \frac{108}{18} = 6$$

Hence the acceleration of the body is 6 m/s²

Problem 18. When three resistors in an electrical circuit are connected in parallel the total resistance R_T is given by:
$$\frac{1}{R_T} = \frac{1}{R_1} + \frac{1}{R_2} + \frac{1}{R_3}$$
Find the total resistance when $R_1 = 5$ Ω, $R_2 = 10$ Ω and $R_3 = 30$ Ω

$$\frac{1}{R_T} = \frac{1}{5} + \frac{1}{10} + \frac{1}{30} = \frac{6 + 3 + 1}{30} = \frac{10}{30} = \frac{1}{3}$$

Taking the reciprocal of both sides gives: $R_T = 3$ Ω
Alternatively, if
$$\frac{1}{R_T} = \frac{1}{5} + \frac{1}{10} + \frac{1}{30}$$

the LCM of the denominators is $30R_T$
Hence

$$30R_T\left(\frac{1}{R_T}\right) = 30R_T\left(\frac{1}{5}\right) + 30R_T\left(\frac{1}{10}\right) + 30R_T\left(\frac{1}{30}\right)$$

Cancelling gives: $30 = 6R_T + 3R_T + R_T$

i.e. $\quad 30 = 10R_T$

$$\boldsymbol{R_T} = \frac{30}{10} = \textbf{3}\Omega, \text{ as above.}$$

THE SOLUTION OF SIMPLE EQUATIONS 7

Problem 19. The extension x m of an aluminium tie bar of length l m and cross-sectional area A m² when carrying a load of F newtons (N) is given by the modulus of elasticity $E = Fl/Ax$. Find the extension of the tie bar (in mm) if $E = 70 \times 10^9$ N/m², $F = 20 \times 10^6$ N, $A = 0.1$ m² and $l = 1.4$ m.

$$E = \frac{Fl}{Ax}$$

Hence

$$70 \times 10^9 \ \frac{N}{m^2} = \frac{(20 \times 10^6 \ N)(1.4 \ m)}{(0.1 \ m^2)(x)}$$

(the unit of x is thus metres).

$$70 \times 10^9 \times 0.1 \times x = 20 \times 10^6 \times 1.4$$

$$x = \frac{20 \times 10^6 \times 1.4}{70 \times 10^9 \times 0.1}$$

Cancelling gives:

$$x = \frac{2 \times 1.4}{7 \times 100} \ m = \frac{2 \times 1.4}{7 \times 100} \times 1000 \ mm$$

Hence the extension of the tie bar, $x = 4$ mm.

Problem 20. Power in a d.c. circuit is given by $P = V^2/R$, where V is the supply voltage and R is the circuit resistance. Find the supply voltage if the circuit resistance is 1.25 Ω and the power measured is 320 W.

Since $P = V^2/R$ then $320 = V^2/1.25$

$$(320)(1.25) = V^2$$

i.e. $\qquad\qquad V^2 = 400$

Supply voltage, $\qquad V = \sqrt{400} = \pm\mathbf{20}$ **V**

Problem 21. A formula relating initial and final states of pressures, P_1 and P_2, volumes V_1 and V_2, and absolute temperatures T_1 and T_2, of an ideal gas is

$$\frac{P_1 V_1}{T_1} = \frac{P_2 V_2}{T_2}$$

Find the value of P_2 given $P_1 = 100 \times 10^3$, $V_1 = 1.0$, $V_2 = 0.266$, $T_1 = 423$ and $T_2 = 293$.

Since

$$\frac{P_1 V_1}{T_1} = \frac{P_2 V_2}{T_2}$$

then

$$\frac{(100 \times 10^3)(1.0)}{423} = \frac{P_2(0.266)}{293}$$

'Cross-multiplying' gives:

$$(100 \times 10^3)(1.0)(293) = P_2(0.266)(423)$$

$$P_2 = \frac{(100 \times 10^3)(1.0)(293)}{(0.266)(423)}$$

Hence $\qquad\qquad P_2 = \mathbf{260 \times 10^3}$ **or** $\mathbf{2.6 \times 10^5}$

Problem 22. The stress f in a material of a thick cylinder can be obtained from

$$\frac{D}{d} = \sqrt{\left(\frac{f+p}{f-p}\right)}$$

Calculate the stress given that $D = 21.5$, $d = 10.75$ and $p = 1800$.

Since

$$\frac{D}{d} = \sqrt{\left(\frac{f+p}{f-p}\right)}$$

then

$$\frac{21.5}{10.75} = \sqrt{\left(\frac{f+1800}{f-1800}\right)}$$

i.e. $\qquad 2 = \sqrt{\left(\frac{f+1800}{f-1800}\right)}$

Squaring both sides gives:

$$4 = \frac{f+1800}{f-1800}$$

$$4(f-1800) = f+1800$$

$$4f - 7200 = f + 1800$$

$$4f - f = 1800 + 7200$$

$$3f = 9000$$

$$f = \frac{9000}{3} = 3000$$

Hence the stress, f, is 3000.

1.4 Further questions on the solution of simple equations

(Answers may be found on page 354)

Solve the equations in questions 1 to 29

1. $2x + 5 = 7$
2. $8 - 3t = 2$
3. $\frac{2}{3}c - 1 = 3$
4. $2x - 1 = 5x + 11$
5. $7 - 4p = 2p - 3$
6. $3x - 2 - 5x = 2x - 4$
7. $20d - 3 + 3d = 11d + 5 - 8$
8. $2(x - 1) = 4$
9. $5(f - 2) - 3(2f + 5) + 15 = 0$
10. $2x = 4(x - 3)$
11. $6(2 - 3y) - 42 = -2(y - 1)$
12. $4(3x + 1) = 7(x + 4) - 2(x + 5)$
13. $8 + 4(x - 1) - 5(x - 3) = 2(5 - 2x)$
14. $2 + \frac{3}{4}y = 1 + \frac{2}{3}y + \frac{5}{6}$
15. $\frac{1}{4}(2x - 1) + 3 = \frac{1}{2}$
16. $\frac{1}{3}(3m - 6) - \frac{1}{4}(5m + 4) + \frac{1}{5}(2m - 9) = -3$
17. $\frac{x}{3} - \frac{x}{5} = 2$
18. $\frac{2}{a} = \frac{3}{8}$
19. $\frac{x + 3}{4} = \frac{x - 3}{5} + 2$
20. $\frac{y - 2}{2y - 3} = \frac{1}{3}$
21. $\frac{x}{4} - \frac{x + 6}{5} = \frac{x + 3}{2}$
22. $\frac{2c - 3}{4} - \frac{1 - c}{5} - 1 = \frac{2c + 3}{3} + \frac{43}{60}$
23. $3\sqrt{t} = 9$
24. $2\sqrt{y} = 5$
25. $4 = \sqrt{\left(\frac{3}{a}\right)} + 3$
26. $10 = 5\sqrt{\left(\frac{x}{2} - 1\right)}$
27. $16 = \frac{t^2}{9}$
28. $\frac{6}{a} = \frac{2a}{3}$
29. $\frac{11}{2} = 5 + \frac{8}{x^2}$

Practical problems involving simple equations

30. A formula used for calculating resistance of a cable is $R = (\rho l)/a$. Given $R = 1.25$, $l = 2500$ and $a = 2 \times 10^{-4}$ find the value of ρ.
31. Force F newtons is given by $F = ma$, where m is the mass in kilograms and a is the acceleration in metres per second squared. Find the acceleration when a force of $4\,kN$ is applied to a mass of $500\,kg$.
32. $PV = mRT$ is the characteristic gas equation. Find the value of m when $P = 100 \times 10^3$, $V = 3.00$, $R = 288$ and $T = 300$.
33. When three resistors, R_1, R_2 and R_3 are connected in parallel the total resistance R_T is determined from
    $$\frac{1}{R_T} = \frac{1}{R_1} + \frac{1}{R_2} + \frac{1}{R_3}$$
    (a) Find the total resistance when $R_1 = 3\,\Omega$, $R_2 = 6\,\Omega$ and $R_3 = 18\,\Omega$.
    (b) Find the value of R_3 given that $R_T = 3\,\Omega$, $R_1 = 5\,\Omega$ and $R_2 = 10\,\Omega$.
34. Ohm's law may be represented by $I = V/R$, where I is the current in amperes, V is the voltage in volts and R is the resistance in ohms. A soldering iron takes a current of $0.30\,A$ from a $240\,V$ supply. Find the resistance of the element.
35. Given $R_2 = R_1(1 + \alpha t)$, find α given $R_1 = 5.0$, $R_2 = 6.03$ and $t = 51.5$.
36. If $v^2 = u^2 + 2as$, find u given $v = 24$, $a = -40$ and $s = 4.05$.
37. The relationship between the temperature on a Fahrenheit scale and that on a Celsius scale is given by $F = (9/5)C + 32$. Express $113°F$ in degrees Celsius.
38. If $t = 2\pi\sqrt{(w/Sg)}$, find the value of S given $w = 1.219$, $g = 9.81$ and $t = 0.3132$.

2

The solution of simultaneous equations

At the end of this chapter you should be able to:

- solve simultaneous equations by substitution
- solve simultaneous equations by elimination
- solve practical engineering problems involving simultaneous equations

2.1 Introduction

When an equation contains **two unknown quantities** it has an infinite number of solutions. When two equations are available connecting the same two unknown values then a unique solution is possible.

Equations which have to be solved together to find the unique values of the unknown quantities, which are true for each of the equations, are called **simultaneous equations**.

There are two methods of solving simultaneous equations analytically:

(a) by **substitution**, and (b) by **elimination**.

Simultaneous equations will be needed to solve problems by Kirchhoff's laws in Chapter 30 and have many other applications in science and engineering as demonstrated in section 2.3 of this chapter.

2.2 Worked problems on the solution of simultaneous equations

Problem 1. Solve the following equations for x and y,
(a) by substitution, and (b) by elimination:

$$x + 2y = -1 \qquad (1)$$
$$4x - 3y = 18 \qquad (2)$$

(a) **By substitution:**

From equation (1): $x = -1 - 2y$

Substituting this expression for x into equation (2) gives:

$$4(-1 - 2y) - 3y = 18$$

This is now a simple equation in y.

Removing the bracket gives:

$$-4 - 8y - 3y = 18$$
$$-11y = 18 + 4 = 22$$
$$y = \frac{22}{-11} = -2$$

Substituting $y = -2$ into equation (1) gives:

$$x + 2(-2) = -1$$
$$x - 4 = -1$$
$$x = -1 + 4 = 3$$

Thus $x = 3$ and $y = -2$ is the solution to the simultaneous equations.

(Check: In equation (2), since $x = 3$ and $y = -2$,

$$\text{LHS} = 4(3) - 3(-2)$$
$$= 12 + 6 = 18 = \text{RHS})$$

(b) **By elimination**

$$x + 2y = -1 \qquad (1)$$
$$4x - 3y = 18 \qquad (2)$$

If equation (1) is multiplied throughout by 4 the coefficient of x will be the same as in equation (2), giving:

$$4x + 8y = -4 \qquad (3)$$

Subtracting equation (3) from equation (2) gives:

$$4x - 3y = 18 \tag{2}$$

$$\underline{4x + 8y = -4} \tag{3}$$

$$0 - 11y = 22$$

Hence $y = 22/-11 = -2$

(Note, in the above subtraction, $18 - (-4) = 18 + 4 = 22$.) Substituting $y = -2$ into either equation (1) or equation (2) will give $x = 3$ as in method (a). The solution **$x = 3$, $y = -2$**, is the only pair of values that satisfies both of the original equations.

Problem 2. Solve, by a substitution method, the simultaneous equations

$$3x - 2y = 12 \tag{1}$$

$$x + 3y = -7 \tag{2}$$

From equation (2), $x = -7 - 3y$.
Substituting for x in equation (1) gives:

$$3(-7 - 3y) - 2y = 12$$

i.e. $-21 - 9y - 2y = 12$

$$-11y = 12 + 21 = 33$$

Hence $y = 33/-11 = -3$

Substituting $y = -3$ in equation (2) gives:

$$x + 3(-3) = -7$$

i.e. $x - 9 = -7$

Hence $x = -7 + 9 = 2$

Thus **$x = 2$, $y = -3$** is the solution of the simultaneous equations. (Such solutions should always be checked by substituting values into each of the original two equations.)

Problem 3. Use an elimination method to solve the simultaneous equations

$$3x + 4y = 5 \tag{1}$$

$$2x - 5y = -12 \tag{2}$$

If equation (1) is multiplied throughout by 2 and equation (2) by 3, then the coefficient of x will be the same in the newly formed equations. Thus

$2 \times$ equation (1) gives: $6x + 8y = 10$ (3)
$3 \times$ equation (2) gives: $6x - 15y = -36$ (4)

Equation (3) – equation (4) gives:

$$0 + 23y = 46$$

i.e. $$y = \frac{46}{23} = 2$$

(Note $+8y - (-15y) = 8y + 15y = 23y$ and $10 - (-36) = 10 + 36 = 46$. (Alternatively, 'change the signs of the bottom line and add'.)
Substituting $y = 2$ in equation (1) gives:

$$3x + 4(2) = 5$$

from which, $3x = 5 - 8 = -3$

and $x = -1$

Checking in equation (2),

$$\text{LHS} = 2(-1) - 5(2) = -2 - 10 = -12 = \text{RHS}$$

Hence **$x = -1$ and $y = 2$** is the solution of the simultaneous equations.
The elimination method is the most common method of solving simultaneous equations.

Problem 4. Solve $7x - 2y = 26$ (1)
                     $6x + 5y = 29$ (2)

When equation (1) is multiplied by 5 and equation (2) by 2 the coefficients of y in each equation are numerically the same, i.e. 10, but are of opposite sign.

$5 \times$ equation (1) gives:

$$35x - 10y = 130 \tag{3}$$

$2 \times$ equation (2) gives:

$$12x + 10y = 58 \tag{4}$$

Adding equations (3) and (4) gives:

$$47x + 0 = 188$$

Hence

$$x = \frac{188}{47} = 4$$

(Note that when the signs of common coefficients are different the two equations are **added**, and when the signs of common coefficients are the same the two equations are **subtracted** (as in Problems 1 and 3).)
Substituting $x = 4$ into equation (1) gives:

$$7(4) - 2y = 26$$

$$28 - 2y = 26$$

$$28 - 26 = 2y$$

$$2 = 2y$$

Hence $\quad y = 1$

Checking, by substituting $x = 4$, $y = 1$ into equation (2) gives:

$$\text{LHS} = 6(4) + 5(1) = 24 + 5 = 29 = \text{RHS}$$

Thus the solution is $x = 4$, $y = 1$, since these values maintain the equality when substituted in both equations.

Problem 5.   Solve $3p = 2q$ $\qquad$ (1)

$\qquad\qquad\qquad$ $4p + q + 11 = 0$ $\qquad$ (2)

Rearranging gives:

$$3p - 2q = 0 \qquad (3)$$
$$4p + q = -11 \qquad (4)$$

Multiplying equation (4) by 2 gives:

$$8p + 2q = -22 \qquad (5)$$

Adding equations (3) and (5) gives:

$$11p + 0 \; -22$$
$$p = \frac{-22}{11} = -2$$

Substituting $p = -2$ into equation (1) gives:

$$3(-2) = 2q$$
$$-6 = 2q$$
$$q = \frac{-6}{2} = -3$$

Checking, by substituting $p = -2$ and $q = -3$ into equation (2), gives:

$$\text{LHS} = 4(-2) + (-3) + 11 = -8 - 3 + 11 = 0$$
$$= \text{RHS}$$

Hence the solution is $p = -2$, $q = -3$

2.3   Practical problems involving simultaneous equations

Problem 6.   The law connecting friction F and load L for an experiment is of the form $F = aL + b$, where a and b are constants. When $F = 5.6$, $L = 8.0$ and when $F = 4.4$, $L = 2.0$. Find the values of a and b and the value of F when $L = 6.5$.

Substituting $F = 5.6$, $L = 8.0$ into $F = aL + b$ gives:

$$5.6 = 8.0\,a + b \qquad (1)$$

Substituting $F = 4.4$, $L = 2.0$ into $F = aL + b$ gives:

$$4.4 = 2.0\,a + b \qquad (2)$$

Subtracting equation (2) from equation (1) gives:

$$1.2 = 6.0a$$
$$a = \frac{1.2}{6.0} = \frac{1}{5}$$

Substituting $a = 1/5$ into equation (1) gives:

$$5.6 = 8.0\left(\frac{1}{5}\right) + b$$
$$5.6 = 1.6 + b$$
$$5.6 - 1.6 = b$$

i.e. $\qquad b = 4$

Checking, substituting $a = 1/5$, $b = 4$ in equation (2) gives:

$$\text{RHS} = 2.0\left(\frac{1}{5}\right) + 4 = 0.4 + 4 = 4.4 = \text{LHS}$$

Hence $a = 1/5$ and $b = 4$.

When $L = 6.5$, $F = aL + b = (1/5)\,(6.5) + 4 = 1.3 + 4$ i.e. **$F = 5.30$.**

Problem 7.   The equation of a straight line, of slope m and intercept on the y-axis c, is $y = mx + c$. If a straight line passes through the point where $x = 1$ and $y = -2$, and also through the point where $x = 3\frac{1}{2}$ and $y = 10\frac{1}{2}$, find the values of the slope and the y-axis intercept.

Substituting $x = 1$ and $y = -2$ into $y = mx + c$ gives:

$$-2 = m + c \qquad (1)$$

Substituting $x = 3\frac{1}{2}$ and $y = 10\frac{1}{2}$ into $y = mx + c$ gives:

$$10\frac{1}{2} = 3\frac{1}{2}\,m + c \qquad (2)$$

Subtracting equation (1) from equation (2) gives:

$$12\frac{1}{2} = 2\frac{1}{2}\,m$$

$$m = \frac{12\frac{1}{2}}{2\frac{1}{2}} = 5$$

Substituting $m = 5$ into equation (1) gives:

$$-2 = 5 + c$$

$$c = -2 - 5 = -7$$

Checking, substituting $m = 5$, $c = -7$ in equation (2), gives:

$$\text{RHS} = \left(3\frac{1}{2}\right)(5) + (-7) = 17\frac{1}{2} - 7 = 10\frac{1}{2} = \text{LHS}$$

Hence the slope, $m = 5$ and the y-axis intercept, $c = -7$.

Problem 8. When Kirchhoff's laws are applied to a particular electrical circuit the currents I_1 and I_2 are connected by the equations:

$$27 = 1.5I_1 + 8(I_1 - I_2) \tag{1}$$
$$-26 = 2I_2 - 8(I_1 - I_2) \tag{2}$$

Solve the equations to find the values of currents I_1 and I_2.

Removing brackets from equation (1) gives:

$$27 = 1.5I_1 + 8I_1 - 8I_2$$

Rearranging gives:

$$9.5I_1 - 8I_2 = 27 \tag{3}$$

Removing brackets from equation (2) gives:

$$-26 = 2I_2 - 8I_1 + 8I_2$$

Rearranging gives:

$$-8I_1 + 10I_2 = -26 \tag{4}$$

Multiplying equation (3) by 5 gives:

$$47.5I_1 - 40I_2 = 135 \tag{5}$$

Multiplying equation (4) by 4 gives:

$$-32I_1 + 40I_2 = -104 \tag{6}$$

Adding equations (5) and (6) gives:

$$15.5I_1 + 0 = 31$$

$$I_1 = \frac{31}{15.5} = 2$$

Substituting $I_1 = 2$ into equation (3) gives:

$$9.5(2) - 8I_2 = 27$$
$$19 - 8I_2 = 27$$
$$19 - 27 = 8I_2$$
$$-8 = 8I_2$$
$$I_2 = -1$$

Hence the solution is $I_1 = 2$ and $I_2 = -1$ (which may be checked in the original equations).

Problem 9. The displacement s metres from a fixed point of a vehicle travelling in a straight line with constant acceleration, a m/s^2, is given by $s = ut + \frac{1}{2}at^2$, where u is the initial velocity in metres per second and t the time in seconds. Determine the initial velocity and the acceleration given that $s = 42$ m when $t = 2$ s and $s = 144$ m when $t = 4$ s. Find also the distance travelled after 3 s.

Substituting $s = 42$, $t = 2$ into $s = ut + \frac{1}{2}at^2$ gives:

$$42 = 2u + \frac{1}{2}a(2)^2$$

i.e. $\quad 42 = 2u + 2a \tag{1}$

Substituting $s = 144$, $t = 4$ into $s = ut + \frac{1}{2}at^2$ gives:

$$144 = 4u + \frac{1}{2}a(4)^2$$

i.e. $\quad 144 = 4u + 8a \tag{2}$

Multiplying equation (1) by 2 gives:

$$84 = 4u + 4a \tag{3}$$

Subtracting equation (3) from equation (2) gives:

$$60 = 0 + 4a$$

$$a = \frac{60}{4} = 15$$

Substituting $a = 15$ into equation (1) gives:

$$42 = 2u + 2(15)$$
$$42 - 30 = 2u$$
$$u = \frac{12}{2} = 6$$

Substituting $a = 15$, $u = 6$ in equation (2) gives:

$$\text{RHS} = 4(6) + 8(15) = 24 + 120 = 144 = \text{LHS}$$

Hence the initial velocity, $u = 6$ m/s and the acceleration, $a = 15$ m/s^2.

Distance travelled after 3 s is given by $s = ut + \frac{1}{2}at^2$, where $t = 3$, $u = 6$ and $a = 15$.
Hence

$$s = 6(3) + \frac{1}{2}(15)(3)^2$$

$$= 18 + 67\tfrac{1}{2}$$

i.e. distance travelled after 3 s = $85\tfrac{1}{2}$ m.

2.4 Further questions on the solution of simultaneous equations

(Answers may be found on page 354.)

In problems 1 to 9, solve the simultaneous equations and verify the results.

1. $a + b = 7$
   $a - b = 3$
2. $2x + 5y = 7$
   $x + 3y = 4$
3. $3s + 2t = 12$
   $4s - t = 5$
4. $3x - 2y = 13$
   $2x + 5y = -4$
5. $5m - 3n = 11$
   $3m + n = 8$
6. $8a - 3b = 51$
   $3a + 4b = 14$
7. $5x = 2y$
   $3x + 7y = 41$
8. $5c = 1 - 3d$
   $2d + c + 4 = 0$
9. $7p + 11 + 2q = 0$
   $-1 = 3q - 5p$

Practical problems involving simultaneous equations

10. In a system of pulleys, the effort P required to raise a load W is given by $P = aW + b$, where a and b are constants. If $W = 40$ when $P = 12$ and $W = 90$ when $P = 22$, find the values of a and b.
11. Applying Kirchhoff's laws to an electrical circuit produces the following equations:

    $5 = 0.2I_1 + 2(I_1 - I_2)$

    $12 = 3I_2 + 0.4I_2 - 2(I_1 - I_2)$

    Determine the values of currents I_1 and I_2.
12. Velocity v is given by the formula $v = u + at$. If $v = 20$ when $t = 2$ and $v = 40$ when $t = 7$ find the values of u and a. Hence find the velocity when $t = 3.5$.
13. $y = mx + c$ is the equation of a straight line of slope m and y-axis intercept c. If the line passes through the point where $x = 2$ and $y = 2$, and also through the point where $x = 5$ and $y = \tfrac{1}{2}$, find the slope and y-axis intercept of the straight line.

3

Evaluation and transposition of formulae

At the end of this chapter you should be able to:

- evaluate given formulae using a calculator
- transpose formulae applicable to engineering applications

3.1 Introduction

The statement $v = u + at$ is said to be a **formula** for v in terms of u, a and t. The letters v, u, a and t are called **symbols**.

The single term on the left-hand side of the equation, v, is called the **subject of the formula**.

Provided values are given for all the symbols in a formula except one, the remaining symbol can be made the subject of the formula and may be evaluated using a calculator. In every chapter of this book, science and engineering formulae need to be evaluated and/or transposed.

3.2 Worked problems on the evaluation of formulae

Problem 1. In an electrical circuit the voltage V is given by Ohm's law, i.e. $V = IR$. Find, correct to 4 significant figures, the voltage when $I = 5.36$ A and $R = 14.76\ \Omega$.

$$V = IR = (5.36)(14.76)$$

Hence **voltage V = 79.11 V** , correct to 4 significant figures.

Problem 2. The surface area A of a hollow cone is given by $A = \pi rl$. Determine the surface area when $r = 3.0$ cm, $l = 8.5$ cm and $\pi = 3.14$.

$$A = \pi rl = (3.14)(3.0)(8.5)\ \text{cm}^2$$

Hence **surface area A = 80.07 cm²**.

Problem 3. Velocity v is given by $v = u + at$. If $u = 9.86$ m/s, $a = 4.25$ m/s² and $t = 6.84$ s, find v, correct to 3 significant figures.

$$v = u + at = 9.86 + (4.25)(6.84)$$
$$= 9.86 + 29.07$$
$$= 38.93$$

Hence **velocity v = 38.9 m/s,** correct to 3 significant figures.

Problem 4. The area, A, of a circle is given by $A = \pi r^2$. Determine the area correct to 2 decimal places, given $\pi = 3.142$ and $r = 5.23$ m.

$$A = \pi r^2 = (3.142)(5.23)^2$$
$$= (3.142)(27.35)$$

Hence **area A = 85.94 m²,** correct to 2 decimal places.

Problem 5. The power P watts dissipated in an electrical circuit may be expressed by the formula $P = V^2/R$. Evaluate the power, correct to 3 significant figures, given that $V = 17.48$ V and $R = 36.12\ \Omega$.

$$P = \frac{V^2}{R} = \frac{(17.48)^2}{36.12} = \frac{305.6}{36.12}$$

Hence **power, P = 8.46 W,** correct to 3 significant figures.

Problem 6.   The volume V cm^3 of a right circular cone is given by $V = \frac{1}{3}\pi r^2 h$. Given that $r = 4.321$ cm, $h = 18.35$ cm and $\pi = 3.142$, find the volume correct to 4 significant figures.

$$V = \tfrac{1}{3}\pi r^2 h = \tfrac{1}{3}(3.142)(4.321)^2(18.35)$$
$$= \tfrac{1}{3}(3.142)(18.67)(18.35)$$

Hence **volume, V = 358.8 cm^3,** correct to 4 significant figures.

Problem 7.   Force F newtons is given by the formula $F = (Gm_1m_2)/d^2$, where m_1 and m_2 are masses, d their distance apart and G is a constant. Find the value of the force given that $G = 6.67 \times 10^{-11}$, $m_1 = 7.36$, $m_2 = 15.5$ and $d = 22.6$. Express the answer in standard form, correct to 3 significant figures.

$$F = \frac{Gm_1m_2}{d^2} = \frac{(6.67 \times 10^{-11})(7.36)(15.5)}{(22.6)^2}$$
$$= \frac{(6.67)(7.36)(15.5)}{(10^{11})(510.8)}$$
$$= \frac{1.490}{10^{11}}$$

Hence **force F = 1.49 $\times$ 10^{-11} newtons,** correct to 3 significant figures.

Problem 8.   The time of swing, t seconds, of a simple pendulum is given by $t = 2\pi\sqrt{(l/g)}$. Determine the time, correct to 3 decimal places, given that $\pi = 3.142$, $l = 12.0$ and $g = 9.81$.

$$t = 2\pi\sqrt{\left(\frac{l}{g}\right)} = (2)(3.142)\sqrt{\left(\frac{12.0}{9.81}\right)}$$
$$= (2)(3.142)\sqrt{(1.223)}$$
$$= (2)(3.142)(1.106)$$

Hence **time t = 6.950 seconds,** correct to 3 decimal places.

Problem 9.   Resistance, R Ω, varies with temperature according to the formula $R = R_0(1 + \alpha t)$. Evaluate R, correct to 3 significant figures, given $R_0 = 14.59$, $\alpha = 0.0043$ and $t = 80$.

$$R = R_0(1 + \alpha t) = 14.59(1 + [(0.0043)(80)])$$
$$= 14.59(1 + 0.344)$$
$$= 14.59(1.344)$$

Hence **resistance, R = 19.6 Ω,** correct to 3 significant figures.

3.3   Transposition of formulae

When a symbol other than the subject is required to be calculated it is usual to rearrange the formula to make a new subject. This rearranging process is called **transposing the formula, or transposition.**

The rules used for transposition of formulae are the same as those used for the solution of simple equations (see Chapter 1), basically, **that the equality of an equation must be maintained.**

3.4   Worked problems on the transposition of formulae

Problem 10.   Transpose $p = q + r + s$ to make r the subject.

The aim is to obtain r on its own on the left-hand side (LHS) of the equation. Changing the equation around so that r is on the LHS gives:

$$q + r + s = p \tag{1}$$

Subtracting $(q + s)$ from both sides of the equation gives:

$$q + r + s - (q + s) = p - (q + s)$$

Thus $q + r + s - q - s = p - q - s$

i.e. $r = p - q - s \tag{2}$

It is shown with simple equations (Chapter 1) that a quantity can be moved from one side of an equation to the other with an appropriate change of sign. Thus equation (2) follows immediately from equation (1) above.

Problem 11. If $a + b = w - x + y$, express x as the subject.

Rearranging gives:

$$w - x + y = a + b \text{ and } -x = a + b - w - y$$

Multiplying both sides by -1 gives:

$$(-1)(-x) = (-1)(a + b - w - y)$$
i.e. $x = -a - b + w + y$

The result of multiplying each side of the equation by -1 is to change all the signs in the equation.

It is conventional to express answers with positive quantities first. Hence rather than $x = -a - b + w + y$,

$$x = w + y - a - b$$

since the order of terms connected by $+$ and $-$ signs is immaterial.

Problem 12. Transpose $v = f \lambda$ to make λ the subject.

Rearranging gives:

$$f \lambda = v$$

Dividing both sides by f gives:

$$\frac{f \lambda}{f} = \frac{v}{f} \text{ i.e. } \lambda = \frac{v}{f}$$

Problem 13. When a body falls freely through a height h, the velocity v is given by $v^2 = 2gh$. Express this formula with h as the subject.

Rearranging gives:

$$2gh = v^2$$

Dividing both sides by $2g$ gives:

$$\frac{2gh}{2g} = \frac{v^2}{2g} \text{ i.e. } h = \frac{v^2}{2g}$$

Problem 14. If $R = V/I$, rearrange to make V the subject.

Rearranging gives:

$$\frac{V}{I} = R$$

Multiplying both sides by I gives:

$$I\left(\frac{V}{I}\right) = I(R)$$

Hence

$$V = IR$$

Problem 15. Transpose $a = F/m$ for m.

Rearranging gives:

$$\frac{F}{m} = a$$

Multiplying both sides by m gives:

$$m\left(\frac{F}{m}\right) = m(a) \text{ i.e. } F = ma$$

Rearranging gives:

$$ma = F$$

Dividing both sides by a gives:

$$\frac{ma}{a} = \frac{F}{a} \text{ i.e. } m = \frac{F}{a}$$

Problem 16. Rearrange the formula $R = (\rho l)/a$ to make (i) a the subject, and (ii) l the subject.

(i) Rearranging gives:

$$\frac{\rho l}{a} = R$$

Multiplying both sides by a gives:

$$a\left(\frac{\rho l}{a}\right) = a(R)$$

i.e. $\rho l = aR$

Rearranging gives:

$$aR = \rho l$$

Dividing both sides by R gives:

$$\frac{aR}{R} = \frac{\rho l}{R}$$

i.e.

$$a = \frac{\rho l}{R}$$

(ii) $\rho l/a = R$

Multiplying both sides by a gives:

$\rho l = aR$

Dividing both sides by ρ gives:

$$\frac{\rho l}{\rho} = \frac{aR}{\rho}$$

i.e.

$$l = \frac{aR}{\rho}$$

Problem 17. Transpose the formula $v = u + (ft)/m$, to make f the subject.

Rearranging gives:

$$u + \frac{ft}{m} = v$$

and

$$\frac{ft}{m} = v - u$$

Multiplying each side by m gives:

$$m\left(\frac{ft}{m}\right) = m(v - u)$$

i.e.

$$ft = m(v - u)$$

Dividing both sides by t gives:

$$\frac{ft}{t} = \frac{m}{t}(v - u)$$

i.e.

$$f = \frac{m}{t}(v - u)$$

Problem 18. The final length, l_2, of a piece of wire heated through $\theta°C$ is given by the formula $l_2 = l_1(1 + \alpha\theta)$. Make the coefficient of expansion, α, the subject.

Rearranging gives:

$$l_1(1 + \alpha\theta) = l_2$$

Removing the bracket gives:

$$l_1 + l_1 \alpha\theta = l_2$$

Rearranging gives:

$$l_1\alpha\theta = l_2 - l_1$$

Dividing both sides by $l_1\theta$ gives:

$$\frac{l_1\alpha\theta}{l_1\theta} = \frac{l_2 - l_1}{l_1\theta}$$

i.e.

$$\alpha = \frac{l_2 - l_1}{l_1\theta}$$

Problem 19. A formula for the distance moved by a body is given by $s = \frac{1}{2}(v + u)t$. Rearrange the formula to make u the subject.

Rearranging gives:

$$\frac{1}{2}(v + u)t = s$$

Multiplying both sides by 2 gives:

$$(v + u)t = 2s$$

Dividing both sides by t gives:

$$\frac{(v + u)t}{t} = \frac{2s}{t}$$

i.e.

$$v + u = \frac{2s}{t}$$

Hence

$$u = \frac{2s}{t} - v \text{ or } \frac{2s - vt}{t}$$

Problem 20. A formula for kinetic energy is $k = \frac{1}{2}mv^2$. Transpose the formula to make v the subject.

Rearranging gives:

$$\frac{1}{2}mv^2 = k$$

Whenever the prospective new subject is a squared term, that term is isolated on the LHS, and then the square root of both sides of the equation is taken.

Multiplying both sides by 2 gives:

$$mv^2 = 2k$$

Dividing both sides by m gives:

$$\frac{mv^2}{m} = \frac{2k}{m}$$

i.e.

$$v^2 = \frac{2k}{m}$$

Taking the square root of both sides gives:

$$\sqrt{v^2} = \sqrt{\left(\frac{2k}{m}\right)}$$

i.e.

$$v = \sqrt{\left(\frac{2k}{m}\right)}$$

Problem 21. In a right-angled triangle having sides x, y and hypotenuse z, Pythagoras' theorem states $z^2 = x^2 + y^2$. Transpose the formula to find x.

Rearranging gives:

$$x^2 + y^2 = z^2$$

and

$$x^2 = z^2 - y^2$$

Taking the square root of both sides gives:

$$x = \sqrt{(z^2 - y^2)}$$

Problem 22. Given $t = 2\pi\sqrt{(l/g)}$, find g in terms of t, l and π.

Whenever the prospective new subject is within a square root sign, it is best to isolate that term on the LHS and then to square both sides of the equation.

Rearranging gives:

$$2\pi\sqrt{\left(\frac{l}{g}\right)} = t$$

Dividing both sides by 2π gives:

$$\sqrt{\left(\frac{l}{g}\right)} = \frac{t}{2\pi}$$

Squaring both sides gives:

$$\frac{l}{g} = \left(\frac{t}{2\pi}\right)^2 = \frac{t^2}{4\pi^2}$$

Cross-multiplying, i.e. multiplying each term by $4\pi^2g$, gives:

$$4\pi^2l = gt^2$$

or

$$gt^2 = 4\pi^2l$$

Dividing both sides by t^2 gives:

$$\frac{gt^2}{t^2} = \frac{4\pi^2l}{t^2}$$

i.e.

$$g = \frac{4\pi^2l}{t^2}$$

Problem 23. Transpose the formula

$$V = \frac{Er}{R + r}$$

to make r the subject.

Rearranging gives:

$$\frac{Er}{R + r} = V$$

Multiplying both sides by $(R + r)$ gives:

$$Er = V(R + r)$$

Removing the bracket gives:

$$Er = VR + Vr$$

Rearranging to obtain terms in r on the LHS side:

$$Er - Vr = VR$$

Factorizing gives:

$$r(E - V) = VR$$

Dividing both sides by $(E - V)$ gives:

$$r = \frac{VR}{E - V}$$

Problem 24. Given that

$$\frac{D}{d} = \sqrt{\left(\frac{f + p}{f - p}\right)}$$

express p in terms of D, d and f.

Rearranging gives:

$$\sqrt{\left(\frac{f + p}{f - p}\right)} = \frac{D}{d}$$

Squaring both sides gives:

$$\frac{f + p}{f - p} = \frac{D^2}{d^2}$$

Cross-multiplying, i.e. multiplying each term by $d^2(f - p)$, gives:

$$d^2(f + p) = D^2(f - p)$$

Removing brackets gives:

$$d^2 f + d^2 p = D^2 f - D^2 p$$

Rearranging, to obtain terms in p on the LHS, gives:

$$d^2 p + D^2 p = D^2 f - d^2 f$$

Factorizing gives:

$$p(d^2 + D^2) = f(D^2 - d^2)$$

Dividing both sides by $(d^2 + D^2)$ gives:

$$p = \frac{f(D^2 - d^2)}{(d^2 + D^2)}$$

3.5 Further questions on evaluation and transposition of formulae

(Answers may be found on page 354.)

Evaluation of formulae

1. The area A of a rectangle is given by the formula $A = lb$. Evaluate the area when $l = 12.4$ cm and $b = 5.37$ cm.
2. The circumference C of a circle is given by the formula $C = 2\pi r$. Determine the circumference given $\pi = 3.14$ and $r = 8.40$ mm.
3. A formula used in connection with gases is $R = (PV)/T$. Evaluate R when $P = 1500$, $V = 5$ and $T = 200$.
4. The potential difference, V volts, available at battery terminals is given by $V = E - Ir$. Evaluate V when $E = 5.62$, $I = 0.70$ and $R = 4.30$.
5. Given force $F = \frac{1}{2}m(v^2 - u^2)$, find F when $m = 18.3$, $v = 12.7$ and $u = 8.24$.
6. The current I amperes flowing in a number of cells is given by $I = (nE)/(R + nr)$. Evaluate the current when $n = 36$, $E = 2.20$, $R = 2.80$ and $r = 0.50$.
7. The time, t seconds, of oscillation for a simple pendulum is given by $t = 2\pi\sqrt{(l/g)}$. Determine the time when $\pi = 3.142$, $l = 54.32$ and $g = 9.81$.

8. Energy, E joules, is given by the formula $E = \frac{1}{2}LI^2$. Evaluate the energy when $L = 5.5$ and $I = 1.2$.
9. Distance s metres is given by the formula $s = ut + \frac{1}{2}at^2$. If $u = 9.50$, $t = 4.60$ and $a = -2.50$, evaluate the distance.

Transposition of formulae

Make the symbol indicated the subject of each of the formulae shown in Problems 10 to 28, and express each in its simplest form.

10. $a + b = c - d - e$ $\hspace{2em}$ (d)
11. $x + 3y = t$ $\hspace{2em}$ (y)
12. $c = 2\pi r$ $\hspace{2em}$ (r)
13. $y = mx + c$ $\hspace{2em}$ (x)
14. $I = PRT$ $\hspace{2em}$ (T)
15. $I = \dfrac{E}{R}$ $\hspace{2em}$ (R)
16. $S = \dfrac{a}{1 - r}$ $\hspace{2em}$ (r)
17. $F = \dfrac{9}{5}C + 32$ $\hspace{2em}$ (C)
18. $y = \dfrac{\lambda(x - d)}{d}$ $\hspace{2em}$ (x)
19. $A = \dfrac{3(F - f)}{L}$ $\hspace{2em}$ (f)
20. $y = \dfrac{Ml^2}{8EI}$ $\hspace{2em}$ (E)
21. $R = R_0(1 + \alpha t)$ $\hspace{2em}$ (t)
22. $\dfrac{1}{R} = \dfrac{1}{R_1} + \dfrac{1}{R_2}$ $\hspace{2em}$ (R_2)
23. $I = \dfrac{E - e}{R + r}$ $\hspace{2em}$ (R)
24. $t = 2\pi\sqrt{\left(\dfrac{l}{g}\right)}$ $\hspace{2em}$ (l)
25. $v^2 = u^2 + 2as$ $\hspace{2em}$ (u)
26. $A = \dfrac{\pi R^2 \theta}{360}$ $\hspace{2em}$ (R)
27. $y = \dfrac{a^2 m - a^2 n}{x}$ $\hspace{2em}$ (a)
28. $m = \dfrac{\mu L}{L + rCR}$ $\hspace{2em}$ (L)
29. A formula for the focal length, f, of a convex lens is

$$\frac{1}{f} = \frac{1}{u} + \frac{1}{v}$$

Transpose the formula to make v the subject and evaluate v when $f = 5$ and $u = 6$.

30. The quantity of heat, Q, is given by the formula $Q = mc(t_2 - t_1)$. Make t_2 the subject

of the formula and evaluate t_2 when $m = 10$, $t_1 = 15$, $c = 4$ and $Q = 1600$.

31. The velocity, v, of water in a pipe appears in the formula

$$h = \frac{0.03Lv^2}{2dg}$$

Express v as the subject of the formula and evaluate v when $h = 0.712$, $L = 150$, $d = 0.30$ and $g = 9.81$.

32. The sag S at the centre of a wire is given by the formula

$$S = \sqrt{\left\{ \frac{3d(l - d)}{8} \right\}}$$

Make l the subject of the formula and evaluate l when $d = 1.75$ and $S = 0.80$.

4

Straight line graphs

At the end of this chapter you should be able to:

- plot a straight line graph from given coordinates

- determine the gradient of a straight line graph

- determine the vertical axis intercept value

- produce a table of coordinates from a given equation to plot a straight line graph

- interpolate and extrapolate values from a straight line graph

- determine the law of a graph

- appreciate where typical practical engineering examples of straight line graphs occur

4.1   Introduction

A graph is a pictorial representation of information showing how one quantity varies with another related quantity.

The most common method of showing a relationship between two sets of data is to use **Cartesian** or **rectangular** axes as shown in Fig. 4.1.

The points on a graph are called **coordinates**. Point A in Fig. 4.1 has the coordinates (3, 2), i.e. 3 units in the x direction and 2 units in the y direction. Similarly, point B has coordinates (−4, 3) and C has coordinates (−3, −2). The origin has coordinates (0, 0).

The horizontal distance of a point from the vertical axis is called the **abscissa** and the vertical distance from the horizontal axis is called the **ordinate**.

Typical examples where the plotting of straight line graphs are used in this book are with distance/time graphs (Chapter 9), speed/time graphs (Chapters 9 and 10) and load/extension graphs (Chapters 17 and 18).

Section 4.2 following explains about gradients and equations of straight line graphs, and section 4.3 demonstrates some practical situations where they are used in science and engineering.

4.2   The straight line graph

Let a relationship between two variables x and y be $y = 3x + 2$.

When $x = 0$, $y = 3(0) + 2 = 2$.
When $x = 1$, $y = 3(1) + 2 = 5$.
When $x = 2$, $y = 3(2) + 2 = 8$, and so on.

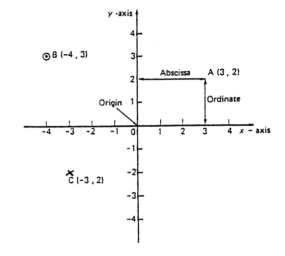

Figure 4.1

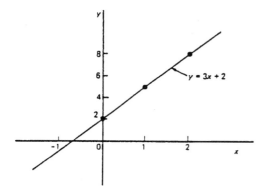

Figure 4.2

Thus coordinates (0, 2), (1, 5) and (2, 8) have been produced from the equation by selecting arbitrary values of x, and are shown plotted in Fig. 4.2. When the points are joined together a **straight-line graph** results.

The **gradient** or **slope** of a straight line is the ratio of the change in the value of y to the change in the value of x between any two points on the line. If, as x increases ($\rightarrow$), y also increases ($\uparrow$), then the gradient is positive.

In Fig. 4.3(a), the gradient of AC is given by

$$\frac{\text{change in } y}{\text{change in } x} = \frac{\text{CB}}{\text{BA}} = \frac{7-3}{3-1}$$

$$= \frac{4}{2} = 2$$

If as x increases ($\rightarrow$), y decreases ($\downarrow$), then the gradient is negative.

In Fig. 4.3(b), the gradient of DF is given by

$$\frac{\text{change in } y}{\text{change in } x} = \frac{\text{FE}}{\text{ED}} = \frac{11-2}{-3-0}$$

$$= \frac{9}{-3} = -3$$

Fig. 3(c) shows a straight line graph $y = 3$. Since the straight line is horizontal the gradient is zero.

The value of y when $x = 0$ is called the **y-axis intercept**. In Fig. 4.3(a) the y-axis intercept is 1 and in Fig. 4.3(b) it is 2. If the equation of a graph is of the form **$y = mx + c$**, where m and c are constants, the graph will always be a straight line, **m representing the gradient** and **c the y-axis intercept**. Thus $y = 5x + 2$ represents a straight line of gradient 5 and y-axis intercept 2. Similarly, $y = -3x - 4$ represents a straight line of gradient -3 and y-axis intercept -4.

Summary of general rules to be applied when drawing graphs

(i)   Give the graph a title clearly explaining what is being illustrated.
(ii) Choose scales such that the graph occupies as much space as possible on the graph paper being used.
(iii) Choose scales so that interpolation is made as easy as possible. Usually scales such as 1 cm = 1 unit, or 1 cm = 2 units, or 1 cm = 10 units are used. Awkward scales such as 1 cm = 3 units or 1 cm = 7 units should not be used.

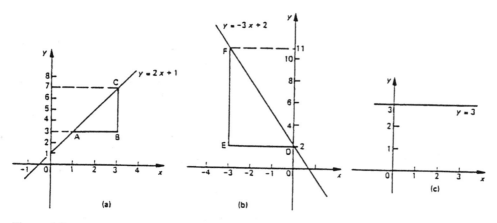

Figure 4.3

(iv) The scales need not start at zero, particularly when starting at zero produces an accumulation of points within a small area of the graph paper.

(v) The coordinates, or points, should be clearly marked. This may be done either by a cross, or by a dot and circle, or just by a dot (see Fig. 4.1).

(vi) A statement should be made next to each axis explaining the numbers represented with their appropriate units.

(vii) Sufficient numbers should be written next to each axis without cramping.

Problem 1. Plot the graph $y = 4x + 3$ in the range $x = -3$ to $x = +4$. From the graph, find (a) the value of y when $x = 2.2$ and (b) the value of x when $y = -3$

Whenever an equation is given and a graph is required, a table giving corresponding values of the variable is necessary. The table is achieved as follows:

When $x = -3$, $y = 4x + 3 = 4(-3) + 3 = -12 + 3 = -9$.
When $x = -2$, $y = 4(-2) + 3 = -8 + 3 = -5$, and so on.

Such a table is shown below:

x	-3	-2	-1	0	1	2	3	4
y	-9	-5	-1	3	7	11	15	19

The coordinates (-3, -9), (-2, -5), (-1, -1), and so on, are plotted and joined together to produce the straight line graph shown in Fig. 4.4. (Note that the scales used on the x and y axes do not have to be the same.) From the graph:

(a) when $x = 2.2$, $y = 11.8$, and
(b) when $y = -3$, $x = -1.5$.

Problem 2. Plot the following graphs on the same axes between the values $x = -3$ to $x = +3$ and determine the gradient and y-axis intercept of each.

(a) $y = 3x$ (b) $y = 3x + 7$
(c) $y = -4x + 4$ (d) $y = -4x - 5$

A table of coordinates is drawn up for each equation.

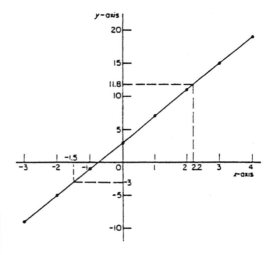

Figure 4.4 Graph of $y = 4x + 3$

(a) $y = 3x$

x	-3	-2	-1	0	1	2	3
y	-9	-6	-3	0	3	6	9

(b) $y = 3x + 7$

x	-3	-2	-1	0	1	2	3
y	-2	1	4	7	10	13	16

(c) $y = -4x + 4$

x	-3	-2	-1	0	1	2	3
y	16	12	8	4	0	-4	-8

(d) $y = -4x - 5$

x	-3	-2	-1	0	1	2	3
y	7	3	-1	-5	-9	-13	-17

Each of the graphs is plotted as shown in Fig. 4.5, and each is a straight line. $y = 3x$ and $y = 3x + 7$ are parallel to each other and thus have the same gradient.
Gradient of AC is given by

$$\frac{BC}{AB} = \frac{16 - 7}{3 - 0} = \frac{9}{3} = 3$$

Hence the gradient of both $y = 3x$ and $y = 3x + 7$ is 3.

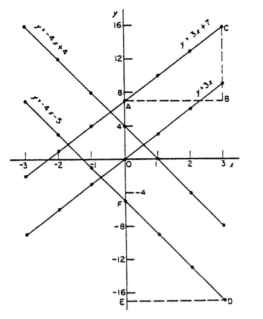

Figure 4.5 Graphs of $y = 3x$, $y = 3x + 7$, $y = -4x + 4$, $y = -4x - 5$

$y = -4x + 4$ and $y = -4x - 5$ are parallel to each other and thus have the same gradient. The gradient of DF is given by

$$DF = \frac{EF}{ED} = \frac{-5 - (-17)}{0 - 3} = \frac{12}{-3} = -4$$

Hence the gradient of both $y = -4x + 4$ and $y = -4x - 5$ is -4.

The y-axis intercept means the value of y where the straight line cuts the y-axis.
From Fig. 4.5,

$$\begin{aligned} y &= 3x \text{ cuts the y-axis at } y = 0 \\ y &= 3x + 7 \text{ cuts the y-axis at } y = +7 \\ y &= -4x + 4 \text{ cuts the y-axis at } y = +4 \\ \text{and} \quad y &= -4x - 5 \text{ cuts the y-axis at } y = -5 \end{aligned}$$

Some general conclusions can be drawn from the graphs shown in Figs. 4.4 and 4.5. When an equation is of the form $y = mx + c$, where m and c are constants, then

(i)   a graph of y against x produces a straight line,
(ii) m represents the slope or gradient of the line, and
(iii) c represents the y-axis intercept.

Thus, given an equation such as $y = 3x + 7$, it may be deduced 'on sight' that its gradient is +3

and its y-axis intercept is +7, as shown in Fig. 4.5. Similarly, if $y = -4x - 5$, then the gradient is -4 and the y-axis intercept is -5, as shown in Fig. 4.5.

When plotting a graph of the form $y = mx + c$, only two coordinates need be determined. When the coordinates are plotted a straight line is drawn between the two points. Normally, three coordinates are determined, the third one acting as a check.

Problem 3.   The following equations represent straight lines. Determine, without plotting graphs, the gradient and y-axis intercept for each.

(a) $y = 3$ (b) $y = 2x$ (c) $y = 5x - 1$
(d) $2x + 3y = 3$

(a) $y = 3$ (which is of the form $y = 0x + 3$) represents a horizontal straight line intercepting the **y-axis at 3**. Since the line is horizontal **its gradient is zero**.
(b) $y = 2x$ is of the form $y = mx + c$, where c is zero. Hence **gradient = 2 and y-axis intercept = 0** (i.e. the origin).
(c) $y = 5x - 1$ is of the form $y = mx + c$. Hence **gradient = 5 and y-axis intercept = -1**.
(d) $2x + 3y = 3$ is not in the form $y = mx + c$ as it stands. Transposing to make y the subject gives

$$3y = 3 - 2x$$

i.e. $$y = \frac{3 - 2x}{3} = \frac{3}{3} - \frac{2x}{3}$$

i.e. $$y = -\frac{2x}{3} + 1$$

which is of the form $y = mx + c$.

Hence **gradient = $-\frac{2}{3}$ and y-axis intercept = +1.**

Problem 4.   Determine the gradient of the straight line graph passing through the coordinates (a) (-2, 5) and (3, 4)
(b) (-2, -3) and (-1, 3).

A straight line graph passing through coordinates (x_1, y_1) and (x_2, y_2) has a gradient given by

$$m = \frac{y_2 - y_1}{x_2 - x_1} \quad \text{(see Fig. 4.6)}$$

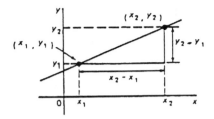

Figure 4.6

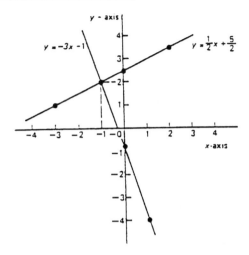

Figure 4.7

(a)   A straight line passes through $(-2, 5)$ and $(3, 4)$, hence $x_1 = -2$, $y_1 = 5$, $x_2 = 3$ and $y_2 = 4$. Hence

$$\textbf{gradient } \boldsymbol{m} = \frac{y_2 - y_1}{x_2 - x_1} = \frac{4 - 5}{3 - (-2)} = -\frac{1}{5}$$

(b)   A straight line passes through $(-2, -3)$ and $(-1, 3)$, hence $x_1 = -2$, $y_1 = -3$, $x_2 = -1$ and $y_2 = 3$. Hence

$$\text{gradient } m = \frac{y_2 - y_1}{x_2 - x_1} = \frac{3 - (-3)}{-1 - (-2)}$$

$$= \frac{3 + 3}{-1 + 2} = \frac{6}{1} = \textbf{6}$$

Problem 5.   Plot the graphs $3x + y + 1 = 0$ and $2y - 5 = x$ on the same axes and find their point of intersection.

Rearranging $3x + y + 1 = 0$ gives $y = -3x - 1$.
Rearranging $2y - 5$ gives $2y = x + 5$ and $y = \frac{1}{2}x + 2\frac{1}{2}$.
Since both equations are of the form $y = mx + c$ both are straight lines.
Knowing an equation is a straight line means that only two coordinates need be plotted and a straight line drawn through them. A third coordinate is usually determined to act as a check. A table of values is produced for each equation as shown below

x	1	0	−1
$-3x - 1$	−4	−1	2

x	2	0	−3
$\frac{1}{2}x + 2\frac{1}{2}$	$3\frac{1}{2}$	$2\frac{1}{2}$	1

The graphs are plotted as shown in Fig. 4.7. **The two straight lines are seen to intersect at $(-1, 2)$.**

4.3   Practical problems involving straight line graphs

When a set of coordinate values are given or are obtained experimentally and it is believed that they follow a law of the form $y = mx + c$, then if a straight line can be drawn reasonably close to most of the coordinate values when plotted, this verifies that a law of the form $y = mx + c$ exists. From the graph, constants m (i.e. gradient) and c (i.e. y-axis intercept) can be determined. This technique is called **determination of law**.

Problem 6.   The temperature in degrees Celsius and the corresponding values in degrees Fahrenheit are shown in the table below. Construct rectangular axes, choose a suitable scale and plot a graph of degrees Celsius (on the horizontal axis) against degrees Fahrenheit (on the vertical axis)

°C	10	20	40	60	80	100
°F	50	68	104	140	176	212

From the graph find (a) the temperature in degrees Fahrenheit at 55°C, (b) the temperature in degrees Celsius at 167°F, (c) the Fahrenheit temperature at 0°C, and (d) the Celsius temperature at 230°F.

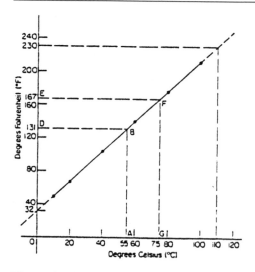

Figure 4.8 Graph of degrees Celsius against degrees Fahrenheit

The coordinates (10, 50), (20, 68), (40, 104), and so on are plotted as shown in Fig. 4.8. When the coordinates are joined, a straight line is produced. Since a straight line results there is a linear relationship between degrees Celsius and degrees Fahrenheit.

(a) To find the Fahrenheit temperature at 55°C a vertical line AB is constructed from the horizontal axis to meet the straight line at B. The point where the horizontal line BD meets the vertical axis indicates the equivalent Fahrenheit temperature.
  Hence 55°C is equivalent to 131°F.
  This process of finding an equivalent value in between the given information in the above table is called **interpolation**.

(b) To find the Celsius temperature at 167°F, a horizontal line EF is constructed as shown in Fig. 4.8. The point where the vertical line FG cuts the horizontal axis indicates the equivalent Celsius temperature.
  Hence 167°F is equivalent to 75°C.

(c) If the graph is assumed to be linear even outside of the given data, then the graph may be extended at both ends (shown by broken lines in Fig. 4.8).
  From Fig. 4.8, 0°C corresponds to 32°F.

(d) 230°F is seen to correspond to 110°C.
  The process of finding equivalent values outside of the given range is called **extrapolation**.

Problem 7.   In an experiment on Charles' law, the value of the volume of gas, V m³, was measured for various temperatures T°C. Results are shown below.

V (m³)	25.0	25.8	26.6	27.4	28.2	29.0
T°C	60	65	70	75	80	85

Plot a graph of volume (vertical) against temperature (horizontal) and from it find (a) the temperature when the volume is 28.6 m³ and (b) the volume when the temperature is 67°C.

If a graph is plotted with both the scales starting at zero then the result is as shown in Fig. 4.9. All of the points lie in the top right-hand corner of the graph, making interpolation difficult. A more accurate graph is obtained if the temperature axis starts at 55°C and the volume axis starts at 24.5 m³. The axes corresponding to these values is shown by the broken lines in Fig. 4.9 and are called **false axes**, since the origin is not now at zero. A magnified version of this relevant part of the graph is shown in Fig. 4.10. From the graph:

(a) when the volume is 28.6 m³, the equivalent temperature is **82.5°C**; and

(b) when the temperature is 67°C, the equivalent volume is **26.1 m³**.

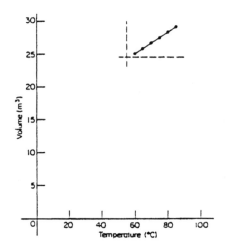

Figure 4.9 Graph of volume against temperature with a zero origin

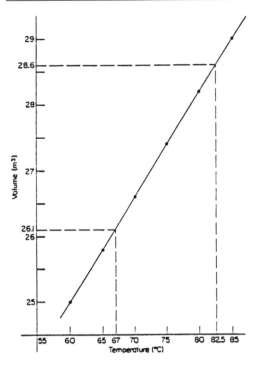

Figure 4.10 Graph of volume against temperature with a non-zero origin

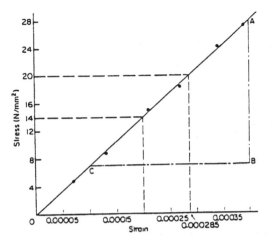

Figure 4.11 Graph of stress against strain for aluminium

The coordinates (0.000 07, 4.9), (0.000 13, 8.7), and so on, are plotted as shown in Fig. 4.11. The graph produced is the best straight line which can be drawn corresponding to these points. (With experimental results it is unlikely that all the points will lie exactly on a straight line.) The graph, and each of its axes, are labelled. Since the straight line passes through the origin, then stress is directly proportional to strain for the given range of values.

(a)   The gradient of the straight line is given by,

$$AC = \frac{AB}{BC} = \frac{28 - 7}{0.000\,40 - 0.000\,10} = \frac{21}{0.000\,30}$$

$$= \frac{21}{3 \times 10^{-4}} = \frac{7}{10^{-4}} \quad = 7 \times 10^{4}$$

$$= 70\,000 \text{ N/mm}^2$$

Thus Young's Modulus of Elasticity for aluminium is 70 000 N/mm².
Since $1\,m^2 = 10^6\,mm^2$, 70 000 N/mm² is equivalent to $70\,000 \times 10^6$ N/m², i.e. **70 × 10⁹ N/m² (or Pascals).**

From Fig. 4.11:

(b)   the value of the strain at a stress of 20 N/mm² is **0.000 285**, and

(b)   the value of the stress when the strain is 0.000 20 is **14 N/mm².**

Problem 8.   In an experiment demonstrating Hooke's law, the strain in an aluminium wire was measured for various stresses. The results were:

Stress (N/mm²)	4.9	8.7	15.0
Strain	0.000 07	0.000 13	0.000 21

Stress (N/mm²)	18.4	24.2	27.3
Strain	0.000 27	0.000 34	0.000 39

Plot a graph of stress (vertically) against strain horizontally.

Find

(a)   Young's Modulus of Elasticity for aluminium, which is given by the gradient of the graph,
(b)   the value of the strain at a stress of 20 N/mm², and
(c)   the value of the stress when the strain is 0.000 20.

Problem 9. The following values of resistance R ohms and corresponding voltage V volts are obtained from a test on a filament lamp.

R (ohms)	30	48.5	73	107	128
V (volts)	16	29	52	76	94

Choose suitable scales and plot a graph with R representing the vertical axis and V the horizontal axis. Determine (a) the slope of the graph, (b) the R-axis intercept value, (c) the equation of the graph, (d) the value of resistance when the voltage is 60 V, and (e) the value of the voltage when the resistance is 40 ohms. (f) If the graph were to continue in the same manner, what value of resistance would be obtained at 110 V?

The coordinates (16, 30), (29, 48.5), and so on, are shown plotted in Fig. 4.12 where the best straight line is drawn through the points.

(a) The slope or gradient of the straight line is given by

$$AC = \frac{AB}{BC} = \frac{135 - 10}{100 - 0} = \frac{125}{100} = \textbf{1.25}$$

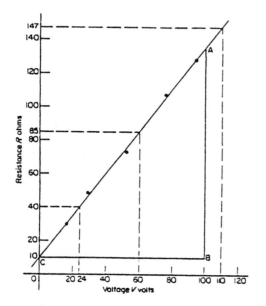

Figure 4.12

(Note that the vertical line AB and the horizontal line BC may be constructed anywhere along the length of the straight line. However, calculations are made easier if the horizontal length of the line BC is carefully chosen, in this case 100.)

(b) The R-axis intercept is at $R = \textbf{10 ohms}$ (by extrapolation).

(c) The equation of a straight line is $y = mx + c$, when y is plotted on the vertical axis and x on the horizontal axis, m represents the gradient and c the y-axis intercept. In this case, R corresponds to y, V corresponds to x, $m = 1.25$ and $c = 10$. Hence the equation of the graph is $R = \textbf{(1.25 } V \textbf{ + 10) } \Omega$.

From Fig. 4.12:

(d) when the voltage is 60 V, the resistance is **85 Ω**,

(e) when the resistance is 40 ohms, the voltage is **24 V**, and

(f) by extrapolation, when the voltage is 110 V, the resistance is **147 Ω**.

Problem 10. Experimental tests to determine the breaking stress σ of rolled copper at various temperatures t gave the following results:

Stress σ (N/cm^2)	8.42	8.02	7.75	7.35	7.06	6.63
Temperature t (°C)	70	200	280	410	500	640

Show that the values obey the law $\sigma = at + b$, where a and b are constants and determine approximate values for a and b. Use the law to determine the stress at 250°C and the temperature when the stress is 7.54 N/cm^2.

The coordinates (70, 8.46), (200, 8.04), and so on, are plotted as shown in Fig. 4.13. Since the graph is a straight line the values obey the law $\sigma = at + b$.

Gradient of the straight line is given by

$$a = \frac{AB}{BC} = \frac{8.36 - 6.76}{100 - 600} = \frac{1.60}{-500}$$

$$= -0.0032$$

Vertical axis intercept, $b = 8.68$.
Hence the law of the graph is $\sigma = \textbf{−0.0032}t + \textbf{8.68}$.

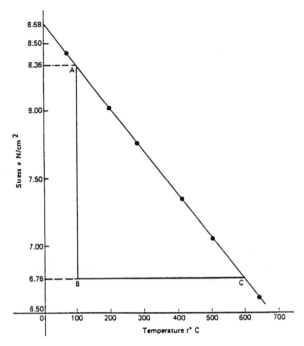

Figure 4.13 Graph of stress against temperature

When the temperature is 250°C, stress $\sigma = -0.0032(250) + 8.68 = \textbf{7.88 N/cm}^2$.
Rearranging $\sigma = -0.0032t + 8.68$ gives

$$0.0032t = 8.68 - \sigma \text{ i.e. } t = \frac{8.68 - \sigma}{0.0032}$$

Hence when the stress $\sigma = 7.54 \text{ N/cm}^2$,

$$\textbf{temperature } t = \frac{8.68 - 7.54}{0.0032}$$

$$= \textbf{356.3°C}$$

4.4 Further questions on straight line graphs

(Answers may be found on page 354.)

The straight line graph

1. Corresponding values obtained experimentally for two quantities are:

x	−2.0	−0.5	0	1.0	2.5	3.0	5.0
y	−13.0	−5.5	−3.0	2.0	9.5	12.0	22.0

Use a horizontal scale for x of 1 cm $= \frac{1}{2}$ unit and a vertical scale for y of 1 cm $= 2$ units and draw a graph of x against y. Label the graph and each of its axes. By interpolation, find from the graph the value of y when x is 3.5.

2. The equation of a line is $4y = 2x + 5$. A table of corresponding values is produced and is shown below. Complete the table and plot a graph of y against x. Find the gradient of the graph.

x	−4	−3	−2	−1	0	1	2	3	4
y		−0.50			1.25				3.25

3. Determine the gradient and intercept on the y-axis for each of the following equations:
   (a) $y = 4x - 2$ (b) $y = -x$ (c) $y = -3x - 4$
   (d) $y = 4$

4. Determine the gradient and y-axis intercept for each of the following equations and sketch the graphs:
   (a) $y = 6x - 3$ (b) $y = -2x + 4$ (c) $y = 3x$
   (d) $y = 7$

5. Determine the gradient of the straight line graphs passing through the coordinates (a) (2, 7) and (−3, 4), (b) (−4, −1) and (−5, 3), (c) $(\frac{1}{4}, -\frac{3}{4})$ and $(-\frac{1}{2}, \frac{5}{8})$.

6. Draw a graph of $y - 3x + 5 = 0$ over a range of $x = -3$ to $x = 4$. Hence determine (a) the value of y when $x = 1.3$ and (b) the value of x when $y = -9.2$.

7. Draw on the same axes the graphs of $y = 3x - 5$ and $3y + 2x = 7$. Find the coordinates of the points of intersection. Check the result obtained by solving the two simultaneous equations algebraically.

Practical problems involving straight line graphs

8. The resistance R ohms of a copper winding is measured at various temperatures $t°C$ and the results are as follows:

R (ohms)	112	120	126	131	136
t (°C)	20	36	48	58	64

Plot a graph of R (vertically) against t (horizontally) and find from it (a) the temperature when the resistance is 122 Ω and (b) the resistance when the temperature is 52°C.

9. The following table gives the force F newtons which, when applied to a lifting machine overcomes a corresponding load of L newtons.

Force F (newtons)	25	47	64	120	149	187
Load L (newtons)	50	140	210	430	550	700

Choose suitable scales and plot a graph of F (vertically) against L (horizontally). Draw the best straight line through the points. Determine from the graph (a) the gradient, (b) the F-axis intercept, (c) the equation of the graph, (d) the force applied when the load is 310 N, and (e) the load that a force of 160 N will overcome. (f) If the graph were to continue in the same manner, what value of force will be needed to overcome a 800 N load?

10. The following table gives the results of tests carried out to determine the breaking stress σ of rolled copper at various temperatures t:

Stress σ (N/cm²)	8.51	8.07	7.80	7.47	7.23	6.78
Temperature t (°C)	75	220	310	420	500	650

Plot a graph of stress (vertically) against temperature (horizontally). Draw the best straight line through the plotted coordinates. Determine the slope of the graph and the vertical axis intercept.

11. The velocity v of a body after varying time intervals t was measured as follows:

t (seconds)	2	5	8	11	15	18
v (m/s)	16.9	19.0	21.1	23.2	26.0	28.1

Plot v vertically and t horizontally and draw a graph of velocity against time. Determine from the graph (a) the velocity after 10 s, (b) the time at 20 m/s and (c) the equation of the graph.

12. The mass m of a steel joist varies with length l as follows:

mass m (kg)	80	100	120	140	160
length l (m)	3.00	3.74	4.48	5.23	5.97

Plot a graph of mass (vertically) against length (horizontally). Determine the equation of the graph.

13. In an experiment demonstrating Hooke's law, the strain in a copper wire was measured for various stresses. The results were:

Stress (pascals)	10.6×10^6	18.2×10^6	24.0×10^6	30.7×10^6	39.4×10^6
Strain	0.00011	0.00019	0.00025	0.00032	0.00041

Plot a graph of stress (vertically) against strain (horizontally). Determine (a) Young's Modulus of Elasticity for copper, which is given by the gradient of the graph (b) the value of strain at a stress of 21×10^6 Pa, (c) the value of stress when the strain is 0.000 30.

14. An experiment with a set of pulley blocks gave the following results:

Effort E (newtons)	9.0	11.0	13.6	17.4	20.8	23.6
Load L (newtons)	15	25	38	57	74	88

Plot a graph of effort (vertically) against load (horizontally) and determine (a) the gradient, (b) the vertical axis intercept, (c) the law of the graph, (d) the effort when the load is 30 N and (e) the load when the effort is 19 N.

15. The variation of pressure p in a vessel with temperature T is believed to follow a law of the form $p = aT + b$, where a and b are constants. Verify this law for the results given below and determine the approximate values of a and b. Hence determine the pressures at temperatures of 285 K and 310 K and the temperature at a pressure of 250 kPa.

Pressure p (kPa)	244	247	252	258	262	267
Temperature T (K)	273	277	282	289	294	300

5

Trigonometry

At the end of this chapter you should be able to:

- define the sine, cosine and tangent of an angle
- evaluate trigonometric ratios for angles of any magnitude
- plot graphs of sine, cosine and tangent over one cycle
- state and use Pythagoras' theorem
- solve right-angled triangles
- use the sine and cosine rules to solve any triangle
- calculate the area of any triangle
- appreciate typical practical science and engineering applications where trigonometry is needed

5.1 Introduction

Trigonometry is the branch of mathematics which deals with the measurement of sides and angles of triangles, and their relationships with each other.

Practical engineering examples where trigonometry is used in this book include the following:

(a) With resolution of forces by calculation and with relative velocities, trigonometric ratios and the theorem of Pythagoras is required (Chapters 8 and 13).
(b) Determining the resultant of coplanar forces by calculation, and with relative velocities requires use of the sine and cosine rules (Chapters 8 and 13).
(c) Finding the area under a speed/time graph gives distance travelled and needs knowledge of the area of triangles (Chapter 9).

5.2 Trigonometric ratios of acute angles

With reference to the right-angled triangle shown in Fig. 5.1:

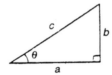

Figure 5.1

(i) sine $\theta = \dfrac{\text{opposite side}}{\text{hypotenuse}}$, i.e. $\sin \theta = \dfrac{b}{c}$

(ii) cosine $\theta = \dfrac{\text{adjacent side}}{\text{hypotenuse}}$, i.e. $\cos \theta = \dfrac{a}{c}$

(iii) tangent $\theta = \dfrac{\text{opposite side}}{\text{adjacent side}}$, i.e. $\tan \theta = \dfrac{b}{a}$

Problem 1. Determine the values of the three trigonometric ratios for angle θ shown in the right-angled triangle ABC of Fig. 5.2.

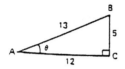

Figure 5.2

By definition:

$$\sin \theta = \frac{\text{opposite site}}{\text{hypotenuse}} = \frac{5}{13} = 0.3846$$

$$\cos \theta = \frac{\text{adjacent side}}{\text{hypotenuse}} = \frac{12}{13} = 0.9231$$

$$\tan \theta = \frac{\text{opposite side}}{\text{adjacent side}} = \frac{5}{12} = 0.4167$$

Problem 2.   From Fig. 5.3, find sin *D*, cos *D* and tan *F*.

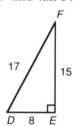

Figure 5.3

$$\sin D = \frac{EF}{DF} = \frac{15}{17} \text{ or } 0.8824$$

$$\cos D = \frac{DE}{DF} = \frac{8}{17} \text{ or } 0.4706$$

$$\tan F = \frac{DE}{EF} = \frac{8}{15} \text{ or } 0.5333$$

5.3   Evaluating trigonometric ratios

The easiest method of evaluating trigonometric functions of any angle is by using a calculator. The following values, correct to 4 decimal places, may be checked:

sine 18°   =0.3090   cosine 56°   = 0.5592
tangent 29° =0.5543   sine 172°   = 0.1392
cosine 115° =−0.4226   tangent 178°   = −0.0349
sine 241.63°=−0.8799   cosine 331.78° = 0.8811
tangent 296.42° = −2.0127

To evaluate, say, sine 42°23′ using a calculator means finding sine $42\frac{23}{60}°$ since there are 60 minutes in 1 degree:

$$\frac{23}{60} = 0.3833, \text{ thus } 42°23' \equiv 42.3833°$$

Thus sine 42°23′ = sine 42.3833° = 0.6741, correct to 4 decimal places.
Similarly, cosine 72°38′ = cosine $72\frac{38}{60}°$ = 0.2985, correct to 4 decimal places.

Problem 3.   Evaluate, correct to 4 decimal places:

(a) sine 11°   (b) sine 121.68°
(c) sine 259°10′

(a) sine 11° = **0.1908**
(b) sine 121.68° = **0.8510**

(c) sine 259°10′ = sine $\left(259\frac{10}{60}\right)°$ = **−0.9822**

Problem 4.   Evaluate, correct to 4 decimal places:

(a) cosine 23°   (b) cosine 159.32°
(c) 321°41′

(a) cosine 23° = **0.9205**
(b) cosine 159.32° = **−0.9356**

(c) cosine 321°41′ = cosine $\left(321\frac{41}{60}\right)°$ = **0.7846**

Problem 5.   Evaluate, correct to 4 significant figures:

(a) tangent 276°   (b) tangent 131.29°
(c) tangent 76°58′

(a) tan 276° = **−9.514**
(b) tan 131.29° = **−1.139**

(c) tan 76°58′ = tan $\left(76\frac{58}{60}\right)°$ = **4.320**

Problem 6. Evaluate, correct to 4 significant figures:

(a) sin 2.162   (b) cos(3π/8)   (c) tan 1.16

(a) sin 2.162 means the sine of 2.162 radians. Hence a calculator needs to be on the radian function. Hence sin 2.162 = **0.8303**
(b) cos (3π/8) = cos 1.178 097 ... = **0.3827**
(c) tan 1.16 = **2.296**

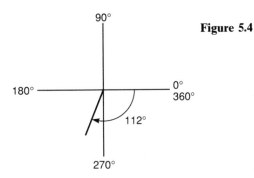

Figure 5.4

Problem 7. Determine the acute angle:

(a) arcsin 0.7321   (b) arccos 0.4174
(c) arctan 1.4695

(a) Note that 'arcsin θ' is an abbreviation for 'the angle whose sine is equal to θ'. 0.7321 is entered into a calculator and then the inverse sine (or sin⁻¹) key is pressed.
Hence arcsin 0.7321 = 47.062 73 ... °
Subtracting 47 leaves 0.062 73 ... °
Multiplying by 60 gives 4' to the nearest minute
Hence arcsin 0.7321 = **47.06°** or **47°4'**.
Alternatively, in radians, arcsin 0.7321 = **0.821 radians**
(b) arccos 0.4174 = **65.33°** or **65°20'** or **1.140 radians**
(c) arctan 1.4695 = **55.76°** or **55°46'** or **0.973 radians**

Problem 8. Evaluate, correct to 4 decimal places:

(a) sine (–112°)   (b) tangent (–217.29°)

(a) By convention, positive angles are shown anticlockwise and negative angles are shown clockwise. From Fig. 5.4, –112° is actually the same as +248° (i.e. 360° – 112°). Hence by calculator, sine (–112°) = sine 248° = **–0.9272**
(b) Tangent (–217.29°) = **–0.7615** (which is the same as tan (360° – 217.29°), i.e. tan 142.71°)

5.4 Graphs of trigonometric functions

By drawing up tables of values from 0° to 360°, graphs of $y = \sin A$, $y = \cos A$ and $y = \tan A$ may be plotted. Values obtained with a calculator (correct to 3 decimal places – which is more than sufficient for plotting graphs), using 30° intervals, are shown below, with the respective graphs shown in Fig. 5.5.

(a) $y = \sin A$

A	0	30°	60°	90°	120°
sin A	0	0.50	0.866	1.00	0.866

A	150°	180°	210°	240°	270°
sin A	0.50	0	–0.50	–0.866	–1.00

A	300°	330°	360°
sin A	–0.866	–0.50	0

(b) $y = \cos A$

A	0	30°	60°	90°	120°
cos A	1.00	0.866	0.50	0	–0.50

A	150°	180°	210°	240°
cos A	–0.866	–1.00	–0.866	–0.50

A	270°	300°	330°	360°
cos A	0	0.50	0.866	1.00

(c) $y = \tan A$

A	0	30°	60°	90°	120°
tan A	0	0.577	1.732	∞	–1.732

A	150°	180°	210°	240°
tan A	–0.577	0	0.577	1.732

A	270°	300°	330°	360°
tan A	∞	–1.732	–0.577	0

(a)

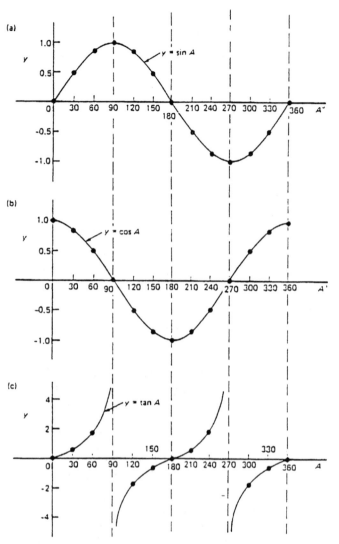

Figure 5.5

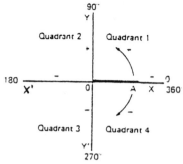

Figure 5.6

5.5   Angles of any magnitude

(i) Figure 5.6 shows rectangular axes XX' and YY' intersecting at origin 0. As with graphical work, measurements made to the right and above 0 are positive while those to the left and downwards are negative. Let OA be free to rotate about 0. By convention, when OA moves anticlockwise angular measurement is considered positive, and vice versa.

(ii) Let OA be rotated anticlockwise so that θ_1 is any angle in the first quadrant and let perpendicular AB be constructed to form the right-angled triangle OAB (see Fig. 5.7). Since all three sides of the triangle are positive, all three trigonometric ratios are positive in the first quadrant. (Note: OA is always positive since it is the radius of a circle.)

(iii) Let OA be further rotated so that θ_2 is any angle in the second quadrant and let AC be

From Fig. 5.5 it is seen that:

(i) Sine and cosine graphs oscillate between peak values of ±1.

(ii) The cosine curve is the same shape as the sine curve but displaced by 90°.

(iii) The sine and cosine curves are continuous and they repeat at intervals of 360°; the tangent curve appears to be discontinuous and repeats at intervals of 180°.

The sine curve in particular has many practical applications in science and engineering, not the least being with alternating currents and voltages.

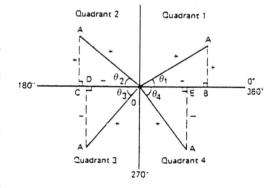

Figure 5.7

constructed to form the right-angled triangle OAC. Then:

$$\sin \theta_2 = \frac{+}{+} = +, \cos \theta_2 = \frac{-}{+} = -,$$

$$\tan \theta_2 = \frac{+}{-} = -$$

(iv) Let OA be further rotated so that θ_3 is any angle in the third quadrant and let AD be constructed to form the right-angled triangle OAD. Then:

$$\sin \theta_3 = \frac{-}{+} = -, \cos \theta_3 = \frac{-}{+} = -,$$

$$\tan \theta_3 = \frac{-}{-} = +$$

(v) Let OA be further rotated so that θ_4 is any angle in the fourth quadrant and let AE be constructed to form the right-angled triangle OAE. Then:

$$\sin \theta_4 = \frac{-}{+} = -, \cos \theta_4 = \frac{+}{+} = +,$$

$$\tan \theta_4 = \frac{-}{+} = -$$

(vi) The results obtained in (ii) to (v) are summarized in Fig. 5.8. The letters underlined spell the word **CAST** when starting in the fourth quadrant and moving in an anticlockwise direction.

(vii) In the first quadrant of Fig. 5.5 all the curves have positive values; in the second only sine is positive; in the third only tangent is positive; in the fourth only cosine is positive (exactly as summarized in Fig. 5.8)

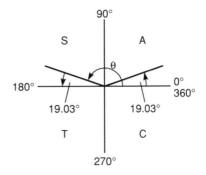

Figure 5.9

A knowledge of angles of any magnitude is needed when finding, for example, all the angles between 0° and 360° whose sine is, say, 0.3261. If 0.3261 is entered into a calculator and then the inverse sine key pressed (or sin⁻¹ key) the answer 19.03° appears. However, there is a second angle between 0° and 360° which the calculator does not give. Sine is also positive in the second quadrant (either from CAST or from Fig. 5.5(a)). The other angle is shown in Fig. 5.9 as angle θ, where $\theta = 180° - 19.03° = 160.97°$. Thus 19.03° **and** 160.97° are the angles between 0° and 360° whose sine is 0.3261 (check that sin 160.97° = 0.3261 on your calculator).

Be careful! Your calculator only gives you one of these answers. The second answer needs to be deduced from a knowledge of angles of any magnitude, as shown in the following problem.

Problem 9. Determine all the angles between 0° and 360° (a) whose sine is −0.4638 and (b) whose tangent is 1.7629.

(a) The angles whose sine is −0.4638 occur in the third and fourth quadrants since sine is negative in these quadrants (see Fig. 5.10(a)).
From Fig. 5.10(b), θ = arcsin 0.4638 = 27° 38′.
Measured from 0°, the two angles between 0° and 360° whose sine is −0.4638 are 180° + 27° 38′, i.e. **207°38′** and 360° − 27°38′, i.e. **332°22′**. (Note that a calculator generally only gives one answer, i.e. −27.632 588°.)

(b) A tangent is positive in the first and third quadrants (see Fig. 5.10(c)).
From Fig. 5.10(d), θ = arctan 1.7629 = 60° 26′.

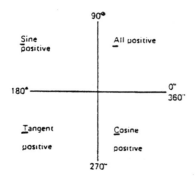

Figure 5.8

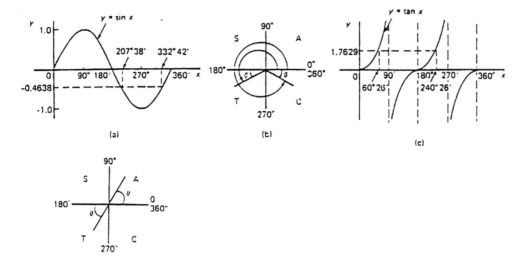

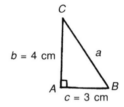

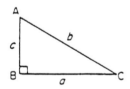

Figure 5.10   (d)

Measured from 0°, the two angles between 0° and 360° whose tangent is 1.7629 are **60°26′** and 180° + 60°26′, i.e. **240°26′**.

5.6   The theorem of Pythagoras

With reference to Fig. 5.11, the side opposite the right angle (side b) is called the **hypotenuse**. The **theorem of Pythagoras** states: '*In any right-angled triangle, the square on the hypotenuse is equal to the sum of the squares on the other two sides*'. Hence $b^2 = a^2 + c^2$

Figure 5.11

Problem 10.   In Fig. 5.12, find the length of BC.

By Pythagoras' theorem:
$$a^2 = b^2 + c^2$$
i.e. $a^2 = 4^2 + 3^2 = 16 + 9 = 25$

Figure 5.12

Hence $a = \sqrt{25} = \pm 5$ (−5 has no meaning in this context and is thus ignored).
Thus **BC = 5 cm**.

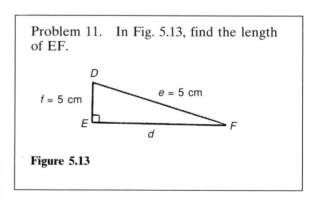

Problem 11.   In Fig. 5.13, find the length of EF.

Figure 5.13

By Pythagoras' theorem:
$$e^2 = d^2 + f^2$$
Hence

$$13^2 = d^2 + 5^2$$

$$169 = d^2 + 25$$

$$d^2 = 169 - 25 = 144$$

Thus

$$d = \sqrt{144} = 12 \text{ cm}$$

Thus **EF = 12 cm**.

5.7 Solution of right-angled triangles

To 'solve a right-angled triangle' means 'to find the unknown sides and angles'. This is achieved by using (i) trigonometric ratios and/or (ii) the theorem of Pythagoras.

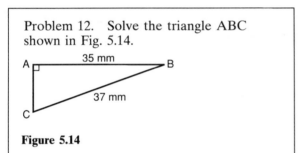

Problem 12.   Solve the triangle ABC shown in Fig. 5.14.

Figure 5.14

To 'solve triangle ABC' means 'to find the length AC and angles B and C'.

$$\sin C = \frac{35}{37} = 0.9459$$

Hence

$$C = \arcsin 0.9459 = \mathbf{71°4'}$$

$B = 180° - 90° - 71°4' = 18°56'$ (since angles in a triangle add up to 180°).

$$\sin B = \frac{AC}{37}$$

hence

$$AC = 37 \sin 18°56' = 37(0.3245) = \mathbf{12.0 \text{ mm}}$$

(Check: Using Pythagoras' theorem $37^2 = 35^2 + 12^2$)

5.8 The sine and cosine rules and areas of triangles

If a triangle is right-angled, trigonometric ratios and the theorem of Pythagoras may be used for

its solution, as shown in section 5.7. However, for a non-right-angled triangle, trigonometric ratios and Pythagoras' theorem **cannot** be used. Instead, two rules, called the **sine rule** and the **cosine rule**, are used.

Sine rule

With reference to triangle ABC of Fig. 5.15, the sine rule states:

$$\frac{a}{\sin A} = \frac{b}{\sin B} = \frac{c}{\sin C}$$

The rule may be used only when:

(i)   1 side and any 2 angles are initially given, or

(ii) 2 sides and an angle (not the included angle) are initially given.

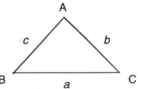

Figure 5.15

Cosine rule

With reference to triangle ABC of Fig. 5.15, the cosine rule states:

$$a^2 = b^2 + c^2 - 2bc \cos A$$
$$\text{or} \quad b^2 = a^2 + c^2 - 2ac \cos B$$
$$\text{or} \quad c^2 = a^2 + b^2 - 2ab \cos C$$

The rule may be used only when:

(i)   2 sides and the included angle are initially given, or

(ii) 3 sides are initially given.

Areas of triangles

The **area of any triangle** such as ABC of Fig. 5.15 is given by:

(i) $\dfrac{1}{2} \times$ base $\times$ perpendicular height

or (ii) $\dfrac{1}{2} ab \sin C$ or $\dfrac{1}{2} ac \sin B$ or $\dfrac{1}{2} bc \sin A$

or (iii) $\sqrt{[s(s-a)(s-b)(s-c)]}$ where

$$s = \dfrac{a+b+c}{2}$$

Problem 13. In the triangle XYZ, X = 51°, Y = 67° and YZ = 15.2 cm. Solve the triangle and find its area.

Figure 5.16

The triangle XYZ is shown in Fig. 5.16. Since the angles in a triangle add up to 180°, then

$$Z = 180° - 51° - 67° = 62°$$

Applying the **sine rule**:

$$\dfrac{15.2}{\sin 15°} = \dfrac{y}{\sin 67°} = \dfrac{z}{\sin 62°}$$

Using

$$\dfrac{15.2}{\sin 51°} = \dfrac{y}{\sin 67°}$$

and transposing gives:

$$y = \dfrac{15.2 \sin 67°}{\sin 51°}$$

$$= 18.00 \text{ cm} = XZ$$

Using

$$\dfrac{15.2}{\sin 51°} = \dfrac{z}{\sin 62°}$$

and transposing gives:

$$z = \dfrac{15.2 \sin 62°}{\sin 51°}$$

$$= 17.27 \text{ cm} = XY$$

Area of triangle XYZ

$$= \dfrac{1}{2} xy \sin Z = \dfrac{1}{2}(15.2)(18.00) \sin 62°$$

$$= 120.8 \text{ cm}^2$$

(or area $= \frac{1}{2} xz \sin Y = \frac{1}{2}(15.2)(17.27) \sin 67° = 120.8 \text{ cm}^2$).

It is always worth checking with triangle problems that the longest side is opposite the largest angle, and vice versa. In this problem, Y is the largest angle and thus XZ should be the longest of the three sides.

Problem 14. Solve triangle DEF and find its area given that EF = 35.0 mm, DE = 25.0 mm and E = 64°.

Triangle DEF is shown in Fig. 5.17. Applying the cosine rule:

$$e^2 = d^2 + f^2 - 2df \cos E$$

i.e.

$$e^2 = (35.0)^2 + (25.0)^2 - \{2(35.0)(25.0) \cos 64°\}$$

$$= 1225 + 625 - 767.1 = 1083$$

$$e = \sqrt{1083} = \textbf{32.91 mm}$$

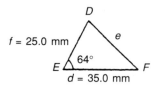

Figure 5.17

Applying the sine rule:

$$\dfrac{32.91}{\sin 64°} = \dfrac{25.0}{\sin F}$$

from which,

$$\sin F = \dfrac{25.0 \sin 64°}{32.91} = 0.6828$$

Thus

$$F = \arcsin 0.6828 = 43°4' \text{ or } 136°56'$$

$F = 136°56'$ is not possible in this case since 136°56' + 64° is greater than 180°. Thus only $\textbf{F = 43°4'}$ is valid.

$D = 180° - 64° - 43°4' = 72°56'$

Area of triangle DEF

$$= \frac{1}{2}df \sin E = \frac{1}{2}(35.0)(25.0) \sin 64°$$

$$= \textbf{393.2 mm}^2$$

Problem 15.   Solve triangle XYZ (Fig. 5.18) and find its area given that $Y = 128°$, $XY = 7.2$ cm and $YZ = 4.5$ cm.

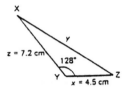

Figure 5.18

Applying the cosine rule:

$$y^2 = x^2 + z^2 - 2xz \cos Y$$
$$= (4.5)^2 + (7.2)^2 - \{2(4.5)(7.2) \cos 128°\}$$
$$= 20.25 + 51.84 - (-39.89)$$
$$= 20.25 + 51.84 + 39.89 = 112.0$$

$y = \sqrt{(112.0)} = \textbf{10.58 cm}$

Applying the sine rule:

$$\frac{10.58}{\sin 128°} = \frac{7.2}{\sin Z}$$

from which,

$$\sin Z = \frac{7.2 \sin 128°}{10.58} = 0.5363$$

Hence $Z = \arcsin 0.5363 = 32°26'$ (or $147°34'$ which, here, is impossible).

$$X = 180° - 128° - 32°26' = \textbf{19°34'}$$

$$\text{Area} = \frac{1}{2}xz \sin Y = \frac{1}{2}(4.5)(7.2) \sin 128°$$

$$= \textbf{12.77 cm}^2$$

Problem 16.   A room 8.0 m wide has a span roof which slopes at 33° on one side and 40° on the other. Find the length of the roof slopes, correct to the nearest centimetre.

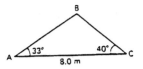

Figure 5.19

A section of the roof is shown in Fig. 5.19. Angle at ridge, $B = 180° - 33° - 40° = 107°$. From the sine rule:

$$\frac{8.0}{\sin 107°} = \frac{a}{\sin 33°}$$

from which,

$$a = \frac{8.0 \sin 33°}{\sin 107°} = 4.556 \text{ m}$$

Also from the sine rule:

$$\frac{8.0}{\sin 107°} = \frac{c}{\sin 40°}$$

from which,

$$c = \frac{8.0 \sin 40°}{\sin 107°} = 5.377 \text{ m}$$

Hence **the roof slopes are 4.56 m and 5.38 m**, correct to the nearest centimetre.

Problem 17.   Two voltage phasors are shown in Fig. 5.20. If $V_1 = 40$ V and $V_2 = 100$ V determine the value of their resultant (i.e. length OA) and the angle the resultant makes with V_1.

Angle OBA $= 180° - 45° = 135°$. Applying the cosine rule:

$$\text{OA}^2 = V_1^2 + V_2^2 - 2V_1V_2 \cos \text{OBA}$$
$$= 40^2 + 100^2 - \{2(40)(100) \cos 135°\}$$
$$= 1600 + 10\,000 - (-5657)$$

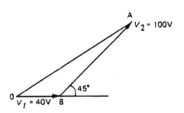

Figure 5.20

$$= 1600 + 10\ 000 + 5657 = 17\ 257$$

The resultant OA

$$= \sqrt{(17257)} = 131.4 \text{ V}$$

Applying the sine rule:

$$\frac{131.4}{\sin 135°} = \frac{100}{\sin \text{AOB}}$$

from which,

$$\sin \text{AOB} = \frac{100 \sin 135°}{131.4} = 0.5381$$

Hence angle AOB = arcsin 0.5381 = 32°33′ (or 147°27′ which is impossible in this case). Hence **the resultant voltage is 131.4 volts at 32°33′ to V_1.**

Problem 18.   In Fig. 5.21, PR represents the inclined jib of a crane and is 10.0 m long. PQ is 4.0 m long. Determine the length of tie QR and the inclination of the jib to the vertical.

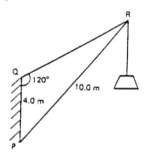

Figure 5.21

Applying the sine rule:

$$\frac{\text{PR}}{\sin 120°} = \frac{\text{PQ}}{\sin R}$$

from which,

$$\sin R = \frac{\text{PQ} \sin 120°}{\text{PR}} = \frac{(4.0) \sin 120°}{10.0}$$

$$= 0.3464$$

Hence R = arcsin 0.3464 = 20°16′ (or 159°44′, which is impossible in this case).
$P = 180° - 120° - 20°16′ = $ **39°44′, which is the inclination of the jib to the vertical**.
Applying the sine rule:

$$\frac{10.0}{\sin 120°} = \frac{\text{QR}}{\sin 39°44′}$$

from which,

$$\text{QR} = \frac{10.0 \sin 39°44′}{\sin 120°} = \textbf{7.38 m = length}$$
$$\textbf{of tie}$$

Problem 19.   The area of a field is in the form of a quadrilateral ABCD as shown in Fig. 5.22. Determine its area.

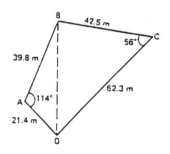

Figure 5.22

A diagonal drawn from B to D divides the quadrilateral into two triangles. Area of quadrilateral ABCD

= area of triangle ABD + area of triangle BCD

$$= \frac{1}{2}(39.8)(21.4) \sin 114° + \frac{1}{2}(42.5)(62.3) \sin 56°$$

$$= 389.04 + 1097.5$$

$$= \textbf{1487 m}^2$$

5.9   Further questions on trigonometry

(Answers may be found on page 354.)

1. In triangle ABC shown in Fig. 5.23, find sin A, cos A, tan A, sin B, cos B and tan B.
2. If tan $\theta = 7/24$ find the other two trigonometric ratios in fraction form.

In Problems 3–6, evaluate correct to 4 decimal places:

3. (a) sine 27° (b) sine 172.41°
   (c) sine 302° 52′
4. (a) cosine 124° (b) cosine 21.46°
   (c) cosine 284°10′
5. (a) tangent 145° (b) tangent 310.59°
   (c) tangent 49°16′

Figure 5.23

6. (a) sine $(2\pi/3)$ (b) cos 1.681   (c) tan 3.672
7. Determine the acute angle in degrees (correct to 2 decimal places), degrees and minutes, and in radians (correct to 3 decimal places):
   (a) arcsin 0.2341   (b) arccos 0.8271
   (c) arctan 0.8106
8. Find all the angles between $0°$ and $360°$ whose sine is -0.7321.
9. Solve for all values of θ between $0°$ and $360°$:
   (a) arccos $-0.5316 = \theta$
   (b) arctan $0.8314 = \theta$
10. Solve the triangles shown in Fig. 5.24.

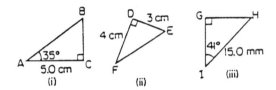

Figure 5.24

11. Use the sine rule to solve the following triangles ABC and find their areas:
    (a) $A = 29°$, $B = 68°$, $b = 27$ mm
    (b) $B = 71°26'$, $C = 56°32'$, $b = 8.60$ cm
    (c) $A = 117°$, $C = 24°30'$, $a = 15.2$ mm

12. Use the cosine and sine rules to solve the following triangles PQR, and find their areas:
    (a) $q = 12$ cm, $r = 16$ cm, $P = 54°$
    (b) $p = 56$ mm, $q = 38$ mm, $R = 64°$
    (c) $q = 3.25$ m, $r = 4.42$ m, $P = 105°$
13. Determine the length of members BF and EB in the roof truss shown in Fig. 5.25.

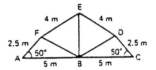

Figure 5.25

14. A laboratory 9.0 m wide has a span roof which slopes at $36°$ on one side and $44°$ on the other. Determine the lengths of the roof slopes.
15. Three forces acting on a fixed point are represented by the sides of a triangle of dimensions 7.2 cm, 9.6 cm and 11.0 cm. Determine the angles between the lines of action and the three forces.
16. A reciprocating engine mechanism is shown in Fig. 5.26. The crank AB is 12.0 cm long and the connecting rod BC is 32.0 cm long. For the position shown determine the length of AC and the angle between the crank and the connecting rod.

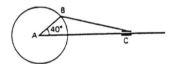

Figure 5.26

6

Areas of irregular shapes

At the end of this chapter you should be able to:

- use the trapezoidal rule to approximately determine an irregular area
- use the mid-ordinate rule to approximately determine an irregular area
- use Simpson's rule to approximately determine an irregular area
- appreciate where typical practical engineering examples of irregular areas occur

6.1 Introduction

Areas of irregular plane surfaces may be approximately determined by using (a) the trapezoidal rule, (b) the mid-ordinate rule, or (c) Simpson's rule. Such methods may be used, for example, by engineers estimating areas of indicator diagrams of steam engines, surveyors estimating areas of plots of land or naval architects estimating areas of water planes or transverse sections of ships.

Examples of the determination of irregular areas in this book are: area under a force/distance graph gives work done (Chapter 15) and the area under a current/time graph is needed to find the mean value of an alternating current, where mean value = area/length of base (Chapter 33).

6.2 The trapezoidal rule

To determine the area PQRS in Fig. 6.1:

(i) Divide base PS into any number of equal intervals, each of width d (the greater the number of intervals, the greater the accuracy).
(ii) Accurately measure ordinates y_1, y_2, y_3, etc.
(iii) Area PQRS

$$= d\left[\frac{y_1 + y_7}{2} + y_2 + y_3 + y_4 + y_5 + y_6\right]$$

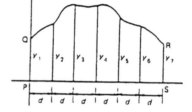

Figure 6.1

In general, the trapezoidal rule states:

> **Area = (width of interval)[½(first + last ordinate) + sum of remaining ordinates]**

6.3 The mid-ordinate rule

To determine the area ABCD of Fig. 6.2:

(i) Divide base AD into any number of equal intervals, each of width d (the greater the number of intervals, the greater the accuracy).
(ii) Erect ordinates in the middle of each interval (shown by broken lines in Fig. 6.2).
(iii) Accurately measure ordinates y_1, y_2, y_3, etc.

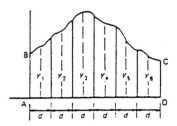

Figure 6.2

(iv)   Area ABCD = $d(y_1 + y_2 + y_3 + y_4 + y_5 + y_6 + y_7)$.

In general, the mid-ordinate rule states:

> **Area = (width of interval)(sum of mid-ordinates)**

6.4   Simpson's rule

To determine the area EFGH of Fig. 6.3:

(i)   Divide base EH into an even number of intervals, each of width d, (the greater the number of intervals, the greater the accuracy).
(ii)   Accurately measure ordinates y_1, y_2, y_3, etc.
(iii)   Area EFGH

$$= \frac{d}{3}[(y_1 + y_7) + 4(y_2 + y_4 + y_6) + 2(y_3 + y_5)]$$

In general, Simpson's rule states:

> **Area = ⅓ (width of interval [(first + last ordinate) + 4(sum of even ordinates) + 2(sum of remaining odd ordinates)]**

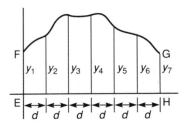

Figure 6.3

With each of the three rules, the greater the number of intervals chosen the more accurate is the estimation of area.

6.5   Worked problems to demonstrate the three rules

> **Problem 1.**   A car starts from rest and its speed is measured every second for 6 s.
>
> | Time t (s) | 0 | 1 | 2 | 3 | 4 | 5 | 6 |
> |---|---|---|---|---|---|---|---|
> | Speed v (m/s) | 0 | 2.5 | 5.5 | 8.75 | 12.5 | 17.5 | 24.0 |
>
> Determine the distance travelled in 6 seconds (i.e. the area under the v/t graph) by (a) the trapezoidal rule, (b) the mid-ordinate rule, and (c) Simpson's rule.

A graph of speed/time is shown in Fig. 6.4.

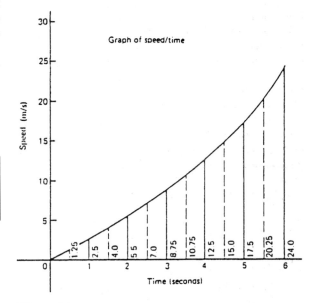

Figure 6.4

(a)   **Trapezoidal rule** (see section 6.2).
The time base is divided into 6 strips each of width 1 s, and the length of the ordinates measured. Thus area

$$= (1)\left[\left(\frac{0 + 24.0}{2}\right) + 2.5 + 5.5 + 8.75 + 12.5 \right.$$
$$\left. + 17.5\right] = \textbf{58.75 m}$$

(b) **Mid-ordinate rule** (see section 6.3)
The time base is divided into 6 strips each of width 1 second. Mid-ordinates are erected as shown in Fig. 6.3 by the broken lines. The length of each mid-ordinate is measured. Thus area

$$= (1)[1.25 + 4.0 + 7.0 + 10.75 + 15.0 + 20.25]$$
$$= \textbf{58.25 m}$$

(c) **Simpson's rule** (see section 6.4)
The time base is divided into 6 strips each of width 1 s, and the length of the ordinates measured. Thus area

$$= \frac{1}{3}(1)[(0 + 24.0) + 4(2.5 + 8.75 + 17.5)$$
$$+ 2(5.5 + 12.5)] = \textbf{58.33 m}$$

Problem 2. An indicator diagram for a steam engine is shown in Fig. 6.5. The base line has been divided into 6 equally spaced intervals and the length of the 7 ordinates measured, with the results shown in centimetres. Determine the area of the indicator diagram using Simpson's rule.

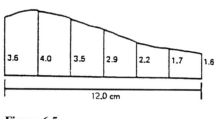

Figure 6.5

6.6 Further questions on areas of irregular shapes

(Answers may be found on page 355.)

1. The velocity of a car at one second intervals is given in the following table:

time t (s)	0	1	2	3	4	5	6
velocity v (m/s)	0	2.0	4.5	8.0	14.0	21.0	29.0

Determine the distance travelled in 6 seconds (i.e. the area under the v/t graph) using an approximate method.

2. The shape of a piece of land is shown in Fig. 6.6. To estimate the area of the land, a surveyor takes measurements at intervals of 50 m, perpendicular to the straight portion with the results shown (the dimensions being in metres). Estimate the area of the land in hectares (1 ha = 10^4 m^2).

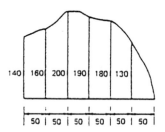

Figure 6.6

3. The deck of a ship is 35 m long. At equal intervals of 5 m the width is given by the following table:

Width (m)	0	2.8	5.2	6.5	5.8	4.1	3.0	2.3

Estimate the area of the deck.

4. An alternating current has the following values at equal intervals of 5 ms:

Time (ms)	0	5	10	15	20	25	30
Current (A)	0	0.9	2.6	4.9	5.8	3.5	0

Plot a graph of current against time and estimate the area under the curve over the 30 ms period using the mid-ordinate rule and determine its mean value.

The width of each interval is 12.0/6 = 2.0 cm. Using Simpson's rule,

$$\text{area} = \frac{1}{3}(2.0)[(3.6 + 1.6) + 4(4.0 + 2.9 + 1.7) + 2(3.5 + 2.2)]$$

$$= \frac{2}{3}[5.2 + 34.4 + 11.4] = \textbf{34 cm}^2$$

Part 2

Science for Mechanical/Thermal Engineering

7

SI units and density

At the end of this chapter you should be able to:

- state the seven basic SI units and their symbols
- understand prefixes to make units larger or smaller
- solve simple problems involving length, area, volume and mass
- define density in terms of mass and volume
- define relative density
- appreciate typical values of densities and relative densities for common materials
- perform calculations involving density, mass, volume and relative density

7.1 SI units

The system of units used in engineering and science is the **Système Internationale d'Unités** (International system of units), usually abbreviated to SI units, and is based on the metric system. This was introduced in 1960 and is now adopted by the majority of countries as the official system of measurement.

The basic units in the SI system are listed below with their symbols:

Quantity	Unit and symbol
length	metre, m
mass	kilogram, kg
time	second, s
electric current	ampere, A
thermodynamic temperature	kelvin, K
luminous intensity	candela, cd
amount of substance	mole, mol

SI units may be made larger or smaller by using **prefixes** which denote multiplication or division by a particular amount. The eight most common multiples, with their meaning, are listed below:

Prefix	Name	Meaning
T	tera	multiply by 1 000 000 000 000 (i.e. $\times 10^{12}$)
G	giga	multiply by 1 000 000 000 (i.e. $\times 10^{9}$)
M	mega	multiply by 1 000 000 (i.e. $\times 10^{6}$)
k	kilo	multiply by 1 000 (i.e. $\times 10^{3}$)
m	milli	divide by 1 000 (i.e. $\times 10^{-3}$)
μ	micro	divide by 1 000 000 (i.e. $\times 10^{-6}$)
n	nano	divide by 1 000 000 000 (i.e. $\times 10^{-9}$)
p	pico	divide by 1 000 000 000 000 (i.e. $\times 10^{-12}$)

Length is the distance between two points. The standard unit of length is the **metre**, although the **centimetre**, **cm**, **millimetre**, **mm** and **kilometre**, **km**, are often used.

1 cm = 10 mm; 1 m = 100 cm = 1000 mm; 1 km = 1000 m.

Area is a measure of the size or extent of a plane surface and is measured by multiplying a length by a length. If the lengths are in metres then the unit of area is the **square metre, m².**

$$1 \text{ m}^2 = 1 \text{ m} \times 1 \text{ m} = 100 \text{ cm} \times 100 \text{ cm} = 10\,000 \text{ cm}^2 \text{ or } 10^4 \text{ cm}^2$$

$$= 1000 \text{ mm} \times 1000 \text{ mm}$$

$$= 1\,000\,000 \text{ mm}^2 \text{ or } 10^6 \text{ mm}^2$$

Conversely, $1 \text{ cm}^2 = 10^{-4} \text{ m}^2$ and $1 \text{ mm}^2 = 10^{-6} \text{ m}^2$.

Volume is a measure of the space occupied by a solid and is measured by multiplying a length by a length by a length. If the lengths are in metres then the unit of volume is in **cubic metres, m³.**

$$1 \text{ m}^3 = 1 \text{ m} \times 1 \text{ m} \times 1 \text{ m}$$

$$= 100 \text{ cm} \times 100 \text{ cm} \times 100 \text{ cm} = 10^6 \text{ cm}^3$$

$$= 1000 \text{ mm} \times 1000 \text{ mm} \times 1000 \text{ mm}$$

$$= 10^9 \text{ mm}^3$$

Conversely, $1 \text{ cm}^3 = 10^{-6} \text{ m}^3$ and $1 \text{ mm}^3 = 10^{-9} \text{ m}^3$.

Another unit used to measure volume, particularly with liquids, is the litre, l, where $1 l = 1000 \text{ cm}^3$.

Mass is the amount of matter in a body and is measured in **kilograms, kg.**

$$1 \text{ kg} = 1000 \text{ g} \text{ (or conversely, } 1 \text{ g} = 10^{-3} \text{ kg)}$$

and 1 tonne (t) = 1000 kg

Problem 1. Express (a) a length of 36 mm in metres, (b) 32 400 mm² in square metres, and (c) 8 540 000 mm³ in cubic metres.

(a) $1 \text{ m} = 10^3 \text{ mm}$ or $1 \text{ mm} = 10^{-3} \text{ m}$. Hence

$$36 \text{ mm} = 36 \times 10^{-3} \text{ m} = \frac{36}{10^3} \text{ m} = \frac{36}{1000} \text{ m}$$

$$= \textbf{0.036 m}$$

(b) $1 \text{ m}^2 = 10^6 \text{ mm}^2$ or $1 \text{ mm}^2 = 10^{-6} \text{ m}^2$. Hence

$$32\,400 \text{ mm}^2 = 32\,400 \times 10^{-6} \text{ m}^2 = \frac{32\,400}{10^6}$$

$$= \textbf{0.0324 m}^2$$

(c) $1 \text{ m}^3 = 10^9 \text{ mm}^3$ or $1 \text{ mm}^3 = 10^{-9} \text{ m}^3$. Hence

$$8\,540\,000 \text{ mm}^3 = 8\,540\,000 \times 10^{-9} \text{ m}^3$$

$$= \frac{8\,540\,000}{10^9} \text{ m}$$

$$= \textbf{8.54} \times \textbf{10}^{-3} \textbf{ m}^3 \text{ or}$$
$$\textbf{0.008 54 m}^3$$

Problem 2. Determine the area of a room 15 m long by 8 m wide in (a) m², (b) cm² and (c) mm².

(a) Area of room = 15 m × 8 m = **120 m²**
(b) 120 m² = 120 × 10⁴ cm², since 1 m² = 10⁴ cm²

$$= \textbf{1 200 000 cm}^2 \text{ or } \textbf{1.2} \times \textbf{10}^6 \textbf{ cm}^2$$

(c) 120 m² = 120 × 10⁶ mm², since 1 m² = 10⁶ mm²

$$= \textbf{120 000 000 mm}^2 \text{ or}$$
$$\textbf{0.12} \times \textbf{10}^9 \textbf{ mm}^2$$

(Note, it is usual to express the power of 10 as a multiple of 3, i.e. × 10³ or × 10⁶ or × 10⁻⁹, and so on.)

Problem 3. A cube has sides each of length 50 mm. Determine the volume of the cube in cubic metres.

Volume of cube = 50 mm × 50 mm × 50 mm = 125 000 mm³.
$1 \text{ mm}^3 = 10^{-9} \text{ m}$, thus volume = 125 000 × 10⁻⁹ m³
$$= \textbf{0.125} \times \textbf{10}^{-3} \textbf{ m}^3$$

Problem 4. A container has a capacity of 2.5 litres. Calculate its volume in (a) m³, (b) mm³.

Since 1 litre = 1000 cm³, 2.5 litres = 2.5 × 1000 cm³ = 2500 cm³.
(a) 2500 cm³ = 2500 × 10⁻⁶ m³ = **2.5 × 10⁻³ m³** or **0.0025 m³**
(b) 2500 cm³ = 2500 × 10³ mm³ = **2 500 000 mm³** or **2.5 × 10⁶ mm³**

7.2 Density

Density is the mass per unit volume of a substance. The symbol used for density is ρ (Greek letter rho) and its units are kg/m³.

$$\text{Density} = \frac{\text{mass}}{\text{volume}}, \text{ i.e.}$$

$$\boxed{\rho = \frac{m}{V}} \quad \text{or} \quad \boxed{m = \rho V} \quad \text{or} \quad \boxed{V = \frac{m}{\rho}}$$

where m is the mass in kg, V is the volume in m³ and ρ is the density in kg/m³.

Some typical values of densities include:

Aluminium	2700 kg/m^3	Steel	7800 kg/m^3
Cast iron	7000 kg/m^3	Petrol	700 kg/m^3
Cork	250 kg/m^3	Lead	11 400 kg/m^3
Copper	8900 kg/m^3	Water	1000 kg/m^3

The **relative density** of a substance is the ratio of the density of the substance to the density of water, i.e.

$$\text{relative density} = \frac{\text{density of substance}}{\text{density of water}}$$

Relative density has no units, since it is the ratio of two similar quantities. Typical values of relative densities can be determined from above (since water has a density of 1000 kg/m³), and include:

Aluminium	2.7	Steel	7.8
Cast iron	7.0	Petrol	0.7
Cork	0.25	Lead	11.4
Copper	8.9		

The relative density of a liquid may be measured using a **hydrometer**.

Problem 5.   Determine the density of 50 cm3 of copper if its mass is 445 g.

Volume = 50 cm^3 = 50 × 10^{-6} m^3; mass = 445 g = 445 × 10^{-3} kg.

$$\text{Density} = \frac{\text{mass}}{\text{volume}} = \frac{445 \times 10^{-3} \text{ kg}}{50 \times 10^{-6} \text{ m}^3}$$

$$= \frac{445}{50} \times 10^3$$

$$= \mathbf{8.9 \times 10^3 \text{ kg/m}^3} \text{ or } \mathbf{8900 \text{ kg/m}^3}$$

Problem 6.   The density of aluminium is 2700 kg/m³. Calculate the mass of a block of aluminium which has a volume of 100 cm³.

Density, ρ = 2700 kg/m^3; volume V = 100 cm^3 = 100 × 10^{-6} m^3.
Since density = mass/volume, then mass = density × volume. Hence

$$\text{mass} = \rho V = 2700 \text{ kg/m}^3 \times 100 \times 10^{-6} \text{ m}^3$$

$$= \frac{2700 \times 100}{10^6} \text{ kg} = \mathbf{0.270 \text{ kg}} \text{ or } \mathbf{270 \text{ g}}$$

Problem 7.   Determine the volume, in litres, of 20 kg of paraffin oil of density 800 kg/m³.

Density = mass/volume   hence   volume = mass/density. Thus

$$\text{volume} = \frac{m}{\rho} = \frac{20 \text{ kg}}{800 \text{ kg/m}^3} = \frac{1}{40} \text{ m}^3$$

$$= \frac{1}{40} \times 10^6 \text{ cm}^3$$

$$= 25\,000 \text{ cm}^3$$

1 litre = 1000 cm^3 hence 25 000 cm^3 = 25 000/1000 = **25 litres**.

Problem 8.   Determine the relative density of a piece of steel of density 7850 kg/m³. Take the density of water as 1000 kg/m³.

$$\text{Relative density} = \frac{\text{density of steel}}{\text{density of water}}$$

$$= \frac{7850}{1000} = \mathbf{7.85}$$

Problem 9.   A piece of metal 200 mm long, 150 mm wide and 10 mm thick has a mass of 2700 g. What is the density of the metal?

Volume of metal = 200 mm × 150 mm × 10 mm = 300 000 mm^3

$$= 3 \times 10^5 \text{ mm}^3 = \frac{3 \times 10^5}{10^6 \text{ m}^3} = 3$$

$$\times 10^{-4} \text{ m}^3$$

Mass = 2700 g = 2.7 kg.

$$\text{Density} = \frac{\text{mass}}{\text{volume}} = \frac{2.7 \text{ kg}}{3 \times 10^{-4} \text{ m}^3}$$

$$= 0.9 \times 10^4 \text{ kg/m}^3 = \mathbf{9000 \text{ kg/m}^3}$$

Problem 10.   Cork has a relative density of 0.25. Calculate (a) the density of cork and (b) the volume in cubic centimetres of 50 g of cork. Take the density of water to be 1000 kg/m³.

(a) Relative density = (density of cork)/(density of water), from which density of cork = relative density × density of water, i.e. density of cork, $\rho = 0.25 \times 1000 = \textbf{2500 kg/m}^3$.

(b) Density = mass/volume, from which volume = mass/density

Mass, $m = 50\,g = 50 \times 10^{-3}$ kg. Hence

$$\text{volume, } V = \frac{m}{\rho} = \frac{50 \times 10^{-3}\ \text{kg}}{250\ \text{kg/m}^3} = \frac{0.05}{250}\text{m}^3$$

$$= \frac{0.05}{250} \times 10^6\ \text{cm}^3 = \textbf{200 cm}^3$$

7.3 Multi-choice questions on SI units and density

(Answers on page 355.)

1. Which of the following statements is true? $1000\ \text{mm}^3$ is equivalent to
   (a) $1\ \text{m}^3$ (b) $10^{-3}\ \text{m}^3$ (c) $10^{-6}\ \text{m}^3$
   (d) $10^{-9}\ \text{m}^3$

2. Which of the following statements is true?
   (a) $1\ \text{mm}^2 = 10^{-4}\ \text{m}^2$ (c) $1\ \text{mm}^3 = 10^{-6}\ \text{m}^3$
   (b) $1\ \text{cm}^3 = 10^{-3}\ \text{m}^3$ (d) $1\ \text{km}^2 = 10^{10}\ \text{cm}^2$

3. Which of the following statements is false? 1000 litres is equivalent to
   (a) $10^3\ \text{m}^3$ (b) $10^6\ \text{cm}^3$ (c) $10^9\ \text{mm}^3$

4. Let mass = A, volume = B and density = C. Which of the following statements is false?

   (a) $A = BC$ (b) $C = \dfrac{A}{B}$ (c) $B = \dfrac{C}{A}$

5. The density of 100 cm³ of a material having a mass of 700 g is:
   (a) $70\,000\ \text{kg/m}^3$ (b) $7000\ \text{kg/m}^3$
   (c) $7\ \text{kg/m}^3$ (d) $70\ \text{kg/m}^3$

6. An alloy has a relative density of 10. If the density of water is $1000\ \text{kg/m}^3$, the density of the alloy is:
   (a) $100\ \text{kg/m}^3$ (b) $0.01\ \text{kg/m}^3$
   (c) $10\,000\ \text{kg/m}^3$ (d) $1010\ \text{kg/m}^3$

7.4 Short answer questions on SI units and density

1. State the SI units for length, mass and time.
2. State the SI units for electric current and thermodynamic temperature.
3. What is the meaning of the following prefixes?

(a) M   (b) m   (c) μ   (d) k

In questions 4 to 8, complete the statements

4. 1 m   = ...mm;   1 km = ...m
5. 1 m²   = ...cm²;   1 cm² = ...mm²
6. 1 l   = ...cm³;   1 m³ = ...mm³
7. 1 kg   = ...g;   1 t   = ...kg
8. 1 mm²   = ...m²;   1 cm³ = ...m³
9. Define density.
10. What is meant by 'relative density'?
11. Relative density of liquids may be measured using a ....

7.5 Further questions on SI units and density

1. Express (a) a length of 52 mm in metres, (b) 20 000 mm² in square metres, and (c) 10 000 000 mm³ in cubic metres.
   [(a) 0.052 m   (b) 0.02 m²   (c) 0.01 m³]

2. A garage measures 5 m by 2.5 m. Determine the area in (a) m² (b) mm².
   [(a) 12.5 m²   (b) 12.5×10^6 mm²]

3. The height of the garage in question 2 is 3 m. Determine the volume in (a) m³ (b) mm³.
   [(a) 37.5 m³   (b) 37.5×10^9 mm³]

4. A bottle contains 6.3 litres of liquid. Determine the volume in (a) m³ (b) cm³ (c) mm³.
   [(a) 0.0063 m³   (b) 6300 cm³ (c) 6.3×10^6 mm³]

5. Determine the density of 200 cm³ of lead which has a mass of 2280 g.
   [11 400 kg/m³]

6. The density of iron is 7500 kg/m³. If the volume of a piece of iron is 200 cm³, determine its mass.
   [1.5 kg]

7. Determine the volume, in litres, of 15 kg of petrol of density 700 kg/m³.
   [20 litres]

8. The density of water is 1000 kg/m³. Determine the relative density of a piece of copper of density 8900 kg/m³.
   [8.9]

9. A piece of metal 100 mm long, 80 mm wide and 20 mm thick has a mass of 1280 g. Determine the density of the metal.
   [8000 kg/m³]

10. Some oil has a relative density of 0.80. Determine (a) the density of the oil, and (b) the volume of 2 kg of oil. Take the density of water as 1000 kg/m³.
    [(a) 800 kg/m³   (b) 0.0025 m³]

8

Forces acting at a point

At the end of this chapter you should be able to:

- distinguish between scalar and vector quantities
- define 'centre of gravity' of an object
- define 'equilibrium' of an object
- understand the terms 'coplanar' and 'concurrent'
- determine the resultant of two coplanar forces using

  (a) the triangle of forces method
  (b) the parallelogram of forces method

- calculate the resultant of two coplanar forces using

  (a) the cosine and sine rules
  (b) resolution of forces

- determine the resultant of more than two coplanar forces using

  (a) the polygon of forces method
  (b) calculation by resolution of forces

- determine unknown forces when three or more coplanar forces are in equilibrium

8.1 Scalar and vector quantities

Quantities used in engineering and science can be divided into two groups:

(a) **Scalar quantities** have a size (or magnitude) only and need no other information to specify them. Thus, 10 centimetres, 50 seconds, 7 litres and 3 kilograms are all examples of scalar quantities.

(b) **Vector quantities** have both a size or magnitude and a direction, called the line of action of the quantity. Thus, a velocity of 50 kilometres per hour due east, an acceleration of 9.81 metres per second squared vertically downwards and a force of 15 newtons at an angle of 30 degrees are all examples of vector quantities.

8.2 Centre of gravity and equilibrium

The **centre of gravity** of an object is a point where the resultant gravitational force acting on the body may be taken to act. For objects of uniform thickness lying in a horizontal plane, the centre of gravity is vertically in line with the point of balance of the object. For a thin uniform rod the point of balance and hence the centre of gravity is halfway along the rod as shown in Fig. 8.1(a).

A thin flat sheet of a material of uniform thickness is called a **lamina** and the centre of gravity of a rectangular lamina lies at the point of intersection of its diagonals, as shown in Fig. 8.1(b). The centre of gravity of a circular lamina is at the centre of the circle, as shown in Fig. 8.1(c).

(a)

(b)

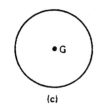

(c)

Figure 8.1

An object is in **equilibrium** when the forces acting on the object are such that there is no tendency for the object to move. The state of equilibrium of an object can be divided into three groups.

(i)   If an object is in **stable equilibrium** and it is slightly disturbed by pushing or pulling (i.e. a disturbing force is applied), the centre of gravity is raised and when the disturbing force is removed, the object returns to its original position. Thus a ball bearing in a hemispherical cup is in stable equilibrium, as shown in Fig. 8.2(a)

(ii) An object is in **unstable equilibrium** if, when a disturbing force is applied, the centre of gravity is lowered and the object moves away from its original position. Thus, a ball bearing balanced on top of a hemispherical cup is in unstable equilibrium, as shown in Fig. 8.2(b)

(iii) When an object in **neutral equilibrium** has a disturbing force applied, the centre of gravity remains at the same height and the object does not move when the disturbing

force is removed. Thus, a ball bearing on a flat horizontal surface is in neutral equilibrium, as shown in Fig. 8.2(c)

8.3   Force

When forces are all acting in the same plane, they are called **coplanar**. When forces act at the same time and at the same point, they are called **concurrent forces**.

Force is a **vector quantity** and thus has both a magnitude and a direction. A vector can be represented graphically by a line drawn to scale in the direction of the line of action of the force. Vector quantities may be shown by using bold, lower case letters, thus **ab** in Fig. 8.3 represents a force of 5 newtons acting in a direction due east.

8.4   The resultant of two coplanar forces

For two forces acting at a point, there are three possibilities.

(a)   For forces acting in the same direction and having the same line of action, the single force having the same effect as both of the forces, called the **resultant force** or just the **resultant**, is the arithmetic sum of the separate forces. Forces of F_1 and F_2 acting at point P, as shown in Fig. 8.4(a), have exactly the same effect on point P as force F shown in Fig. 8.4(b), where $F = F_1 + F_2$

(a)
Stable
equilibrium

(b)
Unstable
equilibrium

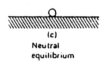

(c)
Neutral
equilibrium

Figure 8.2

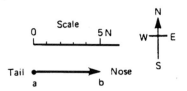

Figure 8.3

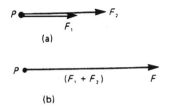

(a)

(b)

Figure 8.4

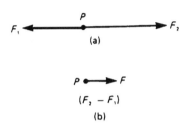

(a)

(b)

Figure 8.5

and acts in the same direction as F_1 and F_2. Thus F is the resultant of F_1 and F_2.

(b) For forces acting in opposite directions along the same line of action, the resultant force is the arithmetic difference between the two forces. Forces of F_1 and F_2 acting at point P as shown in Fig. 8.5(a) have exactly the same effect on point P as force F shown in Fig. 8.5(b), where $F = F_2 - F_1$ and acts in the direction of F_2, since F_2 is greater than F_1. Thus F is the resultant of F_1 and F_2.

(c) When two forces do not have the same line of action, the magnitude and direction of the resultant force may be found by a procedure called vector addition of forces. There are two graphical methods of performing **vector addition**, known as the **triangle of forces** method (see section 8.5) and the **parallelogram of forces** method (see section 8.6).

Problem 1.   Determine the resultant force of two forces of 5 kN and 8 kN,

(a) acting in the same direction and having the same line of action,
(b) acting in opposite directions but having the same line of action.

(a) The vector diagram of the two forces acting in the same direction is shown in Fig. 8.6(a),

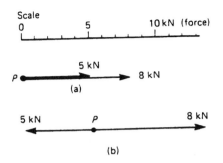

(a)

(b)

Figure 8.6

which assumes that the line of action is horizontal, although since it is not specified, could be in any direction. From above, the resultant force F is given by: $F = F_1 + F_2$, i.e. $F = (5 + 8)$ kN = **13 kN** in the direction of the original forces.

(b) The vector diagram of the two forces acting in opposite directions is shown in Fig. 8.6(b), again assuming that the line of action is in a horizontal direction. From above, the resultant force F is given by: $F = F_2 - F_1$, i.e. $F = (8 - 5)$ kN = **3 kN** in the direction of the 8 kN force.

8.5   Triangle of forces method

A simple procedure for the triangle of forces method of vector addition is as follows:

(i) Draw a vector representing one of the forces, using an appropriate scale and in the direction of its line of action.

(ii) From the **nose** of this vector and using the same scale, draw a vector representing the second force in the direction of its line of action.

(iii) The resultant vector is represented in both magnitude and direction by the vector drawn from the tail of the first vector to the nose of the second vector.

Problem 2.   Determine the magnitude and direction of the resultant of a force of 15 N acting horizontally to the right and a force of 20 N, inclined at an angle of 60° to the 15 N force. Use the triangle of forces method.

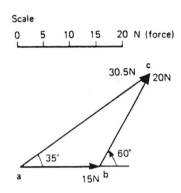

Figure 8.7

Using the procedure given above and with reference to Fig. 8.7

(i)   **ab** is drawn 15 units long horizontally.
(ii)   From b, **bc** is drawn 20 units long, inclined at an angle of 60° to ab. (Note, in angular measure, an angle of 60° from ab means 60° in an anticlockwise direction.)
(iii) By measurement, the resultant **ac** is 30.5 units long inclined at an angle of 35° to **ab**. That is, the resultant force is **30.5 N**, inclined at an angle of **35°** to the 15 N force.

Problem 3.   Find the magnitude and direction of the two forces given, using the triangle of forces method.

First force: 1.5 kN acting at an angle of 30°

Second force: 3.7 kN acting at an angle of −45°

From the above procedure and with reference to Fig. 8.8:

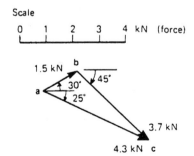

Figure 8.8

(i)   **ab** is drawn at an angle of 30° and 1.5 units in length.
(ii)   From b, **bc** is drawn at an angle of −45° and 3.7 units in length. (Note, an angle of −45° means a clockwise rotation of 45° from a line drawn horizontally to the right.)
(iii) By measurement, the resultant **ac** is 4.3 units long at an angle of −25°. That is, the resultant force is **4.3 kN** at an angle of **−25°**.

8.6   The parallelogram of forces method

A simple procedure for the parallelogram of forces method of vector addition is as follows:

(i)   Draw a vector representing one of the forces, using an appropriate scale and in the direction of its line of action.
(ii)   From the **tail** of this vector and using the same scale draw a vector representing the second force in the direction of its line of action.
(iii) Complete the parallelogram using the two vectors drawn in (i) and (ii) as two sides of the parallelogram.
(iv)   The resultant force is represented in both magnitude and direction by the vector corresponding to the diagonal of the parallelogram drawn from the tail of the vectors in (i) and (ii).

Problem 4.   Use the parallelogram of forces method to find the magnitude and direction of the resultant of a force of 250 N acting at an angle of 135° and a force of 400 N acting at an angle of −120°.

From the procedure given above and with reference to Fig. 8.9:

(i)   **ab** is drawn at an angle of 135° and 250 units in length.
(ii)   **ac** is drawn at an angle of −120° and 400 units in length.
(iii) **bd** and **cd** are drawn to complete the parallelogram.
(iv)   **ad** is drawn. By measurement, **ad** is 413 units long at an angle of −156°. That is, the resultant force is **413 N** at an angle of **−156°**.

0   100 200 300 400 500 N (force)

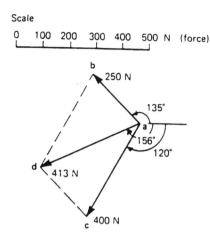

b

250 N

135°

a

156°

120°

d

413 N

c 400 N

Figure 8.9

senting the 8 kN force in magnitude and direction and **ab** representing the 5 kN force in magnitude and direction. The resultant is given by length **ob**. By the cosine rule (refer to Chapter 5),

$$ob^2 = oa^2 + ab^2 - 2(oa)(ab)\cos\angle oab$$
$$= 8^2 + 5^2 - 2(8)(5)\cos 100°$$
$$(\text{since } \angle oab = 180° - 50° - 30° = 100°)$$
$$= 64 + 25 - (-13.892) = 102.892.$$

Hence ob = $\sqrt{(102.892)}$ = 10.14 kN.
By the sine rule,

$$\frac{5}{\sin\angle aob} = \frac{10.14}{\sin 100°}$$

from which,

$$\sin\angle aob = \frac{5\sin 100°}{10.14} = 0.4856$$

Hence $\angle aob$ = arcsin (0.4856) = 29°3'. Thus angle ϕ in Fig. 8.10(b) is 50° − 29°3' = 20°57'.
Hence the resultant of the two forces is 10.14 kN acting at an angle of 20°57' to the horizontal.

8.7 Resultant of coplanar forces by calculation

An alternative to the graphical methods of determining the resultant of two coplanar forces is by **calculation**. This can be achieved by trigonometry using the cosine rule and the sine rule, as shown in Problem 5 following, or by resolution of forces (see section 8.10).

> Problem 5. Use the cosine and sine rules to determine the magnitude and direction of the resultant of a force of 8 kN acting at an angle of 50° to the horizontal and a force of 5 kN acting at an angle of −30° to the horizontal.

The space diagram is shown in Fig. 8.10(a). A sketch is made of the vector diagram, **oa** repre-

8.8 Resultant of more than two coplanar forces

For the three coplanar forces F_1, F_2 and F_3 acting at a point as shown in Fig. 8.11, the vector diagram is drawn using the nose-to-tail method of section 8.5. The procedure is:

(i)   Draw **oa** to scale to represent force F_1 in both magnitude and direction (see Fig. 8.12).

(ii) From the nose of **oa**, draw **ab** to represent force F_2.

(iii) From the nose of **ab**, draw **bc** to represent force F_3.

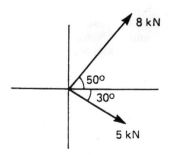

8 kN

50°

30°

5 kN

(a) space diagram

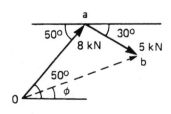

a

50°   30°

8 kN   5 kN

b

50°

0   ϕ

(b) vector diagram

Figure 8.10

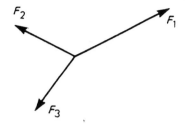

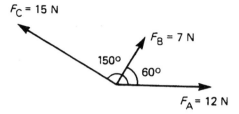

Figure 8.11

Figure 8.13

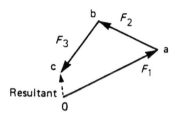

Figure 8.12

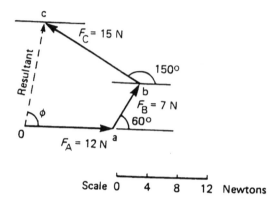

Figure 8.14

(iv) The resultant vector is given by length **oc** in Fig. 8.12. The direction of resultant **oc** is from where we started, i.e. point o, to where we finished, i.e. point c. When acting by itself, the resultant force, given by **oc**, has the same effect on the point as forces F_1, F_2 and F_3 have when acting together. The resulting vector diagram of Fig. 8.12 is called the **polygon of forces**.

Thus the resultant of the three forces, F_A, F_B and F_C is a force of 13.8 N at 80° to the horizontal.

> Problem 6.   Determine graphically the magnitude and direction of the resultant of these three coplanar forces, which may be considered as acting at a point. Force A, 12 N acting horizontally to the right; force B, 7 N inclined at 60° to force A; force C, 15 N inclined at 150° to force A.

> Problem 7.   The following coplanar forces are acting at a point, the given angles being measured from the horizontal: 100 N at 30°, 200 N at 80°, 40 N at –150°, 120 N at –100° and 70 N at –60°. Determine graphically the magnitude and direction of the resultant of the five forces.

The space diagram is shown in Fig. 8.13. The vector diagram (Fig. 8.14) is produced as follows:

(i)   **oa** represents the 12 N force in magnitude and direction.
(ii) From the nose of **oa**, **ab** is drawn inclined at 60° to **oa** and 7 units long.
(iii) From the nose of **ab**, **bc** is drawn 15 units long inclined at 150° to **oa** (i.e. 150° to the horizontal).
(iv) **oc** represents the resultant. By measurement, the resultant is 13.8 N inclined at 80° to the horizontal.

The five forces are shown in the space diagram of Fig. 8.15. Since the 200 N and 120 N forces have the same line of action but are in opposite sense, they can be represented by a single force of 200 – 120, i.e. 80 N acting at 80° to the horizontal. Similarly, the 100 N and 40 N forces can be represented by a force of 100 – 40, i.e. 60 N acting at 30° to the horizontal. Hence the space diagram of Fig. 8.15 may be represented by the space diagram of Fig. 8.16. Such a simplification of the vectors is not essential but it is easier to construct the vector diagram from a space diagram having three forces, than one with five. The vector diagram is shown in Fig. 8.17, **oa** representing the 60 N force, **ab** representing the 80 N force and **bc**

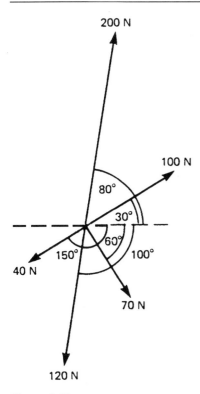

Figure 8.15

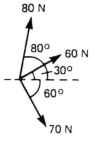

Figure 8.16

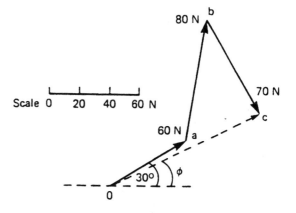

Figure 8.17

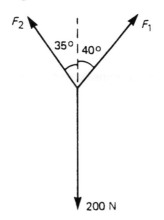

Figure 8.18

the 70 N force. The resultant, **oc**, is found by measurement to represent a force of 112 N and angle ϕ is 25°.

Thus the five forces shown in Fig. 8.15 may be represented by a single force of 112 N at 25° to the horizontal.

8.9 Coplanar forces in equilibrium

When three or more coplanar forces are acting at a point and the vector diagram closes, there is no resultant. The forces acting at the point are in **equilibrium**.

> Problem 8.   A load of 200 N is lifted by two ropes connected to the same point on the load, making angles of 40° and 35° with the vertical. Determine graphically the tensions in each rope when the system is in **equilibrium**.

The space diagram is shown in Fig. 8.18. Since the system is in equilibrium, the vector diagram must close. The vector diagram (Fig. 8.19) is drawn as follows:

(i) The load of 200 N is drawn vertically as shown by **oa**.

(ii) The direction only of force F_1 is known, so from point a, **ad** is drawn at 40° to the vertical.

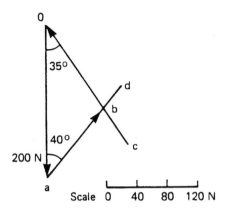

Figure 8.19

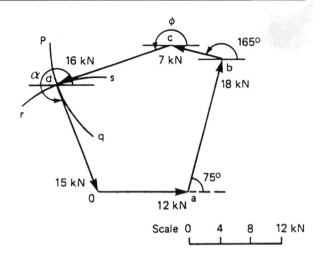

Figure 8.20

(iii)   The direction only of force F_2 is known, so from point o, **oc** is drawn at 35° to the vertical.

(iv)   Lines **ad** and **oc** cross at point b. Hence the vector diagram is given by triangle oab.
By measurement, **ab** is 119 N and **ob** is 133 N.

Thus the tensions in the ropes are $F_1 = 119$ N and $F_2 = 133$ N.

Problem 9.   Five coplanar forces are acting on a body and the body is in equilibrium. The forces are: 12 kN acting horizontally to the right, 18 kN acting at an angle of 75°, 7 kN acting at an angle of 165°, 16 kN acting from the nose of the 7 kN force, and 15 kN acting from the nose of the 16 kN force. Determine the directions of the 16 kN and 15 kN forces relative to the 12 kN force.

With reference to Fig. 8.20, **oa** is drawn 12 units long horizontally to the right. From point a, **ab** is drawn 18 units long at an angle of 75°. From b, **bc** is drawn 7 units long at an angle of 165°. The direction of the 16 kN force is not known, thus arc pq is drawn with a compass, with centre at c, radius 16 units. Since the forces are at equilibrium, the polygon of forces must close. Using a compass with centre at 0, arc rs is drawn having a radius 15 units. The point where the arcs intersect is at d.

By measurement, angle $\phi = 198°$ and $\alpha = 291°$.

Thus the 16 kN force acts at an angle of 198° (or –162°) to the 12 kN force and the 15 kN force acts at an angle of 291° (or –69°) to the 12 kN force.

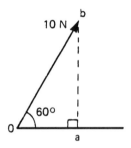

Figure 8.21

8.10   Resolution of forces

A vector quantity may be expressed in terms of its **horizontal** and **vertical components**. For example, a vector representing a force of 10 N at an angle of 60° to the horizontal is shown in Fig. 8.21. If the horizontal line **oa** and the vertical line **ab** are constructed as shown, then **oa** is called the horizontal component of the 10 N force and **ab** the vertical component of the 10 N force.

By trigonometry, $\cos 60° = oa/ob$. Hence the horizontal component, $oa = 10 \cos 60°$. Also, $\sin 60° = ab/ob$. Hence the vertical component, $ab = 10 \sin 60°$.

This process is called **finding the horizontal and vertical components of a vector** or **the resolution of a vector**, and can be used as an alternative to graphical methods for calculating the resultant of two or more coplanar forces acting at a point. For example, to calculate the resultant of a 10 N force

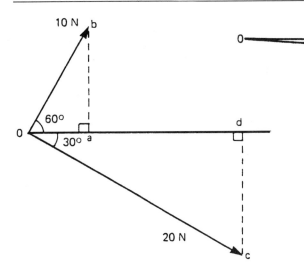

Figure 8.22

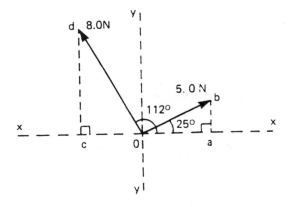

Figure 8.23

This example demonstrates the use of resolution of forces for calculating the resultant of two coplanar forces acting at a point. However the method may be used for more than two forces acting at a point.

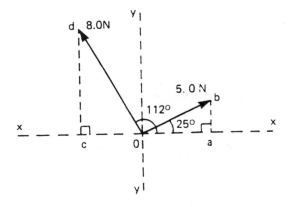

Figure 8.24

acting at 60° to the horizontal and a 20 N force acting at –30° to the horizontal (see Fig. 8.22) the procedure is as follows:

(i) Determine the horizontal and vertical components of the 10 N force, i.e. horizontal component, oa = 10 cos 60° = 5.0 N, and vertical component, ab = 10 sin 60° = 8.66 N.

(ii) Determine the horizontal and vertical components of the 20 N force, i.e. horizontal component, od = 20 cos (–30°) = 17.32 N, and vertical component, cd = 20 sin (–30°) = –10.0 N.

(iii) Determine the total horizontal component, i.e.

oa + od = 5.0 + 17.32 = 22.32 N

(iv) Determine the total vertical component, i.e.

ab + cd = 8.66 + (–10.0) = –1.34 N

(v) Sketch the total horizontal and vertical components as shown in Fig. 8.23. The resultant of the two components is given by length **or** and, by Pythagoras' theorem

or = $\sqrt{[(22.32)^2 + (1.34)^2]}$ = 22.36 N

and using trigonometry,

angle ϕ = arctan $\dfrac{1.34}{22.32}$ = 3°26′

Hence the resultant of the 10 N and 20 N forces shown in Fig. 8.22 is **22.36 N at an angle of –3°26′ to the horizontal**.

Problem 10. Forces of 5.0 N at 25° and 8.0 N at 112° act at a point. By resolving these forces into horizontal and vertical components, determine their resultant.

The space diagram is shown in Fig. 8.24.

(i) The horizontal component of the 5.0 N force, oa = 5.0 cos 25° = 4.532; the vertical component of the 5.0 N force, ab = 5.0 sin 25° = 2.113.

(ii) The horizontal component of the 8.0 N force, oc = 8.0 cos ∠cod = 8.0 cos 68° = 2.997. However, in the second quadrant, the cosine of an acute angle is negative, hence the horizontal component of the 8.0 N force is –2.997.

The vertical component of the 8.0 N force, cd = 8.0 sin ∠cod = 8.0 sin 68° = 7.417. Since, in the second quadrant, the sine of an acute angle is positive, the vertical component of the 8.0 N force is +7.417.

Figure 8.25

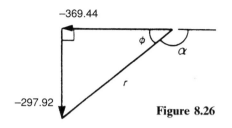

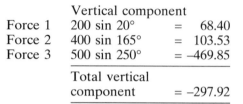

Figure 8.26

(A useful check is that the vertical components ab and cd are both above the XX axis and are thus both positive; horizontal component oa is positive since it is to the right of axis YY and the horizontal component oc is negative since it is to the left of axis YY.)

With a calculator, horizontal and vertical components can be determined easily. For example, component oc in Fig. 8.24 is equivalent to 8.0 cos 112°, which is evaluated as −2.997 directly with a calculator without having to consider the equivalent acute angle (i.e. ∠cod), and whether it is positive or negative in that quadrant.

(iii) Total horizontal component = oa + oc = 4.532 + (−2.997) = +1.535.

(iv) Total vertical component = ab + dc = 2.113 + 7.417 = +9.530.

(v) The components are shown sketched in Fig. 8.25. By Pythagoras' theorem, $r = \sqrt{[(1.535)^2 + (9.530)^2]} = 9.653$, and by trigonometry, angle ϕ = arctan (9.530/1.535) = 80°51′.

Hence the resultant of the two forces shown in Figure 8.24 is a force of 9.653 N acting at 80°51′ to the horizontal.

Problem 11.   Determine by resolution of forces the resultant of the following three coplanar forces acting at a point: 200 N acting at 20° to the horizontal; 400 N acting at 165° to the horizontal; 500 N acting at 250° to the horizontal.

A tabular approach using a calculator may be made as shown below.

	Horizontal component	
Force 1	200 cos 20°	= 187.94
Force 2	400 cos 165°	= −386.37
Force 3	500 cos 250°	= −171.01
Total horizontal component		= −369.44

	Vertical component	
Force 1	200 sin 20°	= 68.40
Force 2	400 sin 165°	= 103.53
Force 3	500 sin 250°	= −469.85
Total vertical component		= −297.92

The total horizontal and vertical components are shown in Fig. 8.26.

Resultant $r = \sqrt{[(369.44)^2 + (297.92)^2]} = 474.60$, and angle ϕ = arctan (297.92/369.44) = 38°53′, from which, α = 180° − 38°53′ = 141°7′.

Thus the resultant of the three forces given is 474.6 N acting at an angle of −141°7′ (or +218°53′) to the horizontal.

Problem 12.   The following coplanar forces act at a point: force A is 18 kN at 15° to the horizontal; force B is 25 kN at 126° to the horizontal; force C is 10 kN at 197° to the horizontal; force D is 15 kN at 246° to the horizontal; force E is 30 kN at 331° to the horizontal. Determine the resultant of the five forces by resolution of forces.

Using a tabular approach:

	Horizontal component	
Force A	18 cos 15°	= 17.39
Force B	25 cos 126°	= −14.69
Force C	10 cos 197°	= −9.56
Force D	15 cos 246°	= −6.10
Force E	30 cos 331°	= 26.24
Total horizontal component		= +13.28

	Vertical component	
Force A	18 sin 15°	= 4.66
Force B	25 sin 126°	= 20.23
Force C	10 sin 197°	= −2.92
Force D	15 sin 246°	= −13.70
Force E	30 sin 331°	= −14.54
Total vertical component		= −6.27

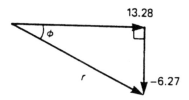

Figure 8.27

The total horizontal and vertical components are shown in Fig. 8.27. By Pythagoras' theorem, resultant $r = \sqrt{[(13.28)^2 + (6.27)^2]} = 14.69$, and angle $\phi = \arctan (6.27/13.28) = 25°16'$.
Hence the resultant of the five forces is 14.69 kN acting at an angle of –25°16'

8.11   Summary

(a)   To determine the **resultant of two coplanar forces** acting at a point, four methods are commonly used. They are:

by drawing: (1) triangle of forces method, and
                (2) parallelogram of forces method, and

by calculation: (3) use of cosine and sine rules, and
                (4) resolution of forces.

(b)   To determine the **resultant of more than two coplanar forces** acting at a point, two methods are commonly used. They are:

by drawing: (1) polygon of forces method, and

by calculation: (2) resolution of forces.

8.12   Multi-choice questions on forces acting at a point

(Answers on page 355.)

1. Which of the following statements is false?
   (a) Scalar quantities have size or magnitude only.
   (b) Vector quantities have both magnitude and direction.
   (c) Mass, length and time are all scalar quantities.
   (d) Distance, velocity and acceleration are all vector quantities.

2. If the centre of gravity of an object which is slightly disturbed is raised and the object returns to its original position when the disturbing force is removed, the object is said to be in
   (a) neutral equilibrium
   (b) stable equilibrium
   (c) static equilibrium
   (d) unstable equilibrium

3. Which of the following statements is false?
   (a) The centre of gravity of a lamina is at its point of balance.
   (b) The centre of gravity of a circular lamina is at its centre.
   (c) The centre of gravity of a rectangular lamina is at the point of intersection of its two sides.
   (d) The centre of gravity of a thin uniform rod is halfway along the rod.

4. The magnitude of the resultant of the vectors shown in Fig. 8.28 is: (a) 2 N   (b) 12 N   (c) 35 N   (d) –2 N

Figure 8.28

5. The magnitude of the resultant of the vectors shown in Fig. 8.29 is: (a) 7 N   (b) 5 N   (c) 1 N   (d) 12 N

Figure 8.29

6. Which of the following statements is false?
   (a) There is always a resultant vector required to close a vector diagram representing a system of coplanar forces acting at a point, which are not in equilibrium.
   (b) A vector quantity has both magnitude and direction.
   (c) A vector diagram representing a system of coplanar forces acting at a point when in equilibrium does not close.
   (d) Concurrent forces are those which act at the same time at the same point.

7. Which of the following statements is false?
   (a) The resultant of coplanar forces of 1 N, 2 N and 3 N acting at a point can be 4 N.
   (b) The resultant of forces of 6 N and 3 N acting in the same line of action but opposite in sense is 3 N.
   (c) The resultant of forces of 6 N and 3 N acting in the same sense and having the same line of action is 9 N.
   (d) The resultant of coplanar forces of 4 N at 0°, 3 N at 90° and 8 N at 180° is 15 N.

8. A space diagram of a force system is shown in Fig. 8.30. Which of the vector diagrams in Fig. 8.31 does **not** represent this force system?

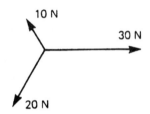

Figure 8.30

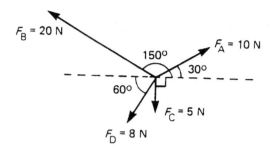

Figure 8.32

9. With reference to Fig. 8.32, which of the following statements is false?
   (a) The horizontal component of F_A is 8.66 N.
   (b) The vertical component of F_B is 10 N.
   (c) The horizontal component of F_C is 0.
   (d) The vertical component of F_D is 4 N.

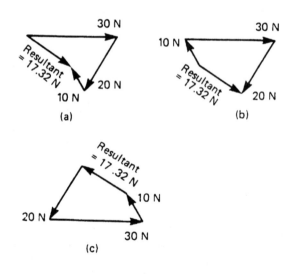

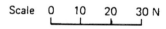

Figure 8.31

8.13   Short answer questions on forces acting at a point

1. Give one example of a scalar quantity and one example of a vector quantity.
2. Explain the difference between a scalar and a vector quantity.
3. What is meant by the centre of gravity of an object?
4. Where is the centre of gravity of a rectangular lamina?
5. What is meant by neutral equilibrium?
6. State the meaning of the term 'coplanar'.
7. What is a concurrent force?
8. State what is meant by a triangle of forces.
9. State what is meant by a parallelogram of forces.
10. State what is meant by a polygon of forces.
11. When a vector diagram is drawn representing coplanar forces acting at a point, and there is no resultant, the forces are in ....
12. Two forces of 6 N and 9 N act horizontally to the right. The resultant is .... N acting ....
13. A force of 10 N acts at an angle of 50° and another force of 20 N acts at an angle of 230°. The resultant is a force .... N acting at an angle of ....°.
14. What is meant by 'resolution of forces'?

15. A coplanar force system comprises a 20 kN force acting horizontally to the right, 30 kN at 45°, 20 kN at 180° and 25 kN at 225°. The resultant is a force of .... N acting at an angle of ....° to the horizontal.

8.14 Further questions on forces acting at a point

In questions 1 to 10, use a graphical method to determine the magnitude and direction of the resultant of the forces given.

1. 1.3 kN and 2.7 kN, having the same line of action and acting in the same direction
   [4.0 kN in the direction of the forces]
2. 470 N and 538 N having the same line of action but acting in opposite directions
   [68 N in the direction of the 538 N force]
3. 13 N at 0° and 25 N at 30°
   [36.8 N at 20°]
4. 5 N at 60° and 8 N at 90°
   [12.6 N at 79°]
5. 1.3 kN at 45° and 2.8 kN at −30°
   [3.4 kN at −8°]
6. 1.7 N at 45° and 2.4 N at −60°
   [2.6 N at −20°]
7. 9 N at 126° and 14 N at 223°
   [15.7 N at −172°]
8. 23.8 N at −50° and 14.4 N at 215°
   [26.7 N at −82°]
9. 0.7 kN at 147° and 1.3 kN at −71°
   [0.86 kN at −100°]
10. 47 N at 79° and 58 N at 247°
    [15.5 N at −152°]
11. Resolve a force of 23.0 N at an angle of 64° into its horizontal and vertical components.
    [10.08 N, 20.67 N]
12. Forces of 7.6 kN at 32° and 11.8 kN at 143° act at a point. Use the cosine and sine rules to calculate the magnitude and direction of their resultant.
    [11.52 kN at 104°59′]
13. In questions 3 to 10, calculate the resultant of the given forces by using the cosine and sine rules.
14. Forces of 5 N at 21° and 9 N at 126° act at a point. By resolving these forces into horizontal and vertical components, determine their resultant.
    [9.09 N at 93°55′]

In questions 15 to 17, determine graphically the magnitude and direction of the resultant of the coplanar forces given which are acting at a point.

15. Force A, 12 N acting horizontally to the right, force B, 20 N acting at 140° to force A; force C, 16 N acting 290° to force A.
    [3.06 N at −45° to force A]
16. Force 1, 23 kN acting at 80° to the horizontal; force 2, 30 kN acting at 37° to force 1; force 3, 15 kN acting at 70° to force 2.
    [53.5 kN at 37° to force 1 (i.e. 117° to the horizontal)]
17. Force P, 50 kN acting horizontally to the right; force Q, 20 kN at 70° to force P; force R, 40 kN at 170° to force P; force S, 80 kN at 300° to force P.
    [72 kN at −37° to force P]
18. Four horizontal wires are attached to a telephone pole and exert tensions of 30 N to the south, 20 N to the east, 50 N to the north-east and 40 N to the north-west. Determine the resultant force on the pole and its direction.
    [43.18 N, 38°49′ east of north]
19. Four coplanar forces acting on a body are such that it is in equilibrium. The vector diagram for the forces is such that the 60 N force acts vertically upwards, the 40 N force acts at 65° to the 60 N force, the 100 N force acts from the nose of the 60 N force and the 90 N force acts from the nose of the 100 N force. Determine the direction of the 100 N and 90 N forces relative to the 60 N force.
    [100 N force at 263° to the 60 N force]
    [90 N force at 132° to the 60 N force]
20. A load of 12.5 N is lifted by two strings connected to the same point on the load, making angles of 22° and 31° on opposite sides of the vertical. Determine the tensions in the strings.
    [5.8 N; 8.0 N]
21. Determine, by resolution of forces, the resultant of the following three coplanar forces acting at a point: 10 kN acting at 32° to the horizontal; 15 kN acting at 170° to the horizontal; 20 kN acting at 240° to the horizontal.
    [18.82 kN at 210°2′ to the horizontal]
22. In questions 15 to 17, calculate the resultant force in each case by resolution of the forces.
23. The following coplanar forces act at a point: force A, 15 N acting horizontally to the right; force B, 23 N at 81° to the horizontal; force C, 7 N at 210° to the horizontal; force D, 9 N at 265° to the horizontal; force E, 28 N at

324° to the horizontal. Determine the resultant of the five forces by resolution of the forces.

[34.96 N at –10°14′ to the horizontal]

24. Forces of 5 kN, 3 kN, 7 kN and F are coplanar and act at a point on a body. Their directions are 63°, 125°, 302° and $\phi°$ respectively from the horizontal. Use resolution of forces to determine the values of F and ϕ when the forces are in equilibrium.

[$F = 4.37$ kN, $\phi = 192°55′$]

25. A two-legged sling and hoist chain used for lifting machine parts is shown in Fig. 8.33. Determine the forces in each leg of the sling if parts exerting a downward force of 15 kN are lifted.

[9.96 kN, 7.77 kN]

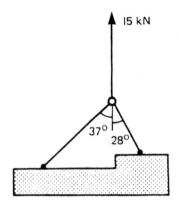

Figure 8.33

9

Speed and velocity

At the end of this chapter you should be able to:

- define speed
- calculate speed given distance and time
- plot a distance/time graph from given data
- determine the average speed from a distance/time graph
- determine the distance travelled from a speed/time graph
- define velocity

9.1. Speed

Speed is the rate of covering distance and is given by:

$$\text{speed} = \frac{\text{distance travelled}}{\text{time taken}}$$

The usual units for speed are metres per second, (m/s or m s^{-1}), or kilometres per hour, (km/h or km h^{-1}). Thus if a person walks 5 kilometres in 1 hour, the speed of the person is 5/1, that is, 5 kilometres per hour.

The symbol for the SI unit of speed (and velocity) is written as m s^{-1}, called the 'index notation'. However, engineers usually use the symbol m/s, called the 'oblique notation', and it is this notation which is largely used in this chapter and other chapters on mechanics. One of the exceptions is when labelling the axes of graphs, when two obliques occur, and in this case the index notation is used. Thus for speed or velocity, the axis markings are speed/m s^{-1} or velocity/m s^{-1}.

Problem 1. A man walks 600 metres in 5 minutes. Determine his speed in (a) metres per second and (b) kilometres per hour.

(a) $\text{Speed} = \dfrac{\text{distance travelled}}{\text{time taken}} = \dfrac{600 \text{ m}}{5 \text{ min}}$

$$= \frac{600 \text{ m}}{5 \text{ min}} \times \frac{1 \text{ min}}{60 \text{ s}} = \textbf{2 m/s}$$

(b) $\dfrac{2 \text{ m}}{1 \text{ s}} = \dfrac{2 \text{ m}}{1 \text{ s}} \times \dfrac{1 \text{ km}}{1000 \text{ m}} \times \dfrac{3600 \text{ s}}{1 \text{ h}} = \textbf{7.2 km/h}$

(Note: to change from m/s to km/h, multiply by 3.6.)

Problem 2. A car travels at 50 kilometres per hour for 24 minutes. Find the distance travelled in this time.

Since

$$\text{speed} = \frac{\text{distance travelled}}{\text{time taken}}$$

then, distance travelled = speed × time taken
Time = 24 minutes = (24/60) hours, hence

$$\text{distance travelled} = 50 \frac{\text{km}}{\text{h}} \times \frac{24}{60} \text{ h} = \textbf{20 km}$$

Problem 3. A train is travelling at a constant speed of 25 metres per second for 16 kilometres. Find the time taken to cover this distance.

Since

$$\text{speed} = \frac{\text{distance travelled}}{\text{time taken}}$$

then

$$\text{time taken} = \frac{\text{distance travelled}}{\text{speed}}$$

16 kilometres = 16 000 metres. Hence,

$$\text{time taken} = \frac{16\,000\,\text{m}}{\dfrac{25\text{m}}{1\,\text{s}}} = 16\,000\,\text{m} \times \frac{1\,\text{s}}{25\,\text{m}}$$

$$= 640\,\text{s}$$

$$640\,\text{s} = 640\,\text{s} \times \frac{1\,\text{min}}{60\,\text{s}} = 10\frac{2}{3}\,\text{min}$$

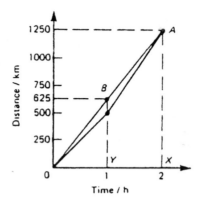

Figure 9.1

9.2 Distance/time graph

One way of giving data on the motion of an object is graphically. A graph of distance travelled (the scale on the vertical axis of the graph) against time (the scale on the horizontal axis of the graph) is called a **distance/time graph**. Thus if an aeroplane travels 500 kilometres in its first hour of flight and 750 kilometres in its second hour of flight, then after 2 hours, the total distance travelled is (500 + 750) kilometres, that is, 1250 kilometres. The distance/time graph for this flight is shown in Fig. 9.1.

The **average speed** is given by

$$\frac{\text{total distance travelled}}{\text{total time taken}}$$

Thus, the average speed of the aeroplane is:

$$\frac{(500 + 750)\,\text{km}}{(1 + 1)\,\text{h}}, \text{ i.e. } \frac{1250}{2} \text{ or } 625\,\text{km/h}$$

If points O and A are joined in Fig. 9.1, the slope of line OA is defined as

$$\frac{\text{change in distance (vertical)}}{\text{change in time (horizontal)}}$$

for any two points on line OA. For point A, the change in distance is AX, that is, 1250 kilometres, and the change in time is OX, that is, 2 hours. Hence the average speed is 1250/2, i.e. 625 kilometres per hour.

Alternatively, for point B on line OA, the change in distance is BY, that is, 625 kilometres and the change in time is OY, that is 1 hour,

hence the average speed is 625/1, i.e. 625 kilometres per hour.

In general, the average speed of an object travelling between points M and N is given by the slope of line MN on the distance/time graph.

Problem 4. A person travels from point O to A, then from A to B and finally from B to C. The distances of A, B and C from O and the times, measured from the start to reach points A, B and C are as shown:

	A	B	C
Distance (m)	100	200	250
Time (s)	40	60	100

Plot the distance/time graph and determine the speed of travel for each of the three parts of the journey.

The vertical scale of the graph is distance travelled and the scale is selected to span 0 to 250 m, the total distance travelled from the start. The horizontal scale is time and spans 0 to 100 seconds, the total time taken to cover the whole journey. Coordinates corresponding to A, B and C are plotted and OA, AB and BC are joined by straight lines. The resulting distance/time graph is shown in Fig. 9.2.

The speed is given by the slope of the distance/time graph. Speed for part OA of the journey = slope of OA = AX/OX

$$= \frac{100\,\text{m}}{40\,\text{s}} = 2\frac{1}{2}\,\text{m/s}$$

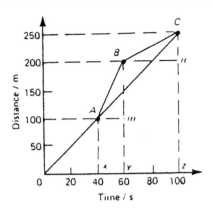

Figure 9.2

Speed for part AB of the journey = slope of AB
= Bm/Am

$$= \frac{(200 - 100) \text{ m}}{(60 - 40) \text{ s}} = \frac{100 \text{ m}}{20 \text{ s}}$$

$$= \textbf{5 m/s}$$

Speed for part BC of the journey = slope of BC
= Cn/Bn

$$= \frac{(250 - 200) \text{ m}}{(100 - 60) \text{ s}}$$

$$= \frac{50 \text{ m}}{40 \text{ s}} = 1\tfrac{1}{4} \textbf{ m/s}$$

Problem 5. Determine the average speed (both in m/s and km/h) for the whole journey for the information given in Problem 4.

Average speed = (total distance travelled)/(total time taken) = slope of line OC. From Fig. 9.2,

$$\text{slope of line OC} = \frac{Cz}{Oz} = \frac{250 \text{ m}}{100 \text{ s}} = \textbf{2.5 m/s}$$

$$2.5 \text{ m/s} = \frac{2.5 \text{ m}}{1 \text{ s}} \times \frac{1 \text{ km}}{1000 \text{ m}} \times \frac{3600 \text{ s}}{1 \text{ h}}$$

$$= 2.5 \times 3.6 \text{ km/h} = \textbf{9 km/h}$$

Problem 6. A coach travels from town A to town B, a distance of 40 kilometres at an average speed of 55 kilometres per hour. It then travels from town B to town C, a distance of 25 kilometres in 35 minutes. Finally, it travels from town C to town D at an average speed of 60 kilometres per hour in 45 minutes. Determine:

(a) the time taken to travel from A to B,
(b) the average speed of the coach from B to C,
(c) the distance from C to D, and
(d) the average speed of the whole journey from A to D.

(a) From town A to town B:
Since

$$\text{speed} = \frac{\text{distance travelled}}{\text{time taken}}$$

then,

$$\text{time taken} = \frac{\text{distance travelled}}{\text{speed}}$$

$$= \frac{40 \text{ km}}{\left(\frac{55 \text{ km}}{1 \text{ h}}\right)} = 40 \text{ km} \times \frac{1 \text{ h}}{55 \text{ km}}$$

$$= \frac{8}{11} \text{ h} \approx \textbf{43.6 min}$$

(b) From town B to town C:
Since

$$\text{speed} = \frac{\text{distance travelled}}{\text{time taken}}$$

and 35 min = $\frac{35}{60}$ h

then,

$$\text{speed} = \frac{25 \text{ km}}{\left(\frac{35}{60} \text{ h}\right)} = \frac{25 \times 60}{35} \text{ km/h}$$

$$\approx \textbf{42.86 km/h}$$

(c) From town C to town D:
Since

$$\text{speed} = \frac{\text{distance travelled}}{\text{time taken}}$$

then, distance travelled = speed × time taken, and 45 min = $\frac{3}{4}$ h, hence,

distance travelled = 60 $\frac{km}{h}$ × $\frac{3}{4}$ h = **45 km**

(d) From town A to town D:

Average speed

$= \dfrac{\text{total distance travelled}}{\text{total time taken}}$

$= \dfrac{(40 + 25 + 45)\ km}{\left(\dfrac{43.6}{60} + \dfrac{35}{60} + \dfrac{45}{60}\ h\right)}$

$= \dfrac{110\ km}{\left(\dfrac{123.6}{60}\ h\right)} = \dfrac{110 \times 60}{123.6}$ km/h

$\approx$ **53.4 km/h**

9.3   Speed/time graph

If a graph is plotted of speed against time, **the area under the graph gives the distance travelled**. This is demonstrated in Problem 7.

Problem 7.   The motion of an object is described by the speed/time graph given in Fig. 9.3. Determine the distance covered by the object when moving from O to B.

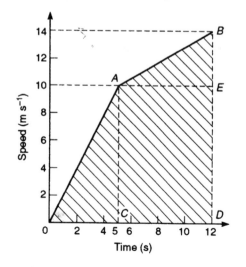

Figure 9.3

The distance travelled is given by the area beneath the speed/time graph, shown shaded in Fig. 9.3.

Area of triangle OAC

$= \dfrac{1}{2}$ × base × perpendicular height

$= \dfrac{1}{2}$ × 5 s × 10 $\dfrac{m}{s}$ = 25 m

Area of rectangle AEDC

= base × height

$= (12 - 5)$ s × $(10 - 0)$ $\dfrac{m}{s}$ = 70 m

Area of triangle ABE

$= \dfrac{1}{2}$ × base × perpendicular height

$= \dfrac{1}{2}$ × $(12 - 5)$ s × $(14 - 10)$ $\dfrac{m}{s}$

$= \dfrac{1}{2}$ × 7 s × 4 $\dfrac{m}{s}$ = 14 m

Hence the distance covered by the object moving from O to B is $(25 + 70 + 14)$ m, i.e. **109 m**.

9.4   Velocity

The **velocity** of an object is the speed of the object **in a specified direction**. Thus, if a plane is flying due south at 500 kilometres per hour, its speed is 500 kilometres per hour, but its velocity is 500 kilometres per hour **due south**. It follows that if the plane had flown in a circular path for one hour at a speed of 500 kilometres per hour, so that one hour after taking off it is again over the airport, its average velocity in the first hour of flight is zero.

The average velocity is given by:

$$\frac{\text{distance travelled in a specific direction}}{\text{time taken}}$$

If a plane flies from place O to place A, a distance of 300 kilometres in one hour, A being due north of O, then OA in Fig. 9.4 represents the first hour of flight. It then flies from A to B,

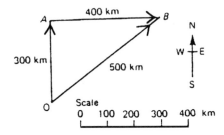

Figure 9.4

a distance of 400 kilometres during the second hour of flight, B being due east of A, thus AB in Fig. 9.4 represents its second hour of flight. Its average velocity for the two hour flight is

$$\frac{\text{distance OB}}{2 \text{ hours}} \text{ i.e. } \frac{500 \text{ km}}{2 \text{ h}}$$

or 250 km/h in direction OB.

A graph of velocity (scale on the vertical axis) against time (scale on the horizontal axis) is called a **velocity/time graph**. The graph shown in Fig. 9.5 represents a plane flying for 3 hours at a constant speed of 600 kilometres per hour in a specified direction. The shaded area represents velocity (vertically) multiplied by time (horizontally), and has units of

$$\frac{\text{kilometres}}{\text{hours}} \times \text{hours}$$

i.e. kilometres, and represents the distance travelled in a specific direction. In this case,

$$\text{distance} = 600 \frac{\text{km}}{\text{h}} \times (3\text{h}) = 1800 \text{ km}$$

Another method of determining the distance travelled is from:

distance travelled = average velocity × time

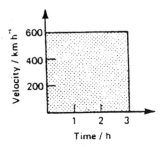

Figure 9.5

Thus if a plane travels due south at 600 kilometres per hour for 20 minutes, the distance covered is

$$\frac{600 \text{ km}}{1 \text{ h}} \times \frac{20}{60} \text{ h, i.e. } 200 \text{ km}$$

9.5 Multi-choice questions on speed and velocity

(Answers on page 355.)

An object travels for 3 s at an average speed of 10 m/s and then for 5 s at an average speed of 15 m/s. In questions 1 to 3, select the correct answers from those given below.

(a) 105 m/s   (b) 3 m   (c) 30 m   (d) $13\frac{1}{8}$ m/s

(e) $3\frac{1}{3}$ m   (f) $\frac{3}{10}$ m   (g) 75 m   (h) $\frac{1}{3}$ m

(i) $12\frac{1}{2}$ m/s

1. The distance travelled in the first 3 s
2. The distance travelled in the latter 5 s
3. The average speed over the 8 s period
4. Which of the following statements is false?
    (a) Speed is the rate of covering distance.
    (b) Speed and velocity are both measured in m/s units.
    (c) Speed is the velocity of travel in a specified direction.
    (d) The area beneath the velocity/time graph gives distance travelled.

In questions 5 to 7, use the table to obtain the quantities stated, selecting the correct answer from (a) to (c) of those given below.

Distance	Time	Speed
20 m	30 s	X
5 km	Y	20 km/h
Z	3 min	10 m/min

(a) 30 m   (b) $\frac{1}{4}$ h   (c) 600 m/s   (d) $\frac{10}{3}$ m

(e) $\frac{2}{3}$ m/s   (f) $\frac{3}{10}$ m   (g) 4 h   (h) $1\frac{1}{4}$ m/s

(i) 100 h

5. Quantity X
6. Quantity Y
7. Quantity Z

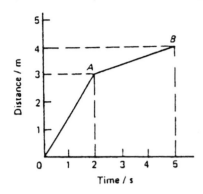

Figure 9.6

In questions 8 to 10, refer to the distance/time graph shown in Fig. 9.6. Select the correct answer from those given below.

(a) $\dfrac{3}{2}$ m/s   (b) 3 m/s   (c) $\dfrac{1}{2}$ m/s

(d) 2 m/s   (e) $\dfrac{4}{5}$ m/s   (f) 6 m/s

(g) $\dfrac{5}{4}$ m/s   (h) $\dfrac{2}{3}$ m/s   (i) $\dfrac{1}{3}$ m/s

8. The average speed when travelling from O to A

9. The average speed when travelling from A to B

10. The average overall speed when travelling from O to B

9.6   Short answer questions on speed and velocity

1. Speed is defined as ....

2. Speed is given by $\dfrac{\text{...............}}{\text{...............}}$

3. The usual units for speed are .... or ....

4. Average speed is given by $\dfrac{\text{...............}}{\text{...............}}$

5. The velocity of an object is ....

6. Average velocity is given by $\dfrac{\text{...............}}{\text{...............}}$

7. The area beneath a velocity/time graph represents the ....

8. Distance travelled = ... × ....

9. The slope of a distance/time graph gives the ....

10. The average speed can be determined from a distance/time graph from ....

9.7   Further questions on speed and velocity

1. A train covers a distance of 96 km in $1\frac{1}{3}$ h. Determine the average speed of the train (a) in km/h and (b) in m/s.
   [(a) 72 km/h   (b) 20 m/s]

2. A horse trots at an average speed of 12 km/h for 18 minutes; determine the distance covered by the horse in this time.
   [3.6 km]

3. A ship covers a distance of 1365 km at an average speed of 15 km/h. How long does it take to cover this distance?
   [3 days 19 hours]

4. Using the information given in the distance/time graph shown in Fig. 9.7, determine the average speed when travelling from O to A, A to B, B to C, O to C and A to C.
   [O to A, 30 km/h; A to B, 40 km/h; B to C, 10 km/h; O to C, 24 km/h; A to C, 20 km/h]

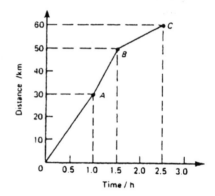

Figure 9.7

5. The distances travelled by an object from point O and the corresponding times taken to reach A, B, C and D, respectively, from the start are as shown:

Points	Start	A	B	C	D
Distance (m)	0	20	40	60	80
Time (s)	0	5	12	18	25

Draw the distance/time graph and hence determine the average speeds from O to A, A to B, B to C, C to D and O to D.
[4 m/s, $2\frac{6}{7}$ m/s, $3\frac{1}{3}$ m/s, $2\frac{6}{7}$ m/s, $3\frac{1}{5}$ m/s]

6. A train leaves station A and travels via stations B and C to station D. The times the train passes the various stations are as shown:

Station A   B   C   D
Times 10.55 am 11.40 am 12.15 pm 12.50 pm
The average speeds are:
A to B, 56 km/h,
B to C, 72 km/h, and
C to D, 60 km/h.

Calculate the total distance from A to D.

[119 km]

7. A gun is fired 5 km north of an observer and the sound takes 15 s to reach him. Determine the average velocity of sound waves in air at this place.

[$333\frac{1}{3}$ m/s or 1200 km/h]

8. The light from a star takes $2\frac{1}{2}$ years to reach an observer. If the velocity of light is 330×10^6 m/s, determine the distance of the star from the observer in kilometres, based on a 365 day year.

[2.6×10^{13} km]

9. The speed/time graph for a car journey is shown in Fig. 9.8. Determine the distance travelled by the car.

[$12\frac{1}{2}$ km]

10. The motion of an object is as follows:

A to B, distance 122 m, time 64 s,
B to C, distance 80 m at an average speed of 20 m/s,
C to D, time 7 s at an average speed of 14 m/s.

Determine the overall average speed of the object when travelling from A to D.

[4 m/s]

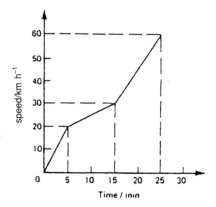

Figure 9.8

10

Acceleration

At the end of this chapter you should be able to:

- define acceleration and know its units
- draw a velocity/time graph
- determine acceleration from a velocity/time graph
- appreciate that 'free fall' has a constant acceleration of 9.8 m/s²
- use the equation of motion $v = u + at$ in calculations

10.1 Introduction to acceleration

Acceleration is the rate of change of velocity with time. The average acceleration, a, is given by:

$$a = \frac{\text{change in velocity}}{\text{time taken}}$$

The usual units are metres per second squared (m/s² or m s⁻²). If u is the initial velocity of an object in metres per second, v is the final velocity in metres per second and t is the time in seconds elapsing between the velocities of u and v, then

$$\textbf{average acceleration, } a = \frac{v - u}{t} \textbf{ m/s}^2$$

10.2 Velocity/time graph

A graph of velocity (scale on the vertical axis) against time (scale on the horizontal axis) is called a **velocity/time graph**, as introduced in Chapter 9. For the velocity/time graph shown in Fig. 10.1, the slope of line OA is given by AX/OX. AX is the change in velocity from an initial velocity u of zero to a final velocity, v, of 4 metres per second. OX is the time taken for this change in velocity, thus

$$\frac{AX}{OX} = \frac{\text{change in velocity}}{\text{time taken}}$$

= the acceleration in the first two seconds

From the graph:

$$\frac{AX}{OX} = \frac{4 \text{ m/s}}{2 \text{ s}} = 2 \text{ m/s}^2$$

i.e. the acceleration is 2 m/s². Similarly, the slope of line AB in Fig. 10.1 is given by BY/AY, i.e. the acceleration between 2 and 5 is

$$\frac{8 - 4}{5 - 2} = \frac{4}{3} = 1\frac{1}{3} \text{ m/s}^2$$

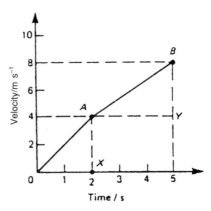

Figure 10.1

In general, the slope of a line on a velocity/time graph gives the acceleration.

The words 'velocity' and 'speed' are commonly interchanged in everyday language. Acceleration is a vector quantity and is correctly defined as the rate of change of velocity with respect to time. However, acceleration is also the rate of change of speed with respect to time in a certain specified direction.

Problem 1.   The speed of a car travelling along a straight road changes uniformly from zero to 50 km/h in 20 s. It then maintains this speed for 30 s and finally reduces speed uniformly to rest in 10 s. Draw the speed/time graph for this journey.

The vertical scale of the speed/time graph is speed (km h⁻¹) and the horizontal scale is time (s). Since the car is initially at rest, then at time 0 seconds, the speed is 0 km/h. After 20 s, the speed if 50 km/h, which corresponds to point A on the speed/time graph shown in Fig. 10.2. Since the change in speed is uniform, a straight line is drawn joining points O and A. The speed is constant at 50 km/h for the next 30 s, hence, horizontal line AB is drawn in Fig. 10.2 for the time period 20 s to 50 s. Finally, the speed falls from 50 km/h at 50 s to zero in 10 s, hence point C on the speed/time graph in Fig. 10.2 corresponds to a speed of zero and a time of 60 s. Since the reduction in speed is uniform, a straight line is drawn joining BC. Thus, the speed/time graph for the journey is as shown in Fig. 10.2.

Problem 2.   For the speed/time graph shown in Fig. 10.2, find the acceleration for each of the three stages of the journey.

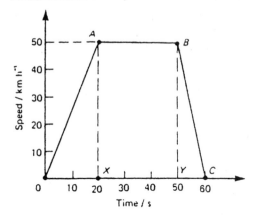

Figure 10.2

From above, the slope of line OA gives the uniform acceleration for the first 20 s of the journey

$$\text{Slope of OA} = \frac{AX}{OX} = \frac{(50 - 0) \text{ km/h}}{(20 - 0) \text{ s}}$$

$$= 50 \frac{\text{km}}{\text{h}} \times \frac{1}{20 \text{ s}}$$

Expressing 50 km/h in metre-second units gives:

$$50 \frac{\text{km}}{\text{h}} = \frac{50 \text{ km}}{1 \text{ h}} \times \frac{1000 \text{ m}}{1 \text{ km}} \times \frac{1 \text{ h}}{3600 \text{ s}}$$

$$= \frac{50}{3.6} \text{ m/s}$$

(Note: to change from km/h to m/s, divide by 3.6.)
Thus,

$$50 \text{ km/h} \times \frac{1}{20 \text{ s}} = \frac{50}{3.6} \text{ m/s} \times \frac{1}{20 \text{ s}}$$

$$= 0.694 \text{ m/s}^2$$

i.e. **the acceleration during the first 20 s is 0.694 m/s².**
Acceleration is defined as

$$\frac{\text{change of velocity}}{\text{time taken}}$$

or

$$\frac{\text{change of speed}}{\text{time taken}}$$

since the car is travelling along a straight road. Since there is no change in speed for the next 30 s (line AB in Fig. 10.2 is horizontal), **then the acceleration for this period is zero.**
From above, the slope of line BC gives the uniform deceleration for the final 10 s of the journey.

$$\text{Slope of BC} = \frac{BY}{YC} = \frac{50 \text{ km/h}}{10 \text{ s}}$$

$$= \frac{50 \text{ m}}{3.6 \text{ s}} \times \frac{1}{10 \text{ s}}$$

$$= 1.39 \text{ m/s}^2$$

i.e. the deceleration during the final 10 s is 1.39 m/s².
Alternatively, **the acceleration is –1.39 m/s².**

10.3   Free-fall and equation of motion

If a dense object such as a stone is dropped from a height, called **free-fall**, it has a constant acceleration of approximately 9.8 m/s². In a vacuum, all objects have this same constant acceleration, vertically downwards, that is, a feather has the same acceleration as a stone. However, if free-fall takes place in air, dense objects have the constant acceleration of 9.8 m/s² over short distances, but objects which have a low density, such as feathers, have little or no acceleration.

For bodies moving with a constant acceleration, the average acceleration is the constant value of the acceleration, and since from section 10.1:

$$a = \frac{v - u}{t}$$

then

$$a \times t = v - u \text{ or } \boldsymbol{v = u + at}$$

where

    u is the initial velocity in m/s,
    v is the final velocity in m/s,
    a is the constant acceleration in m/s²,
    t is the time in s.

When symbol 'a' has a negative value, it is called **deceleration** or **retardation**. The equation $v = u + at$ is called an **equation of motion**.

Problem 3.   A stone is dropped from an aeroplane. Determine (a) its velocity after 2 s and (b) the increase in velocity during the third second, in the absence of all forces except that due to gravity.

The stone is free-falling and thus has an acceleration, a, of approximately 9.8 m/s² (taking downward motion as positive). From above:

    final velocity, $v = u + at$

(a) The initial downward velocity of the stone, u, is zero. The acceleration, a, is 9.8 m/s² downwards and the time during which the stone is accelerating is 2 s. Hence, final velocity, $v = 0 + 9.8 \times 2 = 19.6$ m/s, i.e. **the velocity of the stone after 2 s is approximately 19.6 m/s.**
(b) From part (a) above, the velocity after two seconds, u, is 19.6 m/s. The velocity after 3 s, applying $v = u + at$, is

$$v \approx 19.6 + 9.8 \times 3 \approx 49 \text{ m/s}$$

Thus, **the change in velocity during the third second is**

    (49 – 19.6) m/s, that is, approximately **29.4 m/s**

(Since the value $a = 9.8$ m/s² is only an approximate value, then the answer can only be an approximate value.)

Problem 4.   Determine how long it takes an object, which is free-falling, to change its speed from 100 km/h to 150 km/h, assuming all other forces, except that due to gravity, are neglected.

The initial velocity, u is 100 km/h, i.e. (100/3.6) m/s (see Problem 2). The final velocity, v, is 150 km/h, i.e. (150/3.6) m/s. Since the object is free-falling, the acceleration, a, is approximately 9.8 m/s² downwards (i.e. in a positive direction). From above, $v = u + at$, i.e.

$$\frac{150}{3.6} = \frac{100}{3.6} + 9.8 \times t$$

Transposing, gives

$$9.8 \times t = \frac{150 - 100}{3.6} = \frac{50}{3.6}$$

Hence,

$$t = \frac{50}{3.6 \times 9.8} \approx 1.42 \text{ s}$$

Since the value of a is only approximate, and rounding-off errors have occurred in calculations, then **the approximate time for the velocity to change from 100 km/h to 150 km/h is 1.42 s.**

Problem 5.   A train travelling at 30 km/h accelerates uniformly to 50 km/h in 2 minutes. Determine the acceleration.

$$30 \text{ km/h} = \frac{30}{3.6} \text{ m/s}\quad \text{(see Problem 2)}$$

$$50 \text{ km/h} = \frac{50}{3.6} \text{ m/s}$$

  2 min $= 2 \times 60 = 120$ s

From above, $v = u + at$, i.e.

$$\frac{50}{3.6} = \frac{30}{3.6} + a \times 120$$

Transposing, gives

$$120 \times a = \frac{50 - 30}{3.6}$$

$$a = \frac{20}{3.6 \times 120} = 0.0463 \text{ m/s}^2$$

i.e. **the uniform acceleration of the train is 0.0463 m/s^2**.

Problem 6.   A car travelling at 50 km/h applies its brakes for 6 s and decelerates uniformly at 0.5 m/s. Determine its velocity in km/h after the 6 s braking period.

The initial velocity, u = 50 km/h = (50/3.6) m/s (see Problem 2). From above, $v = u + at$. Since the car is decelerating, i.e. it has a negative acceleration, then $a = -0.5$ m/s^2 and t is 6 s. Thus,

$$\text{final velocity, } v = \frac{50}{3.6} + (-0.5)(6)$$

$$= 13.\dot{8} - 3 = 10.\dot{8} \text{ m/s}$$

$$10.\dot{8} \text{m/s} = 10.\dot{8}\,\frac{\text{m}}{\text{s}} \times \frac{1 \text{ km}}{1000 \text{ m}} \times \frac{3600 \text{ s}}{1 \text{ h}}$$

$$= 10.\dot{8} \times 3.6 \text{ km/h} = 39.2 \text{ km/h}$$

(Note: to convert m/s to km/h, multiply by 3.6.)
Thus, the velocity after braking is 39.2 km/h.

Problem 7.   A cyclist accelerates uniformly at 0.3 m/s^2 for 10 s, and his speed after accelerating is 20 km/h. Find his initial speed.

The final speed, v, is (20/3.6) m/s. Time, t, is 10 s. Acceleration, a, is 0.3 m/s^2. From above, $v = u + at$, where u is the initial speed. Hence,

$$\frac{20}{3.6} = u + 0.3 \times 10$$

$$u = \frac{20}{3.6} - 3 = 2.\dot{5} \text{ m/s}$$

2.5 m/s = 2.5 × 3.6 km/h (see Problem 6) = 9.2 km/h

i.e. **the initial speed of the cyclist is 9.2 km/h**.

10.4   Multi-choice questions on acceleration

(Answers on page 355.)

Six statements, (a) to (f), are given below, some of the statements being true and the remainder false.

(a) Acceleration is the rate of change of velocity with distance.
(b) Average acceleration = (change of velocity)/(time taken).
(c) Average acceleration = $(u - v)/t$, where u is the initial velocity, v is the final velocity and t is the time.
(d) The slope of a velocity/time graph gives the acceleration.
(e) The acceleration of a dense object during free-fall is approximately 9.8 m/s^2 in the absence of all other forces except gravity.
(f) When the initial and final velocities are u and v, respectively, a is the acceleration and t the time, then $u = v + at$.

In problems 1 and 2, select the statements required from those given

1. (b), (c), (d), (e) Which statement is false?
2. (a), (c), (e), (f) Which statement is true?

A car accelerates uniformly from 5 m/s to 15 m/s in 20 s. It stays at the velocity attained at 20 s for 2 min. Finally, the brakes are applied to give a uniform deceleration and it comes to rest in 10 s. Use this data in Problems 3 to 7, selecting the correct answer from (a)–(l) given below.

(a) –1.5 m/s^2 (b) $\frac{2}{15}$ m/s^2 (c) 0
(d) 0.5 m/s^2 (e) 1.3$\dot{8}$ km/h   (f) 7.5 m/s^2
(g) 54 km/h   (h) 2 m/s^2 (i) 18 km/h
(j) $-\frac{1}{10}$ m/s^2 (k) 1.4$\dot{6}$ km/h   (l) $-\frac{2}{3}$ m/s^2

3. The initial speed of the car in km/h
4. The speed of the car after 20 s in km/h
5. The acceleration during the first 20 s period
6. The acceleration during the 2 min period
7. The acceleration during the final 10 s

10.5   Short answer questions on acceleration

1. Acceleration is defined as ....
   ....
2. Acceleration is given by $\dfrac{\text{...............}}{\text{...............}}$

3. The usual units for acceleration are ....
4. The slope of a velocity/time graph gives the ....
5. The value of free-fall acceleration for a dense object is approximately ....
6. The relationship between initial velocity, u, final velocity, v, acceleration, a, and time, t, is ....
7. A negative acceleration is called a .... or a ....

10.6   Further questions on acceleration

1. A coach increases velocity from 4 km/h to 40 km/h at an average acceleration of 0.2 m/s^2. Find the time taken for this increase in velocity.   [50 s]
2. A ship changes velocity from 15 km/h to 20 km/h in 25 min. Determine the average acceleration in m/s^2 of the ship during this time.   [9.26×10^{-4} m/s^2]
3. A cyclist travelling at 15 km/h changes velocity uniformly to 20 km/h in 1 min, maintains this velocity for 5 min and then comes to rest uniformly during the next 15 s. Draw a velocity/time graph and hence determine the accelerations in m/s^2 (a) during the first minute, (b) for the next 5 minutes, and (c) for the last 10 s.
   [(a) 0.0231 m/s^2 (b) 0 (c) −0.370 m/s^2]
4. Assuming uniform accelerations between points, draw the velocity/time graph for the data given below, and hence determine the accelerations from A to B, B to C and C to D:

Point	A	B	C	D
Speed (m/s)	25	5	30	15
Time (s)	15	25	35	45

[A to B −2 m/s^2, B to C 2.5 m/s^2, C to D −1.5 m/s^2]

5. An object is dropped from the third floor of a building. Find its approximate velocity 1.25 s later if all forces except that of gravity are neglected.   [12.25 m/s]
6. During free fall, a ball is dropped from point A and is travelling at 100 m/s when it passes point B. Calculate the time for the ball to travel from A to B if all forces except that of gravity are neglected.   [10.2 s]
7. A piston moves at 10 m/s at the centre of its motion and decelerates uniformly at 0.8 m/s^2. Determine its velocity 3 s after passing the centre of its motion.   [7.6 m/s]
8. The final velocity of a train after applying its brakes for 1.2 min is 24 km/h. If its uniform retardation is 0.06 m/s^2, find its velocity before the brakes are applied.   [39.6 km/h]
9. A plane in level flight at 400 km/h starts to descend at a uniform acceleration of 0.6 m/s^2. It levels off when its velocity is 670 km/h. Calculate the time during which it is losing height.   [2 min 5 s]
10. A lift accelerates from rest uniformly at 0.9 m/s^2 for 1.5 s, travels at constant velocity for 7 s and then comes to rest in 3 s. Determine its velocity when travelling at constant speed and its acceleration during the final 3 s of its travel.

[1.35 m/s, −0.45 m/s^2]

11

Force, mass and acceleration

At the end of this chapter you should be able to:

- define a force and know its unit
- appreciate 'gravitational force'
- state Newton's three laws of motion
- perform calculations involving force $F = ma$
- define 'centripetal acceleration'
- perform calculations involving centripetal force $= mv^2/r$

11.1 Introduction

When an object is pushed or pulled, a force is applied to the object. This force is measured in **newtons (N)**. The effects of pushing or pulling an object are:

(i) to cause a change in the motion of the object, and

(ii) to cause a change in the shape of the object.

If a change occurs in the motion of the object, that is, its velocity changes from u to v, then the object accelerates. Thus, it follows that acceleration results from a force being applied to an object. If a force is applied to an object and it does not move, then the object changes shape, that is, deformation of the object takes place. Usually the change in shape is so small that it cannot be detected by just watching the object. However, when very sensitive measuring instruments are used, very small changes in dimensions can be detected.

A force of attraction exists between all objects. The factors governing the size of this force F are the masses of the objects and the distances between their centres:

$$F \propto \frac{m_1 m_2}{d^2}$$

Thus, if a person is taken as one object and the earth as a second object, a force of attraction exists between the person and the earth. This force is called the **gravitational force** and is the force which gives a person a certain weight when standing on the earth's surface. It is also this force which gives freely falling objects a constant acceleration in the absence of other forces.

11.2 Newton's laws of motion

To make a stationary object move or to change the direction in which the object is moving requires a force to be applied externally to the object. This concept is known as **Newton's first law of motion** and may be stated as:

An object remains in a state of rest, or continues in a state of uniform motion in a straight line, unless it is acted on by an externally applied force

Since a force is necessary to produce a change of motion, an object must have some resistance to a change in its motion. The force necessary to give a stationary pram a given acceleration is far less than the force necessary to give a stationary car the same acceleration. The resistance to a change in motion is called the **inertia** of an object and the amount of inertia depends on the mass of the object. Since a car has a much larger mass than a pram, the inertia of a car is much larger than that of a pram.

Newton's second law of motion may be stated as:

The acceleration of an object acted upon by an external force is proportional to the force and is in the same direction as the force

Thus, force ∝ acceleration, or force = a constant × acceleration, this constant of proportionality being the mass of the object, i.e.

force = mass × acceleration

The unit of force is the newton (N) and is defined in terms of mass and acceleration. One newton is the force required to give a mass of 1 kilogram an acceleration of 1 metre per second squared. Thus

$$F = ma$$

where F is the force in newtons (N), m is the mass in kilograms (kg) and a is the acceleration in metres per second squared (m/s²), i.e.

$$1 \text{ N} = \frac{1 \text{ kg m}}{s^2}$$

It follows that $1 \text{ m/s}^2 = 1 \text{ N/kg}$. Hence a gravitational acceleration of 9.8 m/s² is the same as a gravitational field of 9.8 N/kg.

Newton's third law of motion may be stated as:

For every force, there is an equal and opposite reacting force

Thus, an object on, say, a table, exerts a downward force on the table and the table exerts an equal upward force on the object, known as a **reaction force** or just a **reaction**.

Problem 1.   Calculate the force needed to accelerate a boat of mass 20 tonne uniformly from rest to a speed of 21.6 km/h in 10 minutes.

The mass of the boat, m, is 20 t, that is 20 000 kg. The law of motion, $v = u + at$ can be used to determine the acceleration a. The initial velocity, u, is zero. The final velocity, v, is 21.6 km/h, that is, 21.6/3.6 or 6 m/s. The time, t, is 10 min, that is, 600 s. Thus

$$6 = 0 + a \times 600 \text{ or } a = \frac{6}{600} = \textbf{0.01 m/s}^2$$

From Newton's second law, $F = ma$, i.e.

Force = 20 000 × 0.01 N

= **200 N**

Problem 2.   The moving head of a machine tool requires a force of 1.2 N to bring it to rest in 0.8 s from a cutting speed of 30 m/min. Find the mass of the moving head.

From Newton's second law, $F = ma$, thus $m = F/a$, where force is given as 1.2 N. The law of motion $v = u + at$ can be used to find acceleration a, where $v = 0$, $u = 30$ m/min, that is 30/60 or 0.5 m/s, and $t = 0.8$ s. Thus,

$$0 = 0.5 + a \times 0.8, \text{ i.e.}$$

$$a = -\frac{0.5}{0.8} = -0.625 \text{ m/s}^2$$

or a retardation of 0.625 m/s².
Thus the mass, $m = 1.2/0.625 = \textbf{1.92 kg}$.

Problem 3.   A lorry of mass 1350 kg accelerates uniformly from 9 km/h to reach a velocity of 45 km/h in 18 s. Determine (a) the acceleration of the lorry, (b) the uniform force needed to accelerate the lorry.

(a)   The law of motion $v = u + at$ can be used to determine the acceleration, where final velocity $v = (45/3.6)$ m/s, initial velocity $u = (9/3.6)$ m/s and time t is 18 s. Thus,

$$\frac{45}{3.6} = \frac{9}{3.6} + a \times 18$$

$$a = \frac{1}{18}\left(\frac{45}{3.6} - \frac{9}{3.6}\right) = \frac{1}{18} \times \frac{36}{3.6} = \frac{10}{18} = \frac{5}{9} \text{ m/s}^2$$

(b)   From Newton's second law of motion,

$$F = ma = 1350 \times \frac{5}{9} = \textbf{750 N}$$

Problem 4.   Find the weight of an object of mass 1.6 kg at a point on the earth's surface where the gravitational field is 9.81 N/kg.

The weight of an object is the force acting vertically downwards due to the force of gravity acting on the object. Thus:

Weight = force acting vertically downwards

= mass × gravitational field

= 1.6 × 9.81 = **15.696 N**

Problem 5.   A bucket of cement of mass 40 kg is tied to the end of a rope connected to a hoist. Calculate the tension in the rope when the bucket is suspended but stationary. Take the gravitational field, *g*, as 9.81 N/kg.

The **tension** in the rope is the same as the force acting in the rope. The force acting vertically downwards due to the weight of the bucket must be equal to the force acting upwards in the rope, i.e. the tension.
Weight of bucket of cement, $F = mg = 40 \times 9.81$ = 392.4 N.
Thus, the tension in the rope is also **392.4 N.**

Problem 6.   The bucket of cement in Problem 5 is now hoisted vertically upwards with a uniform acceleration of 0.4 m/s². Calculate the tension in the rope during the period of acceleration.

With reference to Fig. 11.1(a), the forces acting on the bucket are:

(i)   a tension (or force) of *T* acting in the rope;
(ii) a force of *mg* acting vertically downwards, i.e. the weight of the bucket and cement.

The resultant force $F = T - mg$. Hence $ma = T - mg$.

$$40 \times 0.4 = T - 40 \times 9.81 \text{ giving } \boldsymbol{T = 408.4\text{ N}}$$

By comparing this result with that of Problem 5, it can be seen that there is an increase in the tension in the rope when an object is accelerating upwards.

Problem 7.   The bucket of cement in Problem 5 is now lowered vertically downwards with a uniform acceleration of 1.4 m/s². Calculate the tension in the rope during the period of acceleration.

With reference to Fig. 11.1(b), the forces acting on the bucket are:

(i)   a tension (or force) of *T* acting vertically upwards;
(ii) a force of *mg* acting vertically downwards, i.e. the weight of the bucket and cement.

The resultant force $F = mg - T$. Hence, $ma = mg - T$, i.e.

$$T = m(g - a) = 40(9.81 - 1.4) = \boldsymbol{336.4\text{ N}}$$

By comparing this result with that of Problem 5, it can be seen that there is a decrease in the tension in the rope when an object is accelerating downwards.

Problem 8.   Two masses are suspended vertically by a thread over a pulley and are at the same height. One of the masses is 10 kg and the system has an acceleration of 1.2 m/s² when released, the acceleration being towards the 10 kg mass. Determine the value of the other mass, assuming *g* is 9.81 m/s² and all losses and the mass of the pulley are neglected.

Let the unknown mass be *m*, the resultant force be *F*, and the tension in the thread be *T* (see Fig. 11.2). Since the 10 kg mass accelerates downwards, the tension *T* in the thread is less than *mg*. Hence, resultant force = $10g - T$ and the acceleration of 10 kg mass

$$= \frac{10g - T}{10} = 1.2 \tag{11.1}$$

Resultant force on mass, $m = T - mg$ and acceleration of mass,

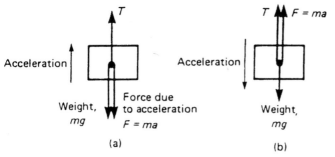

Figure 11.1

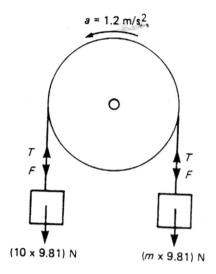

Figure 11.2

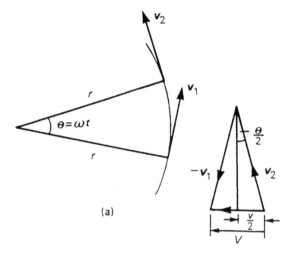

(a)

Figure 11.3

$$\alpha = \frac{T - mg}{m} = 1.2 \qquad (11.2)$$

From equation (11.1), $10g - T = 12$.
From equation (11.2), $T - mg = 1.2\,m$.
Adding gives:

$$10g - mg = 12 + 1.2\,m$$

$$98.1 - 12 = m(1.2 + 9.81)$$

i.e. $$m = \frac{86.1}{11.01} = 7.82 \text{ kg}$$

11.3   Centripetal acceleration

When an object moves in a circular path at
constant speed, its direction of motion is continually changing and hence its velocity (which
depends on both magnitude **and direction**) is also
continually changing. Since acceleration is the
(change in velocity)/(time taken) the object has
an acceleration.

Let the object be moving with a constant
angular velocity of ω and a tangential velocity of
magnitude v and let the change of velocity for a
small change of angle of θ ($= \omega t$) be V (see Fig.
11.3(a)).

Then, $v_2 - v_1 = V$. The vector diagram is shown
in Fig. 11.3(b) and since the magnitudes of v_1 and
v_2 are the same, i.e. v, the vector diagram is also
an isosceles triangle.

Bisecting the angle between v_2 and v_1 gives:

$$\sin \frac{\theta}{2} = \frac{V/2}{v_2} = \frac{V}{2v}$$

i.e.

$$V = 2v \sin \frac{\theta}{2} \qquad (11.1)$$

Since $\theta = \omega t$,

$$t = \theta/\omega \qquad (11.2)$$

Dividing (11.1) by (11.2) gives:

$$\frac{V}{t} = \frac{2v \sin \dfrac{\theta}{2}}{\dfrac{\theta}{\omega}} = \frac{v\omega \sin \dfrac{\theta}{2}}{\dfrac{\theta}{2}}$$

For small angles,

$$\frac{\sin \dfrac{\theta}{2}}{\dfrac{\theta}{2}}$$

is very nearly equal to unity. Hence,

$$\frac{V}{t} = \frac{\text{change of velocity}}{\text{change of time}}$$

$$= \text{acceleration}, \ a = v\omega$$

But, $\omega = v/r$, thus

$$v\omega = v \times \frac{v}{r} = \frac{v^2}{r}$$

That is, the acceleration a is v^2/r and is towards the centre of the circle of motion (along V). It is called the **centripetal acceleration**. If the mass of the rotating object is m, then by Newton's second law, the **centripetal force** is mv^2/r, and its direction is towards the centre of the circle of motion.

Problem 9. A vehicle of mass 750 kg travels round a bend of radius 150 m, at 50.4 km/h. Determine the centripetal force acting on the vehicle.

The centripetal force is given by mv^2/r and its direction is towards the centre of the circle.
$m = 750$ kg, $v = 50.4$ km/h $= (50.4/3.6)$ m/s $= 14$ m/s, $r = 150$ m, thus

$$\text{centripetal force} = \frac{750 \times (14)^2}{150} = \textbf{980 N}$$

Problem 10. An object is suspended by a thread 250 mm long and both object and thread move in a horizontal circle with a constant angular velocity of 2.0 rad/s. If the tension in the thread is 12.5 N, determine the mass of the object.

Centripetal force (i.e. tension in thread) $= mv^2/r$ $= 12.5$ N. The angular velocity $\omega = 2.0$ rad/s and radius $r = 250$ mm $= 0.25$ m. Since linear velocity $v = \omega r$, $v = 2.0 \times 0.25 = 0.5$ m/s. Since $F = mv^2/r$, then $m = Fr/v^2$, i.e.

$$\text{mass of object, } m = \frac{12.5 \times 0.25}{0.5^2} = \textbf{12.5 kg}$$

Problem 11. An aircraft is turning at constant altitude, the turn following the arc of a circle of radius 1.5 km. If the maximum allowable acceleration of the aircraft is 2.5 g, determine the maximum speed of the turn in km/h. Take g as 9.8 m/s².

The acceleration of an object turning in a circle is v^2/r. Thus, to determine the maximum speed of turn $v^2/r = 2.5$ g.

$$v = \sqrt{(2.5\, gr)} = \sqrt{(2.5 \times 9.8 \times 1500)} = \sqrt{36\,750}$$
$$= 191.7 \text{ m/s}$$

191.7 m/s $= 191.7 \times 3.6$ km/h $= \textbf{690 km/h}$

11.4 Multi-choice questions on force, mass and acceleration

(Answers on page 355.)

A man of mass 75 kg is standing in a lift of mass 500 kg. Use this data to determine the answers to questions 1 to 6. Take g as 10 m/s².

1. The tension in a cable when the lift is moving at a constant speed vertically upward is:
   (a) 4250 N (b) 5750 N (c) 4600 N (d) 6900 N
2. The tension in the cable supporting the lift when the lift is moving at a constant speed vertically downwards is:
   (a) 4250 N (b) 5750 N (c) 4600 N (d) 6900 N
3. The reaction force between the man and the floor of the lift when the lift is travelling at a constant speed vertically upwards is:
   (a) 750 N (b) 900 N (c) 600 N (d) 475 N
4. The reaction force between the man and the floor of the lift when the lift is travelling at a constant speed vertically downwards is:
   (a) 750 N (b) 900 N (c) 600 N (d) 475 N
5. The reaction force between the man and the floor of the lift when the lift is accelerating at a rate of 2 m/s² vertically upwards is:
   (a) 750 N (b) 900 N (c) 600 N (d) 475 N
6. The reaction force between the man and the floor of the lift when the lift is accelerating at a rate of 2 m/s² vertically downwards is:
   (a) 750 N (b) 900 N (c) 600 N (d) 475 N

A ball of mass 0.5 kg is tied to a thread and rotated at a constant angular velocity of 10 rad/s in a circle of radius 1 m. Use this data to determine the answers to questions 7 and 8.

7. The centripetal acceleration is:
   (a) 50 m/s² (b) $\frac{100}{2\pi}$ m/s² (c) $\frac{50}{2\pi}$ m/s²
   (d) 100 m/s².
8. The tension in the thread is:
   (a) 25 N (b) $\frac{50}{2\pi}$ N (c) $\frac{25}{2\pi}$ N (d) 50 N
9. Which of the following statements is false?
   (a) An externally applied force is needed to change the direction of a moving object.
   (b) For every force, there is an equal and opposite reaction force.
   (c) A body travelling at a constant velocity in a circle has no acceleration.
   (d) Centripetal acceleration acts towards the centre of the circle of motion.

10. Which of the following statements is true?
   (a) The acceleration of an object acted upon by an external force is proportional to the force and acts in the opposite direction to the force.
   (b) The inertia of an object is the resistance of an object to a change in motion.
   (c) The tension in a cable supporting a lift is greater when the lift is moving with a constant speed vertically upwards than when it is stationary.
   (d) The tension in a cable supporting a lift remains constant irrespective of its motion.

11.5   Short answer questions on force, mass and acceleration

1. Force is measured in ....
2. The two effects of pushing or pulling an object are .... or ....
3. A gravitational force gives free-falling objects a .... in the absence of all other forces.
4. State Newton's first law of motion.
5. Describe what is meant by the inertia of an object.
6. State Newton's second law of motion.
7. Define the newton.
8. State Newton's third law of motion.
9. Explain why an object moving round a circle at a constant angular velocity has an acceleration.
10. Define centripetal acceleration in symbols.
11. Define centripetal force in symbols.

11.6   Further questions on force, mass and acceleration

(Take g as 9.81 m/s^2, and express answers to three significant figure accuracy.)

1. A car initially at rest accelerates uniformly to a speed of 55 km/h in 14 s. Determine the accelerating force required if the mass of the car is 800 kg.   [873 N]
2. The brakes are applied on the car in question 1 when travelling at 55 km/h and it comes to rest uniformly in a distance of 50 m. Calculate the braking force and the time for the car to come to rest.   [1.87 kN, 6.55 s]
3. The tension in a rope lifting a crate vertically upwards is 2.8 kN. Determine its acceleration if the mass of the crate is 270 kg.   [0.560 m/s^2]
4. A ship is travelling at 18 km/h when it stops its engines. It drifts for a distance of 0.6 km and its speed is then 14 km/h. Determine the value of the forces opposing the motion of the ship, assuming the reduction in speed is uniform and the mass of the ship is 2000 t.   [16.5 kN]

A cage having a mass of 2 t is being lowered down a mine shaft. It moves from rest with an acceleration of 4 m/s^2, until it is travelling at 15 m/s. It then travels at constant speed for 700 m and finally comes to rest in 6 s. Use this data to find the answers to questions 5 and 6.

5. Calculate the tension in the cable supporting the cage during (a) the initial period of acceleration, (b) the period of constant speed travel, (c) the final retardation period.   [(a) 11.6 kN (b) 19.6 kN (c) 24.6 kN]
6. A miner having a mass of 80 kg is standing in the cage. Determine the reaction force between the man and the floor of the cage during (a) the initial period of acceleration, (b) the period of constant speed travel, (c) the final retardation period.   [(a) 465 N (b) 785 N (c) 985 N]
7. During an experiment, masses of 4 kg and 5 kg are attached to a thread and the thread is passed over a pulley so that both masses hang vertically downwards and are at the same height. When the system is released, find (a) the acceleration of the system, (b) the tension in the thread, assuming no losses in the system.   [(a) 1.09 m/s^2 (b) 43.6 N]
8. Calculate the centripetal force acting on a vehicle of mass 1 tonne when travelling round a bend of radius 125 m at 40 km/h. If this force should not exceed 750 N, determine the reduction in speed of the vehicle to meet this requirement.   [988 N, 5.1 km/h]
9. A speed-boat negotiates an S-bend consisting of two circular arcs of radii 100 m and 150 m. If the speed of the boat is constant at 34 km/h, determine the change in acceleration when leaving one arc and entering the other.   [1.49 m/s^2]
10. Derive the centripetal force formula $F = mv^2/r$, from first principles for an object of mass m moving at constant angular velocity, ω, in a circle of radius r on a horizontal plane.

12

Simply supported beams

At the end of this chapter you should be able to:

- define a 'moment' of a force and state its unit
- calculate the moment of a force from $M = F \times d$
- understand the conditions for equilibrium of a beam
- state the principle of moments
- perform calculations involving the principle of moments
- recognize typical practical applications of simply supported beams with point loadings
- perform calculations on simply supported beams having point loads

12.1 The moment of a force

When using a spanner to tighten a nut, a force tends to turn the nut in a clockwise direction. This turning effect of a force is called the **moment of a force** or more briefly, just a **moment**. The size of the moment acting on the nut depends on two factors:

(a) the size of the force acting at right angles to the shank of the spanner, and
(b) the perpendicular distance between the point of application of the force and the centre of the nut.

In general, with reference to Fig. 12.1, the moment M of a force acting about a point P is

force × perpendicular distance between the line of action of the force and P, i.e.

$$M = F \times d$$

The unit of a moment is the newton metre (Nm). Thus, if force F in Fig. 12.1 is 7 N and distance d is 3 m, then the moment M is 7 N × 3 m, i.e. 21 Nm.

Problem 1. A force of 15 N is applied to a spanner at an effective length of 140 mm from the centre of a nut. Calculate (a) the moment of the force applied to the nut, (b) the magnitude of the force required to produce the same moment if the effective length is reduced to 100 mm.

From above, $M = F \times d$, where M is the turning moment, F is the force applied at right angles to the spanner and d is the effective length between the force and the centre of the nut. Thus, with reference to Fig. 12.2(a):

(a) Turning moment,

$M = 15\,\text{N} \times 140\,\text{mm} = 2100\,\text{N mm}$

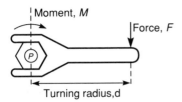

Figure 12.1

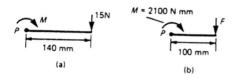

(a)

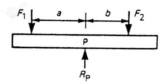

(b)

Figure 12.2

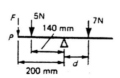

Figure 12.3

$$= 2100 \text{ N mm} \times \frac{1 \text{ m}}{1000 \text{ mm}} = \textbf{2.1 Nm}$$

(b) Turning moment, M is 2100 N mm and the effective length d becomes 100 mm (see Fig. 12.2(b)). Applying $M = F \times d$ gives:

2100 N mm $F \times 100$ mm

from which,

$$\text{force } F = \frac{2100 \text{ N mm}}{100 \text{ mm}} = \textbf{21 N}$$

Problem 2. A moment of 25 Nm is required to operate a lifting jack. Determine the effective length of the handle of the jack if the force applied to it is (a) 125 N, (b) 0.4 kN.

From above, moment $M = F \times d$, where F is the force applied at right angles to the handle and d is the effective length of the handle. Thus:

(a) 25 Nm $= 125$ N $\times d$, from which

$$\text{effective length, } d = \frac{25 \text{ Nm}}{125 \text{ N}} = \frac{1}{5} \text{ m}$$

$$= \frac{1000}{5} \text{ mm}$$

$$= \textbf{200 mm}$$

(b) Turning moment M is 25 Nm and the force F becomes 0.4 kN, i.e. 400 N. Since $M = F \times d$, then 25 Nm $= 400$ N $\times d$. Thus

$$\text{effective length, } d = \frac{25 \text{ Nm}}{400 \text{ N}} = \frac{1}{16} \text{ m}$$

$$= \frac{1000}{16} \text{ mm}$$

$$= \textbf{62.5 mm}$$

12.2 Equilibrium and the principle of moments

If more than one force is acting on an object and the forces do not act at a single point, then the turning effect of the forces, that is, the moment of the forces, must be considered.

Figure 12.3 shows a beam with its support (known as its pivot or fulcrum) at P, acting vertically upwards, and forces F_1 and F_2 acting vertically downwards at distances a and b, respectively, from the fulcrum.

A beam is said to be in **equilibrium** when there is no tendency for it to move. There are two conditions for equilibrium:

(i) The sum of the forces acting vertically downwards must be equal to the sum of the forces acting vertically upwards, i.e. for Fig. 12.3, $R_P = F_1 + F_2$.

(ii) The total moment of the forces acting on a beam must be zero; for the total moment to be zero:

the sum of the clockwise moments about any point must be equal to the sum of the anticlockwise moments about that point.

This statement is known as the **principle of moments**. Hence, taking moments about P in Fig. 3,

$F_2 \times b =$ the clockwise moment, and
$F_1 \times a =$ the anticlockwise moment.

Thus for equilibrium:

$$\boldsymbol{F_1 a = F_2 b}$$

Figure 12.4

Problem 3.   A system of forces is as shown in Fig. 12.4.

(a) If the system is in equilibrium find the distance d.

(b) If the point of application of the 5 N force is moved to point P, distance 200 mm from the support, find the new value of F to replace the 5 N force for the system to be in equilibrium.

(a)   From above, the clockwise moment M_1 is due to a force of 7 N acting at a distance d from the support, called the **fulcrum**, i.e.

$M_1 = 7\,\text{N} \times d$

The anticlockwise moment M_2 is due to a force of 5 N acting at a distance of 140 mm from the fulcrum, i.e.

$M_2 = 5\,\text{N} \times 140\,\text{mm}$

Applying the principle of moments, for the system to be in equilibrium about the fulcrum: clockwise moment = anticlockwise moment, i.e.

$7\,\text{N} \times d = 5 \times 140\,\text{N mm}$

Hence,

$$\text{distance, } d = \frac{5 \times 140\,\text{N mm}}{7\,\text{N}} = \mathbf{100\ mm}$$

(b)   When the 5 N force is replaced by force F at a distance of 200 mm from the fulcrum, the new value of the anticlockwise moment is $F \times 200$. For the system to be in equilibrium: clockwise moment = anticlockwise moment, i.e.

$(7 \times 100)\,\text{N mm} = F \times 200\,\text{mm}$

Hence,

$$\text{new value of force, } F = \frac{700\,\text{N mm}}{200\,\text{mm}} = \mathbf{3.5\ N}$$

Problem 4.   A beam is supported at its centre on a fulcrum and forces act as shown in Fig. 12.5. Calculate (a) force F for the beam to be in equilibrium, (b) the new position of the 23 N force when F is decreased to 21 N, if equilibrium is to be maintained.

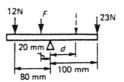

Figure 12.5

(a)   The clockwise moment, M_1, is due to the 23 N force acting at a distance of 100 mm from the fulcrum, i.e.

$M_1 = 23 \times 100 = 2300\,\text{N mm}$

There are two forces giving the anticlockwise moment M_2. One is the force F acting at a distance of 20 mm from the fulcrum and the other a force of 12 N acting at a distance of 80 mm. Thus

$M_2 = (F \times 20) + (12 \times 80)\,\text{N mm}.$

Applying the principle of moments about the fulcrum: clockwise moment = anticlockwise moments, i.e.

$2300 = (F \times 20) + (12 \times 80)$

Hence

$F \times 20 = 2300 - 960$

i.e.

$$\text{force } F = \frac{1340}{20} = \mathbf{67\ N}$$

(b)   The clockwise moment is now due to a force of 23 N acting at a distance of, say, d from the fulcrum. Since the value of F is decreased to 21 N, the anticlockwise moment is $(21 \times 20) + (12 \times 80)\,\text{N mm}$. Applying the principle of moments,

$23 \times d = (21 \times 20) + (12 \times 80)$

i.e.

$$\text{distance } d = \frac{420 + 960}{23} = \frac{1380}{23} = \mathbf{60\ mm}$$

Problem 5.   For the centrally supported uniform beam shown in Fig. 12.6 determine the values of forces F_1 and F_2 when the beam is in equilibrium.

At equilibrium:

    (i) $R = F_1 + F_2$, i.e. $5 = F_1 + F_2$ (1)

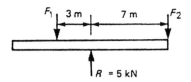

Figure 12.6

and

(ii) $F_1 \times 3 = F_2 \times 7$ (2)

From equation (1), $F_2 = 5 - F_1$.
Substituting for F_2 in equation (2) gives:

$$F_1 \times 3 = (5 - F_1) \times 7$$

i.e.

$$3F_1 = 35 - 7F_1$$
$$10F_1 = 35$$

from which $F_1 = 3.5$ kN. Since $F_2 = 5 - F_1$, $F_2 = 1.5$ kN.
Thus at equilibrium, force F_1 = 3.5 kN and force F_2 = 1.5 kN.

12.3   Simply supported beams having point loads

A **simply supported beam** is one which rests on two supports and is free to move horizontally.
   Two typical simply supported beams having loads acting at given points on the beam (called **point loading**) are shown in Fig. 12.7.

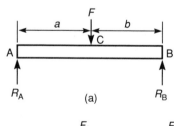

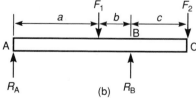

Figure 12.7

A man whose mass exerts a force F vertically downwards, standing on a wooden plank which is simply supported at its ends, may, for example, be represented by the beam diagram of Fig. 12.7(a) if the mass of the plank is neglected. The forces exerted by the supports on the plank, R_A and R_B, act vertically upwards, and are called **reactions**.
   When the forces acting are all in one plane, the algebraic sum of the moments can be taken about **any** point.
   For the beam in Fig. 12.7(a) at equilibrium:

(i) $R_A + R_B = F$, and
(ii) taking moments about A, $Fa = R_B(a + b)$.

(Alternatively, taking moments about C, $R_A a = R_B b$.)
   For the beam in Fig. 12.7(b), at equilibrium

(i) $R_A + R_B = F_1 + F_2$, and
(ii) taking moments about B, $R_A(a + b) + F_2 c = F_1 b$.

Typical **practical applications** of simply supported beams with point loadings include bridges, beams in buildings, and beds of machine tools.

Problem 6.   A beam is loaded as shown in Fig. 12.8.

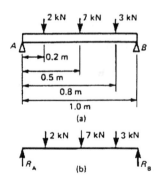

Figure 12.8

Determine (a) the force acting on the beam support at B, (b) the force acting on the beam support at A, neglecting the mass of the beam.

A beam supported as shown in Fig. 12.8 is called a simply supported **beam**.

(a) Taking moments about point A and applying the principle of moments gives: clockwise moments = anticlockwise moments (2×0.2) + (7×0.5) + (3×0.8) kN m = $R_B \times 1.0$ m, where R_B is the force suppporting the beam at B, as shown in Fig. 12.8(b). Thus $(0.4 + 3.5 + 2.4)$ kN m = $R_B \times 1.0$ m, i.e.

$$R_B = \frac{6.3 \text{ kN m}}{1.0 \text{ m}} = \textbf{6.3 kN}$$

(b) For the beam to be in equilibrium, the forces acting upwards must be equal to the forces acting downwards, thus

$$R_A + R_B = (2 + 7 + 3) \text{ kN}$$

$R_B = 6.3$ kN, thus $R_A = 12 - 6.3 = \textbf{5.7 kN}$

Problem 7. For the beam shown in Fig. 12.9 calculate (a) the force acting on support A, (b) distance d, neglecting any forces arising from the mass of the beam.

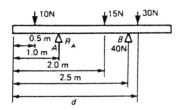

Figure 12.9

(a) From section 12.2,

(the forces acting in an upward direction) = (the forces acting in a downward direction)

Hence

$$(R_A + 40) \text{ N} = (10 + 15 + 30) \text{ N}$$

$$R_A = 10 + 15 + 30 - 40 = \textbf{15 N}$$

(b) Taking moments about the left-hand end of the beam and applying the principle of moments gives:

clockwise moments = anticlockwise moments

$$(10 \times 0.5) + (15 \times 2.0) \text{ N m} + 30 \text{ N} \times d$$
$$= (15 \times 1.0) + (40 \times 2.5) \text{ N m}$$

i.e.

$$35 \text{ Nm} + 30 \text{ N} \times d = 115 \text{ Nm}$$

from which

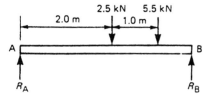

Figure 12.10

$$\text{distance, } d = \frac{(115 - 35) \text{ Nm}}{30 \text{ N}} = \textbf{2.67 m}$$

Problem 8. A metal bar AB is 4.0 m long and is supported at each end in a horizontal position. It carries loads of 2.5 kN and 5.5 kN at distances of 2.0 m and 3.0 m, respectively, from A. Neglecting the mass of the beam, determine the reactions of the supports when the beam is in equilibrium.

The beam and its loads are shown in Fig. 12.10. At equilibrium,

$$R_A + R_B = 2.5 + 5.5 = 8.0 \text{ kN} \qquad (1)$$

Taking moments about A, clockwise moments = anticlockwise moment, i.e.

$$(2.5 \times 2.0) + (5.5 \times 3.0) = 4.0 R_B, \text{ or}$$
$$5.0 + 16.5 = 4.0 R_B$$

from which,

$$R_B = \frac{21.5}{4.0} = 5.375 \text{ kN}$$

From equation (1),

$$R_A = 8.0 - 5.375 = 2.625 \text{ kN}$$

Thus the reactions at the supports at equilibrium are 2.625 kN at A and 5.375 kN at B.

Problem 9. A beam PQ is 5.0 m long and is supported at its ends in a horizontal position as shown in Fig. 12.11. Its mass is equivalent to a force of 400 N acting at its centre as shown. Point loads of 12 kN and 20 kN act on the beam in the positions shown. When the beam is in equilibrium, determine (a) the reactions of the supports, R_P and R_Q, and (b) the position to which the 12 kN load must be moved for the force on the supports to be equal.

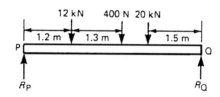

Figure 12.11

(a) At equilibrium,

$$R_P + R_Q = 12 + 0.4 + 20 = 32.4 \text{ kN} \qquad (1)$$

Taking moments about P:

clockwise moments = anticlockwise moments

i.e.

$$(12 \times 1.2) + (0.4 \times 2.5) + (20 \times 3.5)$$
$$= (R_Q \times 5.0)$$

$$14.4 + 1.0 + 70.0 = 5.0 \, R_Q$$

from which,

$$R_Q = \frac{85.4}{5.0} = \textbf{17.08 kN}$$

From equation (1),

$$R_P = 32.4 - R_Q = 32.4 - 17.08 = \textbf{15.32 kN}$$

(b) For the reactions of the supports to be equal,

$$R_P = R_Q = \frac{32.4}{2} = 16.2 \text{ kN}$$

Let the 12 kN load be at a distance d metres from P (instead of at 1.2 m from P). Taking moments about point P gives:

$$12d + (0.4 \times 2.5) + (20 \times 3.5) = 5.0 \, R_Q$$

i.e.

$$12d + 1.0 + 70.0 = 5.0 \times 16.2, \text{ and}$$

$$12d = 81.0 - 71.0$$

from which,

$$d = \frac{10.0}{12} = 0.833 \text{ m}$$

Hence the 12 kN load needs to be moved to a position 833 mm from P for the reactions of the supports to be equal (i.e. 367 mm to the left of its original position)

Problem 10. A uniform steel girder AB which is 6.0 m long has a mass equivalent to 4.0 kN acting at its centre. The girder rests on two supports at C and B as shown in Fig. 12.12. A point load of 20.0 kN is attached to the beam as shown. Determine the value of force F which causes the beam to just lift off the support B.

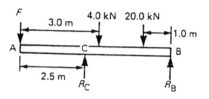

Figure 12.12

At equilibrium,

$$R_C + R_B = F + 4.0 + 20.0$$

When the beam is just lifting off the support B, then $R_B = 0$, hence $R_C = (F + 24.0)$ kN. Taking moments about A:

clockwise moments = anticlockwise moments

i.e.

$$(4.0 \times 3.0) + (5.0 \times 20.0)$$
$$= (R_C \times 2.5) + (R_B \times 6.0)$$

i.e. $12.0 + 100.0 = (F + 24.0) \times 2.5 + 0$

i.e. $\dfrac{112.0}{2.5} = (F + 24.0)$

from which,

$$F = 44.8 - 24.0 = 20.8 \text{ kN}$$

i.e. **the value of force F which causes the beam to just lift off the support B is 20.8 kN.**

12.4 Multi-choice questions on simply supported beams

(Answers on page 355.)

1. A force of 10 N is applied at right angles to the handle of a spanner, 5 m from the centre of a nut. The moment on the nut is:
   (a) 50 Nm   (b) 2 N/m   (c) 0.5 m/N
   (d) 15 Nm

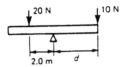

Figure 12.13

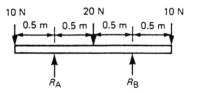

Figure 12.16

2. The distance d in Fig. 12.13 when the beam is in equilibrium is:
   (a) 0.5 m   (b) 1.0 m   (c) 4.0 m   (d) 15 m

3. With reference to Fig. 12.14, the clockwise moment about A is:
   (a) 70 Nm   (b) 10 Nm   (c) 60 Nm
   (d) $5 \times R_B$ Nm

8. With reference to Fig. 12.16, when moments are taken about point A, the sum of the anticlockwise moments is:
   (a) 25 Nm   (b) 20 Nm   (c) 35 Nm
   (d) 30 Nm

9. With reference to Fig. 12.16, when moments are taken about the right-hand end, the sum of the clockwise moments is:
   (a) 10 Nm   (b) 20 Nm   (c) 30 Nm
   (d) 40 Nm

10. With reference to Fig. 12.16, which of the following statements is false?
    (a) $(5 + R_B) = 25$ Nm   (b) $R_A = R_B$
    (c) $(10 \times 0.5) = (10 \times 1) + (10 \times 1.5) + R_A$
    (d) $R_A + R_B = 40$ N

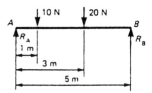

Figure 12.14

4. The force acting at B (i.e. R_B) in Fig. 12.14 is:
   (a) 16 N   (b) 20 N   (c) 5 N   (d) 14 N

5. The force acting at A (i.e. R_A) in Fig. 12.14 is:
   (a) 16 N   (b) 10 N   (c) 15 N   (d) 14 N

6. Which of the following statements is false for the beam shown in Fig. 12.15 if the beam is in equilibrium?
   (a) The anticlockwise moment is 27 N.
   (b) The force F is 9 N
   (c) The reaction at the support R is 18 N.
   (d) The beam cannot be in equilibrium for the given conditions.

12.5 Short answer questions on simply supported beams

1. The moment of a force is the product of .... and ....
2. When a beam has no tendency to move it is in ....
3. State the two conditions for equilibrium of a beam.
4. State the principle of moments.
5. What is meant by a simply supported beam?
6. State two practical applications of simply supported beams.

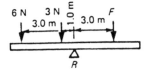

Figure 12.15

7. With reference to Fig. 12.16, the reaction R_A is:
   (a) 10 N   (b) 30 N   (c) 20 N   (d) 40 N

12.6 Further questions on simply supported beams

1. Determine the moment of a force of 25 N applied to a spanner at an effective length of 180 mm from the centre of a nut.   [4.5 Nm]
2. A moment of 7.5 Nm is required to turn a wheel. If a force of 37.5 N is applied to the rim of the wheel, calculate the effective distance from the rim to the hub of the wheel.   [200 mm]

3. Calculate the force required to produce a moment of 27 Nm on a shaft, when the effective distance from the centre of the shaft to the point of application of the force is 180 mm.

[150 N]

4. Determine distance d and the force acting at the support A for the force system shown in Fig. 12.17, when the system is in equilibrium.

[50 mm, 3.8 kN]

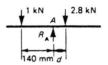

Figure 12.17

5. If the 1 kN force shown in Fig. 12.17 is replaced by a force F at a distance of 250 mm to the left of R_A, find the value of F for the system to be in equilibrium.   [560 N]

6. Determine the values of the forces acting at A and B for the force system shown in Fig. 12.18. $[R_A = R_B = 25\text{ N}]$

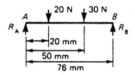

Figure 12.18

7. The forces acting on a beam are as shown in Fig. 12.19. Neglecting the mass of the beam, find the value of R_A and distance d when the beam is in equilibrium.   [5 N, 25 mm]

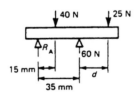

Figure 12.19

8. Calculate the force R_A and distance d for the beam shown in Fig. 12.20. The mass of the beam should be neglected and equilibrium conditions assumed.   [20 kN, 24 mm]

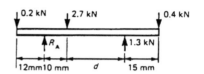

Figure 12.20

9. For the force system shown in Fig. 12.21, find the values of F and d for the system to be in equilibrium.   [1.0 kN, 64 mm]

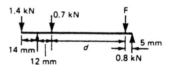

Figure 12.21

10. For the force system shown in Fig. 12.22, determine distance d for the forces R_A and R_B to be equal, assuming equilibrium conditions.   [80 m]

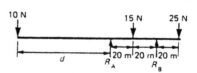

Figure 12.22

11. A simply supported beam AB is loaded as shown in Fig. 12.23. Determine the load F in order that the reaction at A is zero. [36 kN]

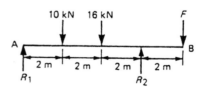

Figure 12.23

12. A uniform wooden beam, 4.8 m long, is supported at its left-hand end and also at 3.2 m from the left-hand end. The mass of the beam is equivalent to 200 N acting vertically downwards at its centre. Determine the reactions at the supports.

[50 N, 150 N]

13. For the simply supported beam PQ shown in Fig. 12.24, determine (a) the reaction at each support, (b) the maximum force which can be applied at Q without losing equilibrium.

[(a) $R_1 = 3$ kN, $R_2 = 12$ kN (b) 15.5 kN]

14. A uniform beam AB is 12 m long and is supported horizontally at distances of 2.0 m and 9.0 m from A. Loads of 60 kN, 104 kN, 50 kN and 40 kN act vertically downwards at A, 5.0 m from A, 7.0 m from A and at B, respectively. Neglecting the mass of the beam, determine the reactions at the supports.

[133.7 kN, 120.3 kN]

15. A uniform girder carrying point loads is shown in Fig. 12.25. Determine the value of load F which causes the beam to just lift off the support B.

[3.25 kN]

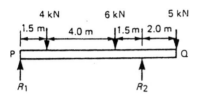

Figure 12.24

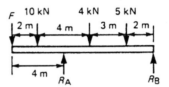

Figure 12.25

13

Linear and angular motion

At the end of this chapter you should be able to:

- appreciate that 2π radians correspond to $360°$
- define linear and angular velocity
- perform calculations on linear and angular velocity using $\omega = 2\pi n$ and $v = \omega r$
- define linear and angular acceleration
- perform calculations on linear and angular acceleration using $\omega_2 = \omega_1 + \alpha t$ and $a = r\alpha$
- select appropriate equations of motion when performing simple calculations
- appreciate the difference between scalar and vector quantities
- use vectors to determine relative velocities, by drawing and by calculation

13.1 The radian

The unit of angular displacement is the **radian**, where one radian is the angle subtended at the centre of a circle by an arc equal in length to the radius, see Fig. 13.1.

The relationship between angle in radians (θ), arc length (s) and radius of a circle (r) is:

$$s = r\theta \qquad (13.1)$$

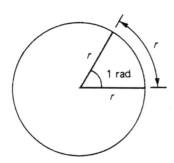

Figure 13.1

Since the arc length of a complete circle is $2\pi r$ and the angle subtended at the centre is $360°$, then from equation (13.1), for a complete circle,

$$2\pi r = r\theta \text{ or } \theta = 2\pi \text{ radians}$$

Thus,

$$2\pi \text{ radians corresponds to } 360° \qquad (13.2)$$

13.2 Linear and angular velocity

Linear velocity v is defined as the rate of change of linear displacement s with respect to time t, and for motion in a straight line:

$$\text{Linear velocity} = \frac{\text{change of displacement}}{\text{change of time}}$$

i.e.

$$v = \frac{s}{t} \qquad (13.3)$$

The unit of linear velocity is metres per second (m/s)

Angular velocity. The speed of revolution of a wheel or a shaft is usually measured in revolutions per minute or revolutions per second but these units do not form part of a coherent system of units. The basis used in SI units is the angle turned through in one second.

Angular velocity is defined as the rate of change of angular displacement θ, with respect to time t, and for an object rotating about a fixed axis at a constant speed:

$$\text{Angular velocity} = \frac{\text{angle turned through}}{\text{time taken}}$$

i.e.

$$\omega = \frac{\theta}{t} \tag{13.4}$$

The unit of angular velocity is radians per second (rad/s).

An object rotating at a constant speed of n revolutions per second subtends an angle of $2\pi n$ radians in one second, that is, its angular velocity

$$\omega = 2\pi n \text{ rad/s} \tag{13.5}$$

From equation (13.1), $s = r\theta$ and from equation (13.4), $\theta = \omega t$, hence $s = r\omega t$, or $s/t = \omega r$. However, from equation (13.3), $v = s/t$, hence

$$v = \omega r \tag{13.6}$$

Equation (13.6) gives the relationship between linear velocity, v, and angular velocity, ω.

Problem 1. A wheel of diameter 540 mm is rotating at $(1500/\pi)$ rev/min. Calculate the angular velocity of the wheel and the linear velocity of a point on the rim of the wheel.

From equation (13.5), angular velocity $\omega = 2\pi n$, where n is the speed of revolution in revolutions per second, i.e. $n = 1500/60\pi$ revolutions per second. Thus

$$\omega = 2\pi\left(\frac{1500}{60\pi}\right) = 50 \text{ rad/s}$$

The linear velocity of a point on the rim, $v = \omega r$, where r is the radius of the wheel, i.e. 0.54/2 or 0.27 m. Thus

$$v = 50 \times 0.27 = 13.5 \text{ m/s}$$

Problem 2. A car is travelling at 64.8 km/h and has wheels of diameter 600 mm.

(a) Find the angular velocity of the wheels in both rad/s and rev/min.
(b) If the speed remains constant for 1.44 km, determine the number of revolutions made by a wheel, assuming no slipping occurs.

(a)   64.8 km/h =

$$64.8 \ \frac{\text{km}}{\text{h}} \times 1000 \ \frac{\text{m}}{\text{km}} \times \frac{1}{3600} \ \frac{\text{h}}{\text{s}}$$

$$= 18 \text{ m/s}$$

That is, the linear velocity, v, is 18 m/s. The radius of a wheel is $(600/2)$ mm = 0.3 m. From equation (13.6), $v = \omega r$, hence $\omega = v/r$, that is, the angular velocity,

$$\omega = \frac{18}{0.3} = 60 \text{ rad/s}$$

From equation (13.5), angular velocity, $\omega = 2\pi n$, where n is in revolutions per second. Hence $n = \omega/2\pi$ and angular speed of a wheel in revolutions per minute is $60\omega/2\pi$. But $\omega = 60$ rad/s, hence

$$\text{angular speed} = \frac{60 \times 60}{2\pi}$$

$$= 573 \text{ revolutions per minute}$$

(b)   From equation (13.3), time taken to travel 1.44 km at a constant speed of 18 m/s is

$$\frac{1440 \text{ m}}{18 \text{ m/s}} = 80 \text{ s}$$

Since a wheel is rotating at 573 revolutions per minute, then in 80/60 minutes it makes $(573 \times 80)/60$, that is, **764 revolutions**.

13.3   Linear and angular acceleration

Linear acceleration, a, is defined as the rate of change of linear velocity with respect to time (as introduced in Chapter 10). For an object whose linear velocity is increasing uniformly:

$$\text{linear acceleration} = \frac{\text{change of linear velocity}}{\text{time taken}}$$

Table 13.1

s = arc length (m)	r = radius of circle (m)
t = time (s)	θ = angle (rad)
v = linear velocity (m/s)	ω = angular velocity (rad/s)
v_1 = initial linear velocity (m/s)	ω_1 = initial angular velocity (rad/s)
v_2 = final linear velocity (m/s)	ω_2 = final angular velocity (rad/s)
a = linear acceleration (m/s^2)	α = angular acceleration (rad/s^2)
n = speed of revolutions (rev/s)	

Equation number	Linear motion	Angular motion
(13.1)	$s = r\theta$ m	
(13.2)		2π rad $= 360°$
(13.3) and (13.4)	$v = \dfrac{s}{t}$ m/s	$\omega = \dfrac{\theta}{t}$ rad/s
(13.5)		$\omega = 2\pi n$ rad/s
(13.6)	$v = \omega r$ m/s	
(13.8) and (13.10)	$v_2 = (v_1 + at)$ m/s	$\omega_2 = (\omega_1 + \alpha t)$ rad/s
(13.11)	$a = r\alpha$ m/s^2	
(13.12) and (13.13)	$s = \left(\dfrac{v_1 + v_2}{2}\right) t$ m	$\theta = \left(\dfrac{\omega_1 + \omega_2}{2}\right) t$ rad
(13.14) and (13.16)	$s = (v_1 t + \frac{1}{2}at^2)$ m	$\theta = (\omega_1 t + \frac{1}{2}\alpha t^2)$ rad
(13.15) and (13.17)	$v_2^2 = (v_1^2 + 2as)$ (m/s)^2	$\omega_2^2 = (\omega_1^2 + 2\alpha\theta)$ (rad/s)^2

i.e.

$$a = \frac{v_2 - v_1}{t} \qquad (13.7)$$

The unit of linear acceleration is metres per second squared (m/s²). Rewriting equation (13.7) with v_2 as the subject of the formula gives:

$$v_2 = v_1 + at \qquad (13.8)$$

Angular acceleration, α, is defined as the rate of change of angular velocity with respect to time. For an object whose angular velocity is increasing uniformly:

Angular acceleration =

$$\frac{\text{change of angular velocity}}{\text{time taken}}$$

that is

$$\alpha = \frac{\omega_2 - \omega_1}{t} \qquad (13.9)$$

The unit of angular acceleration is radians per second squared (rad/s²). Rewriting equation (13.9) with ω_2 as the subject of the formula gives:

$$\omega_2 = \omega_1 + \alpha t \qquad (13.10)$$

From equation (13.6), $v = \omega r$. For motion in a circle having a constant radius r, $v_2 = \omega_2 r$ and $v_1 = \omega_1 r$, hence equation (13.7) can be rewritten as

$$a = \frac{\omega_2 r - \omega_1 r}{t} = \frac{r(\omega_2 - \omega_1)}{t}$$

From equation (13.4), $\theta = \omega t$, and if the angular velocity is changing uniformly from ω_1 to ω_2, then θ = mean angular velocity × time, i.e.

$$\theta = \left(\frac{\omega_1 + \omega_2}{2}\right) t \qquad (13.13)$$

Two further equations of linear motion may be derived from equations (13.8) and (13.11):

$$s = v_1 t + \frac{1}{2} at^2 \qquad (13.14)$$

and

$$v_2{}^2 = v_1{}^2 + 2as \qquad (13.15)$$

Two further equations of angular motion may be derived from equations (13.10) and (13.12):

$$\theta = \omega_1 t + \frac{1}{2}\alpha t^2 \qquad (13.16)$$

and

$$\omega_2{}^2 = \omega_1{}^2 + 2\alpha\theta \qquad (13.17)$$

Table 13.1 summarizes the principal equations of linear and angular motion for uniform changes in velocities and constant accelerations and also gives the relationships between linear and angular quantities.

Problem 5.   Find the number of revolutions made by the shaft in Problem 3 during the 10 s it is accelerating.

From equation (13.13), angle turned through

$$\theta = \left(\frac{\omega_1 + \omega_2}{2}\right) t$$

$$= \frac{1}{2}(\omega_1 + \omega_2)t$$

$$= \frac{1}{2}\left(\frac{300 \times 2\pi}{60} + \frac{800 \times 2\pi}{60}\right)(10) \text{ rad}$$

But there are 2π rad in 1 revolution, hence,

number of revolutions

$$= \frac{1}{2}\left(\frac{300 \times 2\pi}{60} + \frac{800 \times 2\pi}{60}\right)\frac{10}{2\pi}$$

$$= \frac{1}{2}\left(\frac{1100}{60}\right)(10)$$

$$= \frac{1100}{60} \times 5 = 91\frac{2}{3} \text{ revolutions}$$

Problem 6.   The shaft of an electric motor, initially at rest, accelerates uniformly for 0.4 s at 15 rad/s^2. Determine the angle turned through by the shaft, in radians, in this time.

From equation (13.16), $\theta = \omega_1 t + \frac{1}{2}\alpha t_2$. Since the shaft is initially at rest, $\omega_1 = 0$ and $\theta = \frac{1}{2}\alpha t^2$. The angular acceleration, α, is 15 rad/s^2 and time t is 0.4 s, hence angle turned through,

$$\theta = \frac{1}{2} \times 15 \times 0.4^2 = \textbf{1.2 rad}$$

Problem 7.   A flywheel accelerates uniformly at 2.05 rad/s^2 until it is rotating at 1500 revolutions per minute. If it completes 5 revolutions during the time it is accelerating, determine its initial angular velocity in rad/s, correct to four significant figures.

Since the final angular velocity is 1500 revolutions per minute

$$\omega_2 = 1500 \,\frac{\text{rev}}{\text{min}} \times \frac{1 \text{ min}}{60 \text{ s}} \times \frac{2\pi \text{ rad}}{1 \text{ rev}} = 50\pi \text{ rad/s}$$

$$5 \text{ revolutions} = 5 \text{ rev} \times \frac{2\pi \text{ rad}}{1 \text{ rev}} = 10\pi \text{ rad}$$

From equation (13.17), $\omega_2{}^2 = (\omega_1{}^2 + 2\alpha\theta)$ (rad/s)^2, i.e.

$$(50\pi)^2 = \omega_1{}^2 + 2 \times 2.05 \times 10\pi$$

$$\omega_1{}^2 = (50\pi)^2 - (2 \times 2.05 \times 10\pi)$$

$$= (50\pi)^2 - 41\pi = 24\,545$$

i.e.

$$\omega_1 = 156.7 \text{ rad/s}$$

Thus the initial angular velocity is 156.7 rad/s, correct to four significant figures.

13.4   Relative velocity

As stated in Chapter 8, quantities used in engineering and science can be divided into two groups:

(a) **Scalar quantities** have a size or magnitude only and need no other information to specify them. Thus 20 centimetres, 5 seconds, 3 litres and 4 kilograms are all examples of scalar quantities.
(b) **Vector quantities** have both a size (or magnitude), and a direction, called the line of action of the quantity. Thus, a velocity of

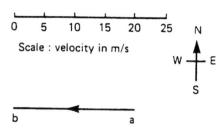

Figure 13.2

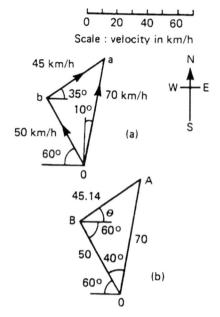

Figure 13.4

30 kilometres per hour due west, and an acceleration of 7 metres per second squared acting vertically downwards, are both vector quantities.

A vector quantity is represented by a straight line lying along the line of action of the quantity and having a length which is proportional to the size of the quantity. Thus **ab** in Fig. 13.2 represents a velocity of 20 m/s, whose line of action is due west. The bold letters, **ab**, indicate a vector quantity and the order of the letters indicate that the time of action is from a to b.

Consider two aircraft A and B flying at a constant altitude, A travelling due north at 200 m/s and B travelling 30° east of north, written N 30° E, at 300 m/s, as shown in Fig. 13.3.

Relative to a fixed point o, **oa** represents the velocity of A and **ob** the velocity of B. The velocity of B relative to A, that is the velocity at which B seems to be travelling to an observer on A, is given by **ab**, and by measurement is 160 m/s in a

direction E 22° N. The velociy of A relative to B, that is, the velocity at which A seems to be travelling to an observer on B, is given by **ba** and by measurement is 160 m/s in a direction W 22° S.

Problem 8.   Two cars are travelling on horizontal roads in straight lines, car A at 70 km/h at N 10° E and car B at 50 km/h at W 60° N. Determine, by drawing a vector diagram to scale, the velocity of car A relative to car B.

With reference to Fig. 13.4(a), **oa** represents the velocity of car A relative to a fixed point 0, and **ob** represents the velocity of car B relative to a fixed point 0. The velocity of car A relative to car B is given by **ba** and by measurement is **45 km/h in a direction of E 35° N**.

Problem 9.   Verify the result obtained in Problem 8 by calculation.

The triangle shown in Fig. 13.4(b) is similar to the vector diagram shown in Fig. 13.4(a). Angle BOA is 180° − (60° + 80°), that is, 40°. Using the cosine rule, from Chapter 5,

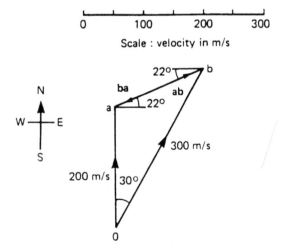

Figure 13.3

$$BA^2 = 50^2 + 70^2 - 2 \times 50 \times 70 \times \cos 40°$$

i.e.

BA = 45.14

Using the sine rule,

$$\frac{50}{\sin\angle BAO} = \frac{45.14}{\sin 40°}$$

$$\sin\angle BAO = \frac{50 \sin 40°}{45.14} = 0.7120$$

Hence angle BAO = 45°24′. Thus angle ABO = 180° − (40° + 45°24′) = 94°36′, and angle θ = 94°36′ − 60° = **34°36′**.
Thus **ba** is **45.14 km/h in a direction E 34°36′ N by calculation**.

Problem 10. A crane is moving in a straight line with a constant horizontal velocity of 2 m/s. At the same time it is lifting a load at a vertical velocity of 5 m/s. Calculate the velocity of the load relative to a fixed point on the earth's surface.

A vector diagram depicting the motion of the crane and load is shown in Fig. 13.5. **oa** represents the velocity of the crane relative to a fixed point on the earth's surface and **ab** represents the velocity of the load relative to the crane. The velocity of the load relative to the fixed point on the earth's surface is **ob**. By Pythagoras' theorem:

$$ob^2 = oa^2 + ab^2$$

$$= 4 + 25 = 29$$

Hence ob = $\sqrt{29}$ = 5.385 m/s.

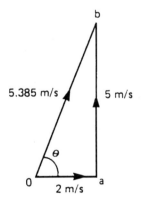

Figure 13.5

$$\text{Tan } \theta = \frac{5}{2} = 2.5$$

Hence θ = arctan 2.5 = 68°12′. That is, the velocity of the load relative to a fixed point on the earth's surface is **5.385 m/s in a direction 68°12′ to the motion of the crane**.

13.5 Multi-choice questions on linear and angular motion

(Answers on page 355.)

1. An angle of 2 rad at the centre of a circle subtends an arc length of 40 mm at the circumference of the circle. The radius of the circle is:
   (a) 40π mm (b) 80 mm (c) 20 mm
   (d) $(40/\pi)$ mm

2. A point on a wheel has a constant angular velocity of 3 rad/s. The angle turned through in 15 seconds is:
   (a) 45 rad (b) 10π rad (c) 5 rad (d) 90π rad

3. An angular velocity of 60 revolutions per minute is the same as
   (a) $(1/2\pi)$ rad/s (b) 120π rad/s
   (c) $(30/\pi)$ rad/s (d) 2π rad/s

4. A wheel of radius 15 mm has an angular velocity of 10 rad/s. A point on the rim of the wheel has a linear velocity of:
   (a) 300π mm/s (b) ⅔ mm/s (c) 150 mm/s
   (d) 1.5 mm/s

5. The shaft of an electric motor is rotating at 20 rad/s and its speed is increased uniformly to 40 rad/s in 5 s. The angular acceleration of the shaft is:
   (a) 4000 rad/s² (b) 4 rad/s² (c) 160 rad/s²
   (d) 12 rad/s²

6. A point on a flywheel of radius 0.5 m has a uniform linear acceleration of 2 m/s². Its angular acceleration is:
   (a) 2.5 rad/s²² (b) 0.25 rad/s² (c) 1 rad/s²
   (d) 4 rad/s²

A car accelerates uniformly from 10 m/s to 20 m/s over a distance of 150 m. The wheels of the car each have a radius of 250 mm. In questions 7 to 10 use this data to determine the quantities stated, selecting the correct answer from those given below.

   (a) 2.5 rad/s   (b) ⅕ s   (c) 1 m/s²
   (d) 20 rad/s   (e) 1 rad/s² (f) ¼ rad/s²
   (g) 10 s   (h) 100 m/s² (i) 40 rad/s
   (j) 3 m/s²   (k) 4 rad/s² (l) 5 s

7. The time the car is accelerating.
8. The initial angular velocity of each of the wheels.
9. The linear acceleration of a point on each of the wheels.
10. The angular acceleration of each of the wheels.

13.6   Short answer questions on linear and angular motion

1. State and define the unit of angular displacement.
2. Write down the formula connecting an angle, arc length and the radius of a circle.
3. Define linear velocity and state its unit.
4. Define angular velocity and state its unit.
5. Write down a formula connecting angular velocity and revolutions per second in coherent units.
6. State the formula connecting linear and angular velocity.
7. Define linear acceleration and state its unit.
8. Define angular acceleration and state its unit.
9. Write down the formula connecting linear and angular acceleration.
10. Define a scalar quantity and give two examples.
11. Define a vector quantity and give two examples.

13.7   Further questions on linear and angular motion

1. A pulley driving a belt has a diameter of 360 mm and is turning at $2700/\pi$ revolutions per minute. Find the angular velocity of the pulley and the linear velocity of the belt assuming that no slip occurs.
   [$\omega = 90$ rad/s, $v = 16.2$ m/s]
2. A bicycle is travelling at 36 km/h and the diameter of the wheels of the bicycle is 500 mm. Determine the angular velocity of the wheels of the bicycle and the linear velocity of a point on the rim of one of the wheels.
   [$\omega = 40$ rad/s, $v = 10$ m/s]
3. A flywheel rotating with an angular velocity of 200 rad/s is uniformly accelerated at a rate of 5 rad/s^2 for 15 s. Find the angular velocity of the flywheel both in rad/s and revolutions per minute.
   [275 rad/s, $8250/\pi$ rev/min]
4. A disc accelerates uniformly from 300 revolutions per minute to 600 revolutions per minute in 25 s. Determine its angular acceleration and the linear acceleration of a point on the rim of the disc, if the radius of the disc is 250 mm.
   [0.4π rad/s^2, 0.1π m/s^2]
5. Calculate the number of revolutions the disc in question 4 makes during the 25 s accelerating period.
   [187.5 revolutions]
6. A grinding wheel makes 300 revolutions when slowing down uniformly from 1000 rad/s to 400 rad/s. Find the time for this reduction in speed.   [2.693 s]
7. Find the angular retardation for the grinding wheel in question 6.   [222.8 rad/s^2]
8. A pulley is accelerated uniformly from rest at a rate of 8 rad/s^2. After 20 s the acceleration stops and the pulley runs at constant speed for 2 min, and then the pulley comes uniformly to rest after a further 40 s. Calculate:
   (a) the angular velocity after the period the acceleration,
   (b) the deceleration,
   (c) the total number of revolutions made by the pulley.
   [(a) 160 rad/s, (b) 4 rad/s^2 (c) $12\,000/\pi$ rev]
9. A ship is sailing due east with a uniform speed of 7.5 m/s relative to the sea. If the tide has a velocity 2 m/s in a north-westerly direction, find the velocity of the ship relative to the sea bed.   [6.248 m/s at E 13°5′ N]
10. A lorry is moving along a straight road at a constant speed of 54 km/h. The tip of its windscreen wiper blade has a linear velocity, when in a vertical position, of 4 m/s. Find the velocity of the tip of the wiper blade relative to the road when in this vertical position.
   [15.52 m/s at 14°56′]
11. A fork-lift truck is moving in a straight line at a constant speed of 5 m/s and at the same time a pallet is being lowered at a constant speed of 2 m/s. Determine the velocity of the pallet relative to the earth.
   [5.385 m/s at −21°48′]

14

Friction

At the end of this chapter you should be able to:

- understand dynamic or sliding friction
- understand static friction or striction
- appreciate factors which affect the size and direction of frictional forces
- define coefficient of friction, μ
- perform calculations involving $F = \mu N$
- state practical applications of friction
- state advantages and disadvantages of frictional forces

14.1 Introduction to friction

When an object, such as a block of wood, is placed on a floor and sufficient force is applied to the block, the force being parallel to the floor, the block slides across the floor. When the force is removed, motion of the block stops; thus there is a force which resists sliding. This force is called **dynamic** or **sliding friction**. A force may be applied to the block which is insufficient to move it. In this case, the force resisting motion is called the **static friction** or **striction**. Thus there are two categories into which a frictional force may be split:

(i) dynamic or sliding friction force which occurs when motion is taking place, and
(ii) static friction force which occurs before motion takes place.

There are three factors which affect the size and direction of frictional forces.

(i) The size of the frictional force depends on the type of surface (a block of wood slides more easily on a polished metal surface than on a rough concrete surface).
(ii) The size of the frictional force depends on the size of the force acting at right angles to the surfaces in contact, called the **normal force**. Thus, if the weight of a block of wood is doubled, the frictional force is doubled when it is sliding on the same surface.
(iii) The direction of the frictional force is always opposite to the direction of motion. Thus the frictional force opposes motion, as shown in Fig. 14.1.

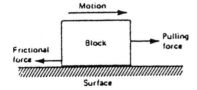
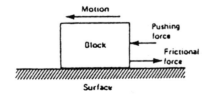

Figure 14.1

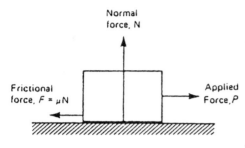

Figure 14.2

14.2   Coefficient of friction

The coefficient of friction, μ, is a measure of the amount of friction existing between two surfaces. A low value of coefficient of friction indicates that the force required for sliding to occur is less than the force required when the coefficient of friction is high. The value of the coefficient of friction is given by

$$\mu = \frac{\text{frictional force } (F)}{\text{normal force } (N)}$$

Transposing gives: frictional force = $\mu \times$ normal force, i.e.

$$\boxed{F = \mu N}$$

The direction of the forces given in this equation is as shown in Fig. 14.2. The coefficient of friction is the ratio of a force to a force, and hence has no units. Typical values for the coefficient of friction when sliding is occurring, i.e. the dynamic coefficient of friction, are:

For polished oiled metal surfaces less than 0.1
For glass on glass   0.4
For rubber on tarmac   close to 1.0

Problem 1.   A block of steel requires a force of 10.4 N applied parallel to a steel plate to keep it moving with constant velocity across the plate. If the normal force between the block and the plate is 40 N, determine the dynamic coefficient of friction.

As the block is moving at constant velocity, the force applied must be that required to overcome frictional forces, i.e.

frictional force, $F = 10.4$ N

The normal force is 40 N, and since $F = \mu N$,

$$\mu = \frac{F}{N} = \frac{10.4}{40} = 0.26$$

i.e. **the dynamic coefficient of friction is 0.26**.

Problem 2.   The surface between the steel block and plate of problem 1 is now lubricated and the dynamic coefficient of friction falls to 0.12. Find the new value of force required to push the block at a constant speed.

The normal force depends on the weight of the block and remains unaltered at 40 N. The new value of the dynamic coefficient of friction is 0.12 and since the frictional force $F = \mu N$, $F = 0.12 \times 40 = 4.8$ N. The block is sliding at constant speed, thus the force required to overcome the frictional force is also 4.8 N, i.e. **the required applied force is 4.8 N**.

Problem 3.   The material of a brake is being tested and it is found that the dynamic coefficient of friction between the material and steel is 0.91. Calculate the normal force when the frictional force is 0.728 kN.

The dynamic coefficient of friction, $\mu = 0.91$. The frictional force, $F = 0.728$ kN $= 728$ N. Since $F = \mu N$, then $N = F/\mu$, i.e.

$$\text{normal force } N = \frac{728}{0.91} = 800 \text{ N}$$

i.e. **the normal force is 800 N**.

14.3   Applications of friction

In some applications, a low coefficient of friction is desirable, for example, in bearings, pistons moving within cylinders, on ski runs, and so on. However, for such applications as force being transmitted by belt drives and braking systems, a high value of coefficient is necessary.

Problem 4.   State three advantages and three disadvantages of frictional forces.

Instances where frictional forces are an advantage include:

(i) Almost all fastening devices rely on frictional forces to keep them in place once secured, examples being screws, nails, nuts, clips and clamps.

(ii) Satisfactory operation of brakes and clutches rely on frictional forces being present.

(iii) In the absence of frictional forces, most accelerations along a horizontal surface are impossible. For example, a person's shoes just slip when walking is attempted and the tyres of a car just rotate with no forward motion of the car being experienced.

Disadvantages of frictional forces include:

(i) Energy is wasted in the bearings associated with shafts, axles and gears due to heat being generated.

(ii) Wear is caused by friction, for example, in shoes, brake lining materials and bearings.

(iii) Energy is wasted when motion through air occurs (it is much easier to cycle with the wind rather than against it).

Problem 5. Discuss briefly two design implications which arise due to frictional forces and how lubrication may or may not help.

(i) Bearings are made of an alloy called white metal, which has a relatively low melting point. When the rotating shaft rubs on the white metal bearing, heat is generated by friction, often in one spot and the white metal may melt in this area, rendering the bearing useless. Adequate lubrication (oil or grease) separates the shaft from the white metal, keeps the coefficient of friction small and prevents damage to the bearing. For very large bearings, oil is pumped under pressure into the bearing and the oil is used to remove the heat generated, often passing through oil coolers before being recirculated. Designers should ensure that the heat generated by friction can be dissipated.

(ii) Wheels driving belts, to transmit force from one place to another, are used in many workshops. The coefficient of friction between the wheel and the belt must be high, and it may be increased by dressing the belt with a tar-like substance. Since frictional force is proportional to the normal force, a slipping belt is made more efficient by tightening it, thus increasing the normal and hence the frictional force. Designers should incorporate some belt tension mechanism into the design of such a system.

Problem 6. Explain what is meant by the terms (a) the limiting or static coefficient of friction and (b) the sliding or dynamic coefficient of friction.

(a) When an object is placed on a surface and a force is applied to it in a direction parallel to the surface, if no movement takes place, then the applied force is balanced exactly by the frictional force. As the size of the applied force is increased, a value is reached such that the object is just on the point of moving. The limiting or static coefficient of friction is given by the ratio of this applied force to the normal force, where the normal force is the force acting at right angles to the surfaces in contact.

(b) Once the applied force is sufficient to overcome the striction its value can be reduced slightly and the object moves across the surface. A particular value of the applied force is then sufficient to keep the object moving at a constant velocity. The sliding or dynamic coefficient of friction is the ratio of the applied force, to maintain constant velocity, to the normal force.

14.4 Multi-choice questions on friction

(Answers on page 355.)

Questions 1 to 5 refer to the statements given below. Select the statement required from each group given.

(a) The coefficient of friction depends on the type of surfaces in contact.

(b) The coefficient of friction depends on the force acting at right angles to the surfaces in contact.

(c) The coefficient of friction depends on the area of the surfaces in contact.

(d) Frictional force acts in the opposite direction to the direction of motion.

(e) Frictional force acts in the direction of motion.

(f) A low value of coefficient of friction is required between the belt and the wheel in a belt drive system.

(g) A low value of coefficient of friction is required for the materials of a bearing.

(h) The dynamic coefficient of friction is given by (normal force)/(frictional force) at constant speed.

(i) The coefficient of static friction is given by (applied force)/(frictional force) as sliding is just about to start.

(j) Lubrication results in a reduction in the coefficient of friction.

1. Which statement is false from (a), (b), (f) and (i)?

2. Which statement is false from (b), (e), (g) and (j)?

3. Which statement is true from (c), (f), (h) and (i)?

4. Which statement is false from (b), (c), (e) and (j)?

5. Which statement is false from (a), (d), (g) and (h)?

6. The normal force between two surfaces is 100 N and the dynamic coefficient of friction is 0.4. The force required to maintain a constant speed of sliding is:
   (a) 100.4 N   (b) 40 N   (c) 99.6 N   (d) 250 N

7. The normal force between two surfaces is 50 N and the force required to maintain a constant speed of sliding is 25 N. The dynamic coefficient of friction is:
   (a) 25   (b) 2   (c) 75   (d) 0.5

8. The maximum force which can be applied to an object without sliding occurring is 60 N, and the static coefficient of friction is 0.3. The normal force between the two surfaces is:
   (a) 200 N   (b) 18 N   (c) 60.3 N   (d) 59.7 N

14.5 Short answer questions on friction

1. The .... of frictional force depends on the .... of surfaces in contact.

2. The .... of frictional force depends on the size of the .... to the surfaces in contact.

3. The .... of frictional force is always .... to the direction of motion.

4. The coefficient of friction between surfaces should be a .... value for materials concerned with bearings.

5. The coefficient of friction should have a .... value for materials concerned with braking systems.

6. The coefficient of dynamic or sliding friction is given by
   ....
   ....

7. The coefficient of static or limiting friction is given by
   ....
   ....
   when .... is just about to take place.

8. Lubricating surfaces in contact result in a .... of the coefficient of friction.

14.6 Further questions on friction

1. Briefly discuss the factors affecting the size and direction of frictional forces.

2. Name three practical applications where a low value of coefficient of friction is desirable and state briefly how this is achieved in each case.

3. Give three practical applications where a high value of coefficient of friction is required when transmitting forces and discuss how this is achieved.

4. For an object on a surface, two different values of coefficient of friction are possible. Give the names of these two coefficients of friction and state how their values may be obtained.

5. Discuss briefly the effects of frictional force on the design of (a) a hovercraft, (b) a screw and (c) a braking system.

6. The coefficient of friction of a brake pad and a steel disc is 0.82. Determine the normal force between the pad and the disc if the frictional force required is 1025 N. [1250 N]

7. A force of 0.12 kN is needed to push a bale of cloth along a chute at a constant speed. If the normal force between the bale and the chute is 500 N, determine the dynamic coefficient of friction. [0.24]

8. The normal force between a belt and its driver wheel is 750 N. If the static coefficient of friction is 0.9 and the dynamic coefficient of friction is 0.87, calculate (a) the maximum force which can be transmitted and (b) maximum force which can be transmitted when the belt is running at a constant speed.
   [(a) 675 N (b) 652.5 N]

15

Work, energy and power

At the end of this chapter you should be able to:

- define work and state its unit
- perform simple calculations on work done
- appreciate that the area under a force/distance graph gives work done
- perform calculations on a force/distance graph to determine work done
- define energy and state its unit
- state several forms of energy
- state the principle of conservation of energy
- state examples of energy conversions
- define efficiency
- calculate efficiency of systems
- define power and state its unit
- understand that power = force × velocity
- perform calculations involving power, work done, energy and efficiency
- define potential energy
- perform calculations involving potential energy = mgh
- define kinetic energy
- perform calculations involving kinetic energy = $\frac{1}{2} mv^2$
- distinguish between elastic and inelastic collisions

15.1 Work

If a body moves as a result of a force being applied to it, the force is said to do work on the body. The amount of work done is the product of the applied force and the distance, i.e.

work done = force × distance moved in the direction of the force

The unit of work is the **joule, J**, which is defined as the amount of work done when a force of 1 newton acts for a distance of 1 m in the direction of the force. Thus,

$$1 \, J = 1 \, N \, m$$

If a graph is plotted of experimental values of force (on the vertical axis) against distance moved (on the horizontal axis) a force/distance graph or work diagram is produced. **The area under the graph represents the work done.**

For example, a constant force of 20 N used to raise a load a height of 8 m may be represented on a force/distance graph as shown in Fig. 15.1(a). The area under the graph shown shaded represents the work done. Hence

$$\text{work done} = 20 \, N \times 8 \, m = 160 \, J$$

Similarly, a spring extended by 20 mm by a force of 500 N may be represented by the work diagram shown in Fig. 15.1(b)

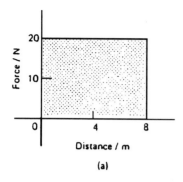

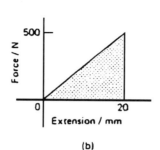

(a) (b) **Figure 15.1**

work done = shaded area

$$= \frac{1}{2} \times \text{base} \times \text{height}$$

$$= \frac{1}{2} \times (20 \times 10^{-3}) \text{ m} \times 500 \text{ N} = \mathbf{5\,J}$$

The work done by a variable force may be found by determining the area enclosed by the force/distance graph using the trapezoidal rule, the mid-ordinate rule or Simpson's rule (see Chapter 6).

Problem 1.   Calculate the work done when a force of 40 N pushes an object a distance of 500 m in the same direction as the force.

---

work done = force × distance moved in the direction of the force.

$$= 40 \text{ N} \times 500 \text{ m} = 20\,000 \text{ J}$$
$$(\text{since } 1 \text{ J} = 1 \text{ N m})$$

i.e.

work done = **20 kJ**

Problem 2.   Calculate the work done when a mass is lifted vertically by a crane to a height of 5 m, the force required to lift the mass being 98 N.

When work is done in lifting then:

work done = (weight of the body) × (vertical distance moved)

Weight is the downward force due to the mass of an object. Hence

work done = 98 N × 5 m = **490 J**

Problem 3.   A motor supplies a constant force of 1 kN which is used to move a load a distance of 5 m. The force is then changed to a constant 500 N and the load is moved a further 15 m. Draw the force/distance graph for the operation and from the graph determine the work done by the motor.

The force/distance graph or work diagram is shown in Fig. 15.2. Between points A and B a constant force of 1000 N moves the load 5 m; between points C and D a constant force of 500 N moves the load from 5 m to 20 m

Total work done

= area under the force/distance graph

= area ABFE + area CDGF

= (1000 N × 5 m) + (500 N × 15 m)

= 5000 J + 7500 J = 12 500 J = **12.5 kJ**

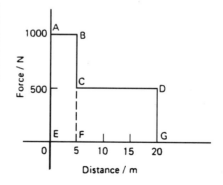

Figure 15.2

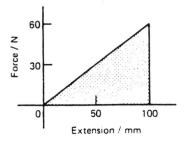

Figure 15.3

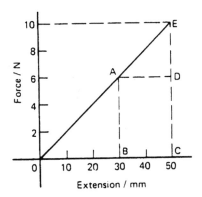

Figure 15.4

Problem 4.   A spring, initially in a relaxed
state, is extended by 100 mm. Determine
the work done by using a work diagram if
the spring requires a force of 0.6 N per
mm of stretch.

Force required for a 100 mm extension = 100 mm
× 0.6 N mm^{-1} = 60 N. Figure 15.3 shows the
force/extension graph or work diagram representing the increase in extension in proportion to
the force, as the force is increased from 0 to 60
N. The work done is the area under the graph
(shown shaded). Hence

$$\text{work done} = \frac{1}{2} \times \text{base} \times \text{height}$$

$$= \frac{1}{2} \times 100 \text{ mm} \times 60 \text{ N}$$

$$= \frac{1}{2} \times 100 \times 100^{-3} \text{ m} \times 60 \text{ N} = \textbf{3 J}$$

(Alternatively, average force during extension =
(60 – 0)/2 = 30 N and total extension = 100 mm =
0.1 m. Hence work done = average force × extension = 30 N × 0.1 m = **3 J**.)

Problem 5.   A spring requires a force of
10 N to cause an extension of 50 mm.
Determine the work done in extending the
spring (a) from zero to 30 mm, and (b)
from 30 mm to 50 mm.

Figure 15.4 shows the force/extension graph for
the spring.

(a)   Work done in extending the spring from
zero to 30 mm is given by area ABO of Fig.
15.4, i.e.

$$\text{work done} = \frac{1}{2} \times \text{base} \times \text{height}$$

$$= \frac{1}{2} \times 30 \times 10^{-3} \text{ m} \times 6 \text{ N}$$

$$= 90 \times 10^{-3} \text{ J} = \textbf{0.09 J}$$

(b)   Work done in extending the spring from
30 mm to 50 mm is given by area ABCE of
Figure 15.4, i.e.

$$\text{work done} = \text{area ABCD} + \text{area ADE}$$

$$= (20 \times 10^{-3} \text{ m} \times 6 \text{ N})$$

$$+ \frac{1}{2} (20 \times 10^{-3} \text{ m})(4 \text{ N})$$

$$= 0.12 \text{ J} + 0.04 \text{ J} = \textbf{0.16 J}$$

Problem 6.   Calculate the work done
when a mass of 20 kg is lifted vertically
through a distance of 5.0 m.

The force to be overcome when lifting a mass of
20 kg vertically upwards is mg, i.e. 20 × 9.81 =
196.2 N.
Work done = force × distance = 196.2 × 5.0 =
981 J.

Problem 7.   Water is pumped vertically
upwards through a distance of 50.0 m and
the work done is 294.3 kJ. Determine the
number of litres of water pumped. (1 litre
of water has a mass of 1 kg.)

Work done = force × distance, i.e. 294 300 =
force × 50.0, from which force = 294 300/50.0 =

5886 N. The force to be overcome when lifting a mass m kg vertically upwards is mg, i.e. ($m \times$ 9.81) N. Thus $5886 = m \times 9.81$, from which mass, $m = 5886/9.81 = 600$ kg. Since 1 litre of water has a mass of 1 kg, **600 litres of water are pumped**.

Problem 8.   The force on a cutting tool of a shaping machine varies over the length of cut as follows:

Distance (mm)	0	20	40	60	80	100
Force (kN)	60	72	65	53	44	50

Determine the work done as the tool moves through a distance of 100 mm.

The force/distance graph for the given data is shown in Fig. 15.5. The work done is given by the area under the graph. This area may be determined by an approximate method. Using the mid-ordinate rule (see Chapter 6), with each strip of width 20 mm, mid-ordinates y_1, y_2, y_3, y_4 and y_5 are erected as shown, and each is measured.

Area under curve

$= $ (width of each strip)(sum of mid-ordinate values)

$= (20)(69 + 69.5 + 59 + 48 + 45.5)$
$= (20)(291)$

$= 5820$ kN mm $= 5820$ Nm $= 5820$ J

Hence the work done as the tool moves through 100 mm is **5.82 kJ**.

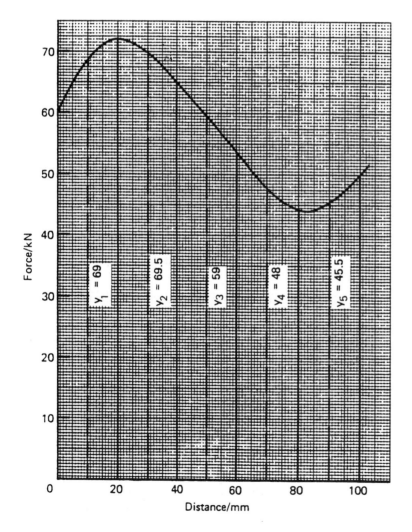

Figure 15.5

15.2 Energy

Energy is the capacity, or ability, to do work. The unit of energy is the joule, the same as for work. Energy is expended when work is done. There are several forms of energy and these include:

(i) Mechanical energy
(ii) Heat or thermal energy
(iii) Electrical energy
(iv) Chemical energy
(v) Nuclear energy
(vi) Light energy
(vii) Sound energy

Energy may be converted from one form to another. The **principle of conservation of energy** states that the total amount of energy remains the same in such conversions, i.e. energy cannot be created or destroyed. Some examples of energy conversions include:

(i) Mechanical energy is converted to electrical energy by a generator.
(ii) Electrical energy is converted to mechanical energy by a motor.
(iii) Heat energy is converted to mechanical energy by a steam engine.
(iv) Mechanical energy is converted to heat energy by friction.
(v) Heat energy is converted to electrical energy by a solar cell.
(vi) Electrical energy is converted to heat energy by an electric fire.
(vii) Heat energy is converted to chemical energy by living plants.
(viii) Chemical energy is converted to heat energy by burning fuels.
(ix) Heat energy is converted to electrical energy by a thermocouple.
(x) Chemical energy is converted to electrical energy by batteries.
(xi) Electrical energy is converted to light energy by a light bulb.
(xii) Sound energy is converted to electrical energy by a microphone.
(xiii) Electrical energy is converted to chemical energy by electrolysis.

Efficiency is defined as the ratio of the useful output energy to the input energy. The symbol for efficiency is η (Greek letter eta). Hence

$$\text{efficiency, } \eta = \frac{\text{useful output energy}}{\text{input energy}}$$

Efficiency has no units and is often stated as a percentage. A perfect machine would have an efficiency of 100%. However, all machines have an efficiency lower than this due to friction and other losses. Thus, if the input energy to a motor is 1000 J and the output energy is 800 J then the efficiency is

$$\frac{800}{1000} \times 100\%, \text{ i.e. } 80\%$$

Problem 9. A machine exerts a force of 200 N in lifting a mass through a height of 6 m. If 2 kJ of energy are supplied to it, what is the efficiency of the machine?

Work done in lifting mass

= force × distance moved

= weight of body × distance moved

= 200 N × 6 m = 1200 J
= useful energy output

Energy input = 2 kJ = 2000 J.

$$\text{Efficiency, } \eta = \frac{\text{useful output energy}}{\text{input energy}}$$

$$= \frac{1200}{2000} = \textbf{0.6 or 60\%}$$

Problem 10. Calculate the useful output energy of an electric motor which is 70% efficient if it uses 600 J of electrical energy.

$$\text{Efficiency, } \eta = \frac{\text{useful output energy}}{\text{input energy}}$$

thus

$$\frac{70}{100} = \frac{\text{output energy}}{600 \text{ J}}$$

from which,

$$\text{output energy} = \frac{70}{100} \times 600 = \textbf{420 J}$$

Problem 11. 4 kJ of energy are supplied to a machine used for lifting a mass. The force required is 800 N. If the machine has an efficiency of 50%, to what height will it lift the mass?

Efficiency, $\eta = \dfrac{\text{output energy}}{\text{input energy}}$

i.e.

$\dfrac{50}{100} = \dfrac{\text{output energy}}{4000 \text{ J}}$

from which,

output energy $= \dfrac{50}{100} \times 4000 = 2000 \text{ J}$

Work done = force × distance moved, hence
2000 J = 800 N × height, from which,

height $= \dfrac{2000 \text{ J}}{800 \text{ N}} = \mathbf{2.5 \text{ m}}$

Problem 12.   A hoist exerts a force of
500 N in raising a load through a height of
20 m. The efficiency of the hoist gears is
75% and the efficiency of the motor is 80%.
Calculate the input energy to the hoist.

The hoist system is shown diagrammatically in
Fig. 15.6.

Output energy = work done

$= \text{force} \times \text{distance} = 500 \text{ N} \times 20 \text{ m}$

$= 10\,000 \text{ J}$

For the gearing,

efficiency $= \dfrac{\text{output energy}}{\text{input energy}}$

i.e.

$\dfrac{75}{100} = \dfrac{10\,000}{\text{input energy}}$

from which, the input energy to the gears =
10 000 × (100/75) = 13 333 J. The input energy to
the gears is the same as the output energy of the
motor. Thus, for the motor,

efficiency $= \dfrac{\text{output energy}}{\text{input energy}}$

$= \dfrac{80}{100} = \dfrac{13\,333}{\text{input energy}}$

Hence

input energy to the system

$= 13\,333 \times \dfrac{100}{80}$

$= 16\,670 \text{ J}$

$= \mathbf{16.67 \text{ kJ}}$

15.3   Power

Power is a measure of the rate at which work is
done or at which energy is converted from one
form to another.

$$\boxed{\text{Power } P = \dfrac{\text{energy used}}{\text{time taken}}}$$

$$\left(\text{or } P = \dfrac{\text{work done}}{\text{time taken}}\right)$$

The unit of power is the **watt, W**, where 1 watt is
equal to 1 joule per second. The watt is a small
unit for many purposes and a larger unit called
the kilowatt, kW, is used, where 1 kW = 1000 W.
The power output of a motor which does 120 kJ
of work in 30 s is thus given by

$$P = \dfrac{120 \text{ kJ}}{30 \text{ s}} = 4 \text{ kW}$$

(For electrical power, see Chapter 26.)
Since work done = force × distance, then

$$\text{Power} = \dfrac{\text{work done}}{\text{time taken}} = \dfrac{\text{force} \times \text{distance}}{\text{time taken}}$$

$$= \text{force} \times \dfrac{\text{distance}}{\text{time taken}}$$

However,

$$\dfrac{\text{distance}}{\text{time taken}} = \text{velocity}$$

Hence

$$\boxed{\textbf{power = force} \times \textbf{velocity}}$$

Problem 13.   The output power of a
motor is 8 kW. How much work does it do
in 30 s?

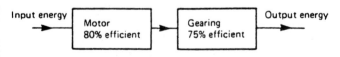

Figure 15.6

Power = (work done)/(time taken), from which,

work done = power × time

$$= 8000 \text{ W} \times 30 \text{ s} = 240\,000 \text{ J}$$

$$= \textbf{240 kJ}$$

Problem 14. Calculate the power required to lift a mass through a height of 10 m in 20 s if the force required is 3924 N.

Work done = force × distance moved = 3924 N × 10 m = 39 240 J.

$$\text{Power} = \frac{\text{work done}}{\text{time taken}} = \frac{39\,240 \text{ J}}{20 \text{ s}}$$

$$= \textbf{1962 W or 1.962 kW}$$

Problem 15. 10 kJ of work is done by a force in moving a body uniformly through 125 m in 50 s. Determine (a) the value of the force and (b) the power.

(a) Work done = force × distance. Hence 10 000 J = force × 125 m, from which,

$$\text{force} = \frac{10\,000 \text{ J}}{125 \text{ m}} = 80 \text{ N}$$

(b) Power $= \dfrac{\text{work done}}{\text{time taken}} = \dfrac{10\,000 \text{ J}}{50 \text{ s}} = \textbf{200 W}$

Problem 16. A car hauls a trailer at 90 km/h when exerting a steady pull of 600 N. Calculate (a) the work done in 30 minutes and (b) the power required.

(a) Work done = force × distance moved. Distance moved in 30 min, i.e. $\frac{1}{2}$ h, at 90 km/h = 45 km. Hence

$$\text{work done} = 600 \text{ N} \times 45\,000 \text{ m}$$
$$= \textbf{27 000 kJ or 27 MJ}$$

(b) Power required $= \dfrac{\text{work done}}{\text{time taken}} = \dfrac{27 \times 10^6 \text{ J}}{30 \times 60 \text{ s}}$

$$= \textbf{15 000 W or 15 kW}$$

Problem 17. To what height will a mass of weight 981 N be raised in 40 s by a machine using a power of 2 kW?

Work done = force × distance. Hence, work done = 981 N × height.
Power = (work done)/(time taken), from which,

work done = power × time taken

$$= 2000 \text{ W} \times 40 \text{ s} = 80\,000 \text{ J}$$

Hence 80 000 = 981 N × height, from which,

$$\text{height} = \frac{80\,000 \text{ J}}{981 \text{ N}} = \textbf{81.55 m}$$

Problem 18. A planing machine has a cutting stroke of 2 m and the stroke takes 4 seconds. If the constant resistance to the cutting tool is 900 N calculate for each cutting stroke (a) the power consumed at the tool point, and (b) the power input to the system if the efficiency of the system is 75%.

(a) Work done in each cutting stroke = force × distance = 900 N × 2 m = 1800 J. Power consumed at tool point is given by

$$\frac{\text{work done}}{\text{time taken}} = \frac{1800 \text{ J}}{4 \text{ s}}$$

$$= \textbf{450 W}$$

(b) Efficiency $= \dfrac{\text{output energy}}{\text{input energy}} = \dfrac{\text{output power}}{\text{input power}}$

Hence

$$\frac{75}{100} = \frac{450}{\text{input power}}$$

from which,

$$\text{input power} = 450 \times \frac{100}{75} = \textbf{600 W}$$

Problem 19. An electric motor provides power to a winding machine. The input power to the motor is 2.5 kW and the overall efficiency is 60%. Calculate (a) the output power of the machine, (b) the rate at which it can raise a 300 kg load vertically upwards.

(a)   Efficiency, $\eta = \dfrac{\text{power output}}{\text{power input}}$

i.e.

$$\frac{60}{100} = \frac{\text{power output}}{2500}$$

from which

power output $= \dfrac{60}{100} \times 2500 = $ **1500 W**

(b)   Power output = force $\times$ velocity, from which, velocity = (power output)/(force). Force acting on the 300 kg load due to gravity = 300 kg $\times$ 9.81 m/s^2 = 2943 N. Hence

velocity $= \dfrac{1500}{2943} = $ **0.510 m/s** or **510 mm/s**

Problem 20.   A lorry is travelling at a constant velocity of 72 km/h. The force resisting motion is 800 N. Calculate the tractive power necessary to keep the lorry moving at this speed.

Power = force $\times$ velocity. The force necessary to keep the lorry moving at constant speed is equal and opposite to the force resisting motion, i.e. 800 N. Velocity = 72 km/h = (72/36) m/s = 20 m/s. Hence power = 800 N $\times$ 20 m/s = 16 000 N m/s = 16 000 J/s = 16 000 W or 16 kW.

Thus the tractive power needed to keep the lorry moving at a constant speed of 72 km/h is 16 kW.

Problem 21.   The variation of tractive force with distance for a vehicle which is accelerating from rest is:

force (kN)	8.0	7.4	5.8	4.5	3.7	3.0
distance (m)	0	10	20	30	40	50

Determine the average power necessary if the time taken to travel the 50 m from rest is 25 s.

The force/distance diagram is shown in Fig. 15.7. The work done is determined from the area under the curve. Using the mid-ordinate rule with five intervals (see Chapter 6) gives:

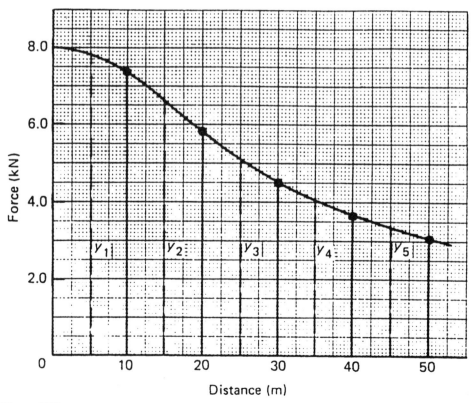

Figure 15.7

area = (width of interval)(sum of mid-ordinate)

$= (10) (y_1 + y_2 + y_3 + y_4 + y_5)$

$= 10(7.8 + 6.6 + 5.1 + 4.0 + 3.3)$

$= 10(26.8) = 268$ kN m,

i.e. work done = 268 kJ

Average power $= \dfrac{\text{work done}}{\text{time taken}} = \dfrac{268\,000 \text{ J}}{25 \text{ s}}$

$= \mathbf{10\,720\ W}$ or $\mathbf{10.72\ kW}$

15.4 Potential and kinetic energy

Mechanical engineering is concerned principally with two kinds of energy, potential energy and kinetic energy.

Potential energy is energy due to the position of the body. The force exerted on a mass of m kg is mg N (where $g = 9.81$ m/s², the acceleration due to gravity). When the mass is lifted vertically through a height h m above some datum level, the work done is given by: force × distance = $(mg)(h)$ J. This work done is stored as potential energy in the mass.

Hence,

> **potential energy = *mgh* joules**

(the potential energy at the datum level being taken as zero).

Kinetic energy is the energy due to the motion of a body. Suppose a force F acts on an object of mass m originally at rest (i.e. $u = 0$) and accelerates it to a velocity v in a distance s:

work done = force × distance

$= Fs = (ma)(s)$
(if no energy is lost)

where a is the acceleration.
Since $v^2 = u^2 + 2as$ and $u = 0$, $v^2 = 2as$, from which $a = v^2/2s$, hence

work done $= m\left(\dfrac{v^2}{2s}\right)s = \dfrac{1}{2}\,mv^2$

This energy is called the kinetic energy of the mass m, i.e.

> **kinetic energy $= \dfrac{1}{2}\,mv^2$ joules**

As stated in section 15.2, energy may be converted from one form to another. The **principle of conservation of energy** states that the total amount of energy remains the same in such conversions, i.e. energy cannot be created or destroyed.

In mechanics, the potential energy possessed by a body is frequently converted into kinetic energy, and vice versa. When a mass is falling freely, its potential energy decreases as it loses height, and its kinetic energy increases as its velocity increases. Ignoring air frictional losses, at all times:

> **Potential energy + kinetic energy**
> **= a constant**

If friction is present, then work is done overcoming the resistance due to friction and this is dissipated as heat. Then,

> **Initial energy = final energy + work done**
> **overcoming frictional resistance**

Kinetic energy is not always conserved in collisions. Collisions in which kinetic energy is conserved (i.e. stays the same) are called **elastic collisions**, and those in which it is not conserved are termed **inelastic collisions**.

> Problem 22. A car of mass 800 kg is climbing an incline at 10° to the horizontal. Determine the increase in potential energy of the car as it moves a distance of 50 m up the incline.

With reference to Fig. 15.8, $\sin 10° = h/50$, from which,

$h = 50 \sin 10°$

$= 8.682$ m

Hence

increase in potential energy

$= mgh$

$= 800$ kg × 9.81 m/s² × 8.682 m

$= \mathbf{68\,140\ J}$ or $\mathbf{68.14\ kJ}$

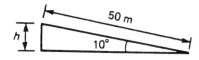

Figure 15.8

Problem 23. At the instant of striking, a hammer of mass 30 kg has a velocity of 15 m/s. Determine the kinetic energy in the hammer.

Kinetic energy $= \frac{1}{2} mv^2 = \frac{1}{2} (30 \text{ kg}) (15 \text{ m/s})^2$, i.e.

kinetic energy in hammer = **3375 J**

Problem 24. A lorry having a mass of 1.5 t is travelling along a level road at 72 km/h. When the brakes are applied, the speed decreases to 18 km/h. Determine how much the kinetic energy of the lorry is reduced.

Initial velocity of lorry

$$v_1 = 72 \text{ km/h}$$

$$= 72 \frac{\text{km}}{\text{h}} \times 1000 \frac{\text{km}}{\text{h}} \times \frac{1 \text{ h}}{3600 \text{ s}}$$

$$= \frac{72}{3.6} = 20 \text{ m/s}$$

Final velocity of lorry $v_2 = (18/3.6) = 5$ m/s. Mass of lorry, $m = 1.5$ t $= 1500$ kg.

Initial kinetic energy of the lorry

$$= \frac{1}{2} mv_1{}^2$$

$$= \frac{1}{2} (1500)(20)^2 = 300 \text{ kJ}$$

Final kinetic energy of the lorry

$$= \frac{1}{2} mv_2{}^2$$

$$= \frac{1}{2} (1500)(5)^2 = 18.75 \text{ kJ}$$

Hence the change in kinetic energy = 300 − 18.75 = **281.25 kJ**.
(Part of this reduction in kinetic energy is converted into heat energy in the brakes of the lorry and is hence dissipated in overcoming frictional forces and air friction.)

Problem 25. A canister containing a meteorology balloon of mass 4 kg is fired vertically upwards from a gun with an initial velocity of 400 m/s. Neglecting the air resistance, calculate (a) its initial kinetic energy, (b) its velocity at a height of 1 km, (c) the maximum height reached.

(a) Initial kinetic energy $= \frac{1}{2} mv^2 = \frac{1}{2} (4)(400)^2$ = **320 kJ**.

(b) At a height of 1 km, potential energy = mgh = $4 \times 9.81 \times 1000 = 39.24$ kJ. By the principle of conservation of energy: potential energy + kinetic energy at 1 km = initial kinetic energy. Hence

$$39\,240 + \frac{1}{2} mv^2 = 320\,000$$

from which,

$$\frac{1}{2} (4) v^2 = 320\,000 - 39\,240 = 280\,760$$

Hence

$$v = \sqrt{\left(\frac{2 \times 280\,760}{4}\right)} = 374.7 \text{ m/s}$$

i.e. **the velocity of the canister at a height of 1 km is 374.7 m/s**.

(c) At the maximum height, the velocity of the canister is zero and all the kinetic energy has been converted into potential energy. Hence potential energy = initial kinetic energy = 320 000 J (from part (a)). Then $320\,000 = mgh = (4)(9.81) h$, from which,

$$\text{height } h = \frac{320\,000}{(4)(9.81)} = 8155 \text{ m}$$

i.e. **the maximum height reached is 8155 m**.

Problem 26. A pile-driver of mass 500 kg falls freely through a height of 1.5 m on to a pile of mass 200 kg. Determine the velocity with which the driver hits the pile. If, at impact, 3 kJ of energy are lost due to heat and sound, the remaining energy being possessed by the pile and driver as they are driven together into the ground a distance of 200 mm, determine (a) the common velocity immediately after impact, (b) the average resistance of the ground.

The potential energy of the pile-driver is converted into kinetic energy. Thus potential energy = kinetic energy, i.e. $mgh = \frac{1}{2} mv^2$, from which, velocity $v = \sqrt{(2gh)} = \sqrt{[(2)(9.81)(1.5)]} = 5.42$ m/s. Hence the pile-driver hits the pile at a velocity of **5.42 m/s**.

(a) Before impact, kinetic energy of pile driver $= \frac{1}{2} mv^2 = \frac{1}{2} (500)(5.42)^2 = 7.34$ kJ.

Kinetic energy after impact = 7.34 − 3 = 4.34 kJ. Thus the pile-driver and pile together have a mass of 500 + 200 = 700 kg and possess kinetic energy of 4.34 kJ. Hence

$$4.34 \times 10^3 = \frac{1}{2} mv^2 = \frac{1}{2} (700) v^2$$

from which,

$$\text{velocity } v = \sqrt{\left(\frac{2 \times 4.34 \times 10^3}{700}\right)} = 3.52 \text{ m/s}$$

Thus the common velocity after impact is **3.52 m/s**.

(b) The kinetic energy after impact is absorbed in overcoming the resistance of the ground, in a distance of 200 mm. Kinetic energy = work done = resistance × distance, i.e. 4.34×10^3 = resistance × 0.200, from which,

$$\text{resistance} = \frac{4.34 \times 10^3}{0.200} = 21\,700 \text{ N}$$

Hence the average resistance of the ground is **21.7 kN**.

Problem 27.   A car of mass 600 kg reduces speed from 90 km/h to 54 km/h in 15 s. Determine the braking power required to give this change of speed.

Change in kinetic energy of car = $\frac{1}{2} mv_1^2 - \frac{1}{2} mv_2^2$, where m = mass of car = 600 kg. v_1 = initial velocity = 90 km/h = (90/3.6) m/s = 25 m/s. v_2 = final velocity = 54 km/h = (54/3.6) m/s = 15 m/s. Hence

$$\text{change in kinetic energy} = \frac{1}{2} m(v_1^2 - v_2^2)$$

$$= \frac{1}{2} (600)(25^2 - 15^2) = 120\,000 \text{ J}$$

$$\textbf{Braking power} = \frac{\text{change in energy}}{\text{time taken}}$$

$$= \frac{120\,000 \text{ J}}{15 \text{ s}} = \textbf{8000 W or 8 kW}$$

15.5 Multi-choice questions on work, energy and power

(Answers on page 355.)

1. State which of the following is incorrect:
(a) $1 \text{ W} = 1 \text{ J s}^{-1}$
(b) $1 \text{ J} = 1 \text{ N/m}$
(c) η = (output energy)/(input energy)
(d) energy = power × time
2. An object is lifted 2000 mm by a crane. If the force required is 100 N, the work done is:
(a) ½₀ N (b) 200 kN (c) 200 N (d) 20 N
3. A motor having an efficiency of 0.8 uses 800 J of electrical energy. The output energy of the motor is
(a) 800 J (b) 1000 J (c) 640 J
4. 6 kJ of work is done by a force in moving an object uniformly through 120 m in 1 minute. The force applied is
(a) 50 N (b) 20 N (c) 720 N (d) 12 N
5. For the object in question 4, the power developed is:
(a) 6 kW (b) 12 kW (c) ⅚ W (d) 0.1 kW
6. Which of the following statements is false?
(a) The unit of energy and work is the same.
(b) The area under a force/distance graph gives the work done.
(c) Electrical energy is converted to mechanical energy by a generator.
(d) Efficiency is the ratio of the useful output energy to the input energy.
7. A machine using a power of 1 kW requires a force of 100 N to raise a mass in 10 s. The height the mass is raised in this time is:
(a) 100 m (b) 1 km (c) 10 m (d) 1 m
8. A force/extension graph for a spring is shown in Fig. 15.9.

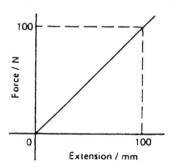

Figure 15.9

Which of the following statements is false? The work done in extending the spring:
(a) from 0 to 100 mm is 5 N
(b) from 0 to 50 mm is 1.25 N
(c) from 20 mm to 60 mm is 1.6 N
(d) from 60 mm to 100 mm is 3.75 N

9. A vehicle of mass 1 tonne climbs an incline of 30° to the horizontal. Taking the acceleration due to gravity as 10 m/s², the increase in potential energy of the vehicle as it moves a distance of 200 m up the incline is:
   (a) 1 kJ   (b) 2 MJ   (c) 1 MJ   (d) 2 kJ

10. A bullet of mass 100 g is fired from a gun with an initial velocity of 360 km/h. Neglecting air resistance, the initial kinetic energy possessed by the bullet is:
    (a) 6.48 kJ (b) 500 J (c) 500 kJ (d) 6.48 MJ

15.6 Short answer questions on work, energy and power

1. Define work in terms of force applied and distance moved.
2. Define energy, and state its unit.
3. Define the joule.
4. The area under a force/distance graph represents ....
5. Name five forms of energy.
6. State the principle of conservation of energy.
7. Give two examples of conversion of heat energy to other forms of energy.
8. Give two examples of conversion of electrical energy to other forms of energy.
9. Give two examples of conversion of chemical energy to other forms of energy.
10. Give two examples of conversion of mechanical energy to other forms of energy.
11. (a) Define efficiency in terms of energy input and energy output.
    (b) State the symbol used for efficiency.
12. Define power and state its unit.
13. Define potential energy.
14. The change in potential energy of a body of mass m kg when lifted vertically upwards to a height h m is given by ....
15. What is kinetic energy?
16. The kinetic energy of a body of mass m kg and moving at a velocity of v m/s is given by ....
17. State the principle of conservation of energy.
18. Distinguish between elastic and inelastic collisions.

15.7 Further questions on work, energy and power

1. Determine the work done when a force of 50 N pushes an object 1.5 km in the same direction as the force.   [75 kJ]

2. Calculate the work done when a mass of weight 200 N is lifted vertically by a crane to a height of 100 m.   [20 kJ]

3. A motor supplies a constant force of 2 kN to move a load 10 m. The force is then changed to a constant 1.5 kN and the load is moved a further 20 m. Draw the force/distance graph for the complete operation, and, from the graph, determine the total work done by the motor.   [50 kJ]

4. A spring, initially relaxed, is extended 80 mm. Draw a work diagram and hence determine the work done if the spring requires a force of 0.5 N/mm of stretch.   [1.6 J]

5. A spring requires a force of 50 N to cause an extension of 100 mm. Determine the work done in extending the spring (a) from 0 to 100 mm, and (b) from 40 mm to 100 mm.   [(a) 2.5 J (b) 2.1 J]

6. The resistance to a cutting tool varies during the cutting stroke of 800 mm as follows:
   (i) The resistance increases uniformly from an initial 5000 N to 10 000 N as the tool moves 500 mm.
   (ii) The resistance falls uniformly from 10 000 N to 6000 N as the tool moves 300 mm.
   Draw the work diagram and calculate the work done in one cutting stroke.   [6.15 kJ]

7. A machine lifts a mass of weight 490.5 N through a height of 12 m when 7.85 kJ of energy is supplied to it. Determine the efficiency of the machine.   [75%]

8. Determine the output energy of an electric motor which is 60% efficient if it uses 2 kJ of electrical energy.   [1.2 kJ]

9. State five possible energy conversions for a motor car.

10. A machine which is used for lifting a particular mass is supplied with 5 kJ of energy. If the machine has an efficiency of 65% and exerts a force of 812.5 N to what height will it lift the mass?   [4 m]

11. A load is hoisted 42 m and requires a force of 100 N. The efficiency of the hoist gear is 60% and that of the motor is 70%. Determine the input energy to the hoist.   [10 kJ]

12. The output power of a motor is 10 kW. How much work does it do in 1 minute? [600 kJ]

13. Determine the power required to lift a load through a height of 20 m in 12.5 s if the force required is 2.5 kN.   [4 kW]

14. 25 kJ of work is done by a force in moving an object uniformly through 50 m in 40 s. Calculate (a) the value of the force, and (b) the power.   [(a) 500 N (b) 625 W]

15. A car towing another at 54 km/h exerts a steady pull of 800 N. Determine (a) the work done in 1/4 hr, and (b) the power required.   [(a) 10.8 MJ (b) 12 kW]

16. To what height will a mass of weight 500 N be raised in 20 s by a motor using 4 kW of power?   [160 m]

17. The output power of a motor is 10 kW. Determine (a) the work done by the motor in 2 hours, and (b) the energy used by the motor if it is 72% efficient.   [(a) 72 MJ (b) 100 MJ]

18. A car is travelling at a constant speed of 81 km/h. The frictional resistance to motion is 0.60 kN. Determine the power required to keep the car moving at this speed.   [13.5 kN]

19. A constant force of 2.0 kN is required to move the table of a shaping machine when a cut is being made. Determine the power required if the stroke of 1.2 m is completed in 5.0 s.   [480 W]

20. A body of mass 15 kg has its speed reduced from 30 km/h to 18 km/h in 4.0 s. Calculate the power required to effect this change of speed.   [83.33 W]

21. The variation of force with distance for a vehicle which is decelerating is as follows:

Distance (m) 600 500 400 300 200 100 0
Force (kN)   24 20 16 12   8   4 0

If the vehicle covers the 600 m in 1.2 minutes, find the power needed to bring the vehicle to rest.   [100 kJ]

22. A cylindrical bar of steel is turned in a lathe. The tangential cutting force on the tool is 0.5 kN and the cutting speed is 180 mm/s. Determine the power absorbed in cutting the steel.   [90 W]

23. An object of mass 400 g is thrown vertically upwards and its maximum increase in potential energy is 32.6 J. Determine the maximum height reached, neglecting air resistance.   [8.31 m]

24. A ball bearing of mass 100 g rolls down from the top of a chute of length 400 m inclined at an angle of 30° to the horizontal. Determine the decrease in potential energy of the ball bearing as it reaches the bottom of the chute.   [196.2 J]

25. A vehicle of mass 800 kg is travelling at 54 km/h when its brakes are applied. Find the kinetic energy lost when the car comes to rest.   [90 kJ]

26. Supplies of mass 300 kg are dropped from a helicopter flying at an altitude of 60 m. Determine the potential energy of the supplies relative to the ground at the instant of release, and its kinetic energy as it strikes the ground.   [176.6 kJ, 176.6 kJ]

27. A shell of mass 10 kg is fired vertically upwards with an initial velocity of 200 m/s. Determine its initial kinetic energy and the maximum height reached, correct to the nearest metre, neglecting air resistance.   [200 kJ, 2039 m]

28. The potential energy of a mass is increased by 20.0 kJ when it is lifted vertically through a height of 25.0 m. It is now released and allowed to fall freely. Neglecting air resistance, find its kinetic energy and its velocity after it has fallen 10.0 m.   [8 kJ, 14.0 m/s]

29. A pile-driver of mass 400 kg falls freely through a height of 1.2 m on to a pile of mass 150 kg. Determine the velocity with which the driver hits the pile. If, at impact, 2.5 kJ of energy are lost due to heat and sound, the remaining energy being possessed by the pile and driver as they are driven together into the ground a distance of 150 mm, determine (a) the common velocity after impact, (b) the average resistance of the ground.   [4.85 m/s (a) 2.83 m/s (b) 14.70 kN]

16

Simple machines

At the end of this chapter you should be able to:

- define a simple machine
- define force ratio, movement ratio, efficiency and limiting efficiency
- perform simple calculations involving force ratio, movement ratio, efficiency and limiting efficiency
- understand pulley systems
- perform calculations involving pulley systems
- understand a simple screw-jack
- perform calculations involving screw-jacks
- understand gear trains
- perform calculations involving gear trains
- understand levers
- perform calculations involving levers

16.1 Machines

A machine is a device which can change the magnitude or line of action, or both magnitude and line of action of a force. A simple machine usually amplifies an input force, called the **effort**, to give a larger output force, called the **load**. Some typical examples of simple machines include pulley systems, screw-jacks, gear systems and lever systems.

16.2 Force ratio, movement ratio and efficiency

The **force ratio** or **mechanical advantage** is defined as the ratio of load to effort, i.e.

$$\boxed{\text{Force ratio} = \frac{\text{load}}{\text{effort}}} \qquad (16.1)$$

Since both load and effort are measured in newtons, force ratio is a ratio of the same units and thus is a dimensionless quantity.

The **movement ratio** or **velocity ratio** is defined as the ratio of the distance moved by the effort to the distance moved by the load, i.e.

$$\boxed{\begin{array}{l}\text{Movement ratio} = \\[4pt] \dfrac{\textbf{distance moved by the effort}}{\textbf{distance moved by the load}}\end{array}} \quad (16.2)$$

Since the numerator and denominator are both measured in metres, movement ratio is a ratio of the same units and thus is a dimensionless quantity.

The **efficiency of a simple machine** is defined as the ratio of the force ratio to the movement ratio, i.e.

$$\text{Efficiency} = \frac{\text{force ratio}}{\text{movement ratio}}$$

Since the numerator and denominator are both dimensionless quantities, efficiency is a dimensionless quantity. It is usually expressed as a percentage, thus:

> **Efficiency =**
>
> $$\frac{\textbf{force ratio}}{\textbf{movement ratio}} \times \textbf{100 per cent} \quad (16.3)$$

Due to the effects of friction and inertia associated with the movement of any object, some of the input energy to a machine is converted into heat and losses occur. Since losses occur, the energy output of a machine is less than the energy input, thus the mechanical efficiency of any machine cannot reach 100%.

For simple machines, the relationship between effort and load is of the form: $F_e = aF_l + b$, where F_e is the effort, F_l is the load and a and b are constants. From equation (16.1),

$$\text{Force ratio} = \frac{\text{load}}{\text{effort}} = \frac{F_l}{F_e} = \frac{F_l}{aF_l + b}$$

Dividing both numerator and denominator by F_l gives:

$$\frac{F_l}{aF_l + b} = \frac{1}{a + \dfrac{b}{F_l}}$$

When the load is large, F_l is large and b/F_l is small compared with a. The force ratio then becomes approximately equal to $1/a$ and is called the **limiting force ratio**.

The **limiting efficiency** of a simple machine is defined as the ratio of the limiting force ratio to the movement ratio, i.e.

Limiting efficiency =

$$\frac{1}{a \times \text{movement ratio}} \times 100\%$$

where a is the constant for the law of the machine: $F_e = aF_l + b$. Due to friction and inertia, the limiting efficiency of simple machines is usually well below 100%.

Problem 1.   A simple machine raises a load of 160 kg through a distance of 1.6 m. The effort applied to the machine is 200 N and moves through a distance of 16 m. Taking g as 9.8 m/s^2, determine the force ratio, movement ratio and efficiency of the machine.

From equation (16.1),

$$\text{force ratio} = \frac{\text{load}}{\text{effort}} = \frac{160 \text{ kg}}{200 \text{ N}}$$

$$= \frac{160 \times 9.8 \text{ N}}{200 \text{ N}} = \textbf{7.84}$$

From equation (16.2),

movement ratio =

$$\frac{\text{distance moved by effort}}{\text{distance moved by load}} = \frac{16 \text{ m}}{1.6 \text{ m}} = \textbf{10}$$

From equation (16.3),

$$\text{efficiency} = \frac{\text{force ratio}}{\text{movement ratio}} \times 100$$

$$= \frac{7.84}{10} \times 100 = \textbf{78.4\%}$$

Problem 2.   For the simple machine of Problem 1, determine:

(a) the distance moved by the effort to move the load through a distance of 0.9 m,

(b) the effort which would be required to raise a load of 200 kg, assuming the same efficiency,

(c) the efficiency if, due to lubrication, the effort to raise the 160 kg load is reduced to 180 N.

(a)   Since the movement ratio is 10, then from equation (16.2), Distance moved by the effort = 10 × distance moved by the load = 10 × 0.9 = **9 m**.

(b)   Since the force ratio is 7.84, then from equation (16.1)

$$\text{Effort} = \frac{\text{load}}{7.84} = \frac{200 \times 9.8}{7.84} = \textbf{250 N}$$

(c)   The new force ratio is given by

$$\frac{\text{load}}{\text{effort}} = \frac{160 \times 9.8}{180} = 8.711$$

Hence the new efficiency after lubrication is

$$\frac{8.711}{10} \times 100$$

$$= \textbf{87.11\%}$$

Problem 3. In a test on a simple machine, the effort/load graph was a straight line of the form $F_e = aF_1 + b$. Two values lying on the graph were at $F_e = 10$ N, $F_1 = 30$ N, and at $F_e = 74$ N, $F_1 = 350$ N. The movement ratio of the machine was 17. Determine: (a) the limiting force ratio, (b) the limiting efficiency of the machine.

(a)   The equation $F_e = aF_1 + b$ is of the form $y = mx + c$, where m is the gradient of the graph. The slope of the line passing through points (x_1, y_1) and (x_2, y_2) of the graph $y = mx + c$ is given by:

$$m = \frac{y_2 - y_1}{x_2 - x_1} \text{ (see Chapter 4)}$$

Thus for $F_e = aF_1 + b$, the slope a is given by

$$a = \frac{74 - 10}{350 - 30} = \frac{64}{320} = 0.2$$

The **limiting force ratio** is $\dfrac{1}{a}$, that is $\dfrac{1}{0.2} = \textbf{5}$

(b)   The **limiting efficiency**

$$= \frac{1}{a \times \text{movement ratio}} \times 100$$

$$= \frac{1}{0.2 \times 17} \times 100 = \textbf{29.4\%}$$

16.3   Pulleys

A **pulley system** is a simple machine. A single-pulley system, shown in Fig. 16.1(a), changes the line of action of the effort, but does not change the magnitude of the force. A two-pulley system, shown in Fig. 16.1(b), changes both the line of action and the magnitude of the force.

Theoretically, each of the ropes marked (i) and (ii) share the load equally, thus the theoretical effort is only half of the load, i.e. the theoretical

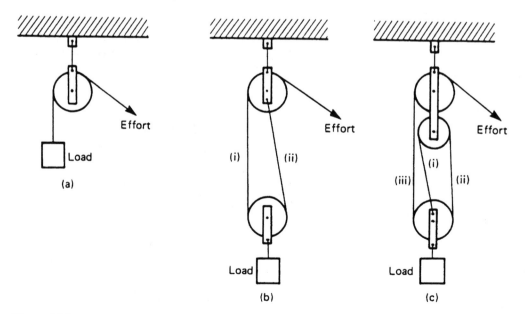

Figure 16.1

force ratio is 2. In practice the actual force ratio is less than 2 due to losses. A three-pulley system is shown in Fig. 16.1(c). Each of the ropes marked (i), (ii) and (iii) carry one-third of the load, thus the theoretical force ratio is 3. In general, for a multiple pulley system having a total of n pulleys, the theoretical force ratio is n. Since the theoretical efficiency of a pulley system (neglecting losses) is 100 and since from equation (16.3),

$$\text{efficiency} = \frac{\text{force ratio}}{\text{movement ratio}} \times 100\%$$

it follows that when the force ratio is n,

$$100 = \frac{n}{\text{movement ratio}} \times 100$$

that is, the movement ratio is also n

Problem 4. A load of 80 kg is lifted by a three-pulley system similar to that shown in Fig. 16.1(c) and the applied effort is 392 N. Calculate (a) the force ratio, (b) the movement ratio, (c) the efficiency of the system. Take g to be 9.8 m/s².

(a) From equation (16.1), the force ratio is given by load/effort. The load is 80 kg, i.e. (80 × 9.8) N. Hence

$$\textbf{force ratio} = \frac{80 \times 9.8}{392} = \textbf{2}$$

(b) From above, for a system having n pulleys, the movement ratio is n. Thus for a three-pulley system, the **movement ratio** is 3.

(c) From equation (16.3),

$$\textbf{efficiency} = \frac{\text{force ratio}}{\text{movement ratio}} \times 100$$

$$= \frac{2}{3} \times 100 = \textbf{66.67\%}$$

Problem 5. A pulley system consists of two blocks, each containing three pulleys and connected as shown in Fig. 16.2. An effort of 400 N is required to raise a load of 1500 N. Determine (a) the force ratio, (b) the movement ratio, (c) the efficiency of the pulley system.

Figure 16.2

(a) From equation (16.1),

$$\text{force ratio} = \frac{\text{load}}{\text{effort}} = \frac{1500}{400} = \textbf{3.75}$$

(b) An n-pulley system has a movement ratio of n, hence this 6-pulley system has a movement ratio of **6**.

(c) From equation (16.3),

$$\text{efficiency} = \frac{\text{force ratio}}{\text{movement ratio}} \times 100$$

$$= \frac{3.75}{6} \times 100 = \textbf{62.5\%}$$

16.4 The screw-jack

A simple screw-jack is shown in Fig. 16.3 and is a simple machine since it changes both the magnitude and the line of action of a force.

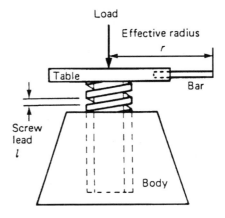

Figure 16.3

The screw of the table of the jack is located in a fixed nut in the body of the jack. As the table is rotated by means of a bar, it raises or lowers a load placed on the table. For a single-start thread, as shown, for one complete revolution of the table, the effort moves through a distance $2\pi r$, and the load moves through a distance equal to the lead of the screw, say, l.

$$\text{Movement ratio} = \frac{2\pi r}{l} \qquad (16.4)$$

Problem 6. A screw-jack is being used to support the axle of a car, the load on it being 2.4 kN. The screw jack has an effort of effective radius 200 mm and a single-start square thread, having a lead of 5 mm. Determine the efficiency of the jack if an effort of 60 N is required to raise the car axle.

From equation (16.3),

$$\text{efficiency} = \frac{\text{force ratio}}{\text{movement ratio}} \times 100\%$$

$$\text{force ratio} = \frac{\text{load}}{\text{effort}} = \frac{2400\text{ N}}{60\text{ N}} = 40$$

From equation (16.4),

$$\text{movement ratio} = \frac{2\pi r}{l} = \frac{2\pi\,(200)\text{ mm}}{5\text{ mm}} = 251.3$$

Hence

$$\text{efficiency} = \frac{40}{251.3} \times 100 = \mathbf{15.9\%}$$

16.5   Gear trains

A simple gear train is used to transmit rotary motion and can change both the magnitude and the line of action of a force, hence is a simple machine. The gear train shown in Fig. 16.4 consists of **spur gears** and has an effort applied to one gear, called the **driver**, and a load applied to the other gear, called the **follower**.

In such a system, the teeth on the wheels are so spaced that they exactly fill the circumference with a whole number of identical teeth, and the teeth on the driver and follower mesh without interference. Under these conditions, the number of teeth on the driver and follower are in direct

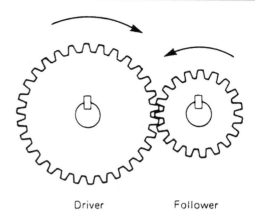

Driver   Follower

Figure 16.4

proportion to the circumference of these wheels, i.e.

$$\frac{\textbf{number of teeth on driver}}{\textbf{number of teeth on follower}}$$

$$= \frac{\textbf{circumference of driver}}{\textbf{circumference of follower}} \qquad (16.5)$$

If there are, say, 40 teeth on the driver and 20 teeth on the follower then the follower makes two revolutions for each revolution of the driver. In general

$$\frac{\text{number of revolutions made by driver}}{\text{number of revolutions made by follower}}$$

$$= \frac{\text{number of teeth on follower}}{\text{number of teeth on driver}} \qquad (16.6)$$

It follows from equation (16.6) that the speeds of the wheels in a gear train are inversely proportional to the number of teeth. The ratio of the speed of the driver wheel to that of the follower is the movement ratio, i.e.

$$\textbf{Movement ratio} = \frac{\textbf{speed of driver}}{\textbf{speed of follower}}$$

$$= \frac{\textbf{teeth on follower}}{\textbf{teeth on driver}} \qquad (16.7)$$

When the same direction of rotation is required on both the driver and the follower an **idler wheel** is used as shown in Fig. 16.5.

Let the driver, idler, and follower be A, B and C, respectively, and let N be the speed of rotation and T be the number of teeth. Then from equation (16.7),

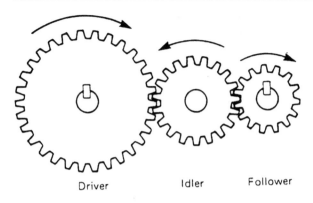

Figure 16.5

$$\frac{N_B}{N_A} = \frac{T_A}{T_B} \text{ and } \frac{N_C}{N_B} = \frac{T_B}{T_C}$$

Thus

$$\frac{\text{speed of A}}{\text{speed of C}} = \frac{N_A}{N_C} = \frac{N_B \dfrac{T_B}{T_A}}{N_B \dfrac{T_B}{T_C}}$$

$$= \frac{T_B}{T_A} \times \frac{T_C}{T_B} = \frac{T_C}{T_A}$$

This shows that the movement ratio is independent of the idler, only the direction of the follower being altered.

A compound gear train is shown in Fig. 16.6, in which gear wheels B and C are fixed to the same shaft and hence $N_B = N_C$. From equation (16.7),

$$\frac{N_A}{N_B} = \frac{T_B}{T_A}, \text{ i.e. } N_B = N_A \times \frac{T_A}{T_B}$$

Also,

$$\frac{N_D}{N_C} = \frac{T_C}{T_D}, \text{ i.e. } N_D = N_C \times \frac{T_C}{T_D}$$

But $N_B = N_C$, hence

$$N_D = N_A \times \frac{T_A}{T_B} \times \frac{T_C}{T_D} \tag{16.8}$$

For compound gear trains having, say, P gear wheels,

$$N_P = N_A \times \frac{T_A}{T_B} \times \frac{T_C}{T_D} \times \frac{T_E}{T_F} \cdots \times \frac{T_O}{T_P}$$

from which

$$\textbf{movement ratio} = \frac{N_A}{N_P}$$

$$= \frac{T_B}{T_A} \times \frac{T_D}{T_o} \cdots \times \frac{T_P}{T_C}$$

Problem 7. A driver gear on a shaft of a motor has 35 teeth and meshes with a follower having 98 teeth. If the speed of the motor is 1400 revolutions per minute, find the speed of rotation of the follower.

From equation (16.7),

$$\frac{\text{speed of driver}}{\text{speed of follower}} = \frac{\text{teeth on follower}}{\text{teeth on driver}}$$

i.e.

$$\frac{1400}{\text{speed of follower}} = \frac{98}{35}$$

Hence

$$\text{speed of follower} = \frac{1400 \times 35}{98} = \textbf{500 rev/min}$$

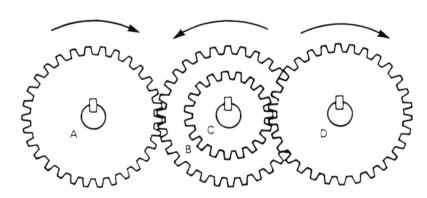

Figure 16.6

Problem 8.   A compound gear train similar to that shown in Fig. 16.6 consists of a driver gear A, having 40 teeth, engaging with gear B, having 160 teeth. Attached to the same shaft as B, gear C has 48 teeth and meshes with gear D on the output shaft, having 96 teeth. Determine (a) the movement ratio of this gear system and (b) the efficiency when the force ratio is 6.

(a)   From equation (16.8),

the speed of D = speed of A $\times \dfrac{T_A}{T_B} \times \dfrac{T_C}{T_D}$

From equation (16.7),

$$\text{movement ratio} = \frac{\text{speed of A}}{\text{speed of D}}$$

$$= \frac{T_B}{T_A} \times \frac{T_D}{T_C}$$

$$= \frac{160}{40} \times \frac{96}{48} = \mathbf{8}$$

(b)   The efficiency of any simple machine is

$$\frac{\text{force ratio}}{\text{movement ratio}} \times 100\%$$

Thus

$$\text{efficiency} = \frac{6}{8} \times 100 = \mathbf{75\%}$$

16.6   Levers

A **lever** can alter both the magnitude and the line of action of a force and is thus classed as a simple machine. There are three types or orders of levers, as shown in Fig. 16.7.

A lever of the first order has the fulcrum placed between the effort and the load, as shown in Fig. 16.7(a).

A lever of the second order has the load placed between the effort and the fulcrum, as shown in Fig. 16.7(b).

A lever of the third order has the effort applied between the load and the fulcrum, as shown in Fig. 16.7(c).

Problems on levers can largely be solved by applying the principle of moments (see Chapter 12). Thus for the lever shown in Fig. 16.7(a), when the lever is in equilibrium,

anticlockwise moment = clockwise moment

i.e. $a \times F_1 = b \times F_e$. Thus

$$\mathbf{force\ ratio} = \frac{F_1}{F_e} = \frac{b}{a}$$

$$= \frac{\textbf{distance of effort from fulcrum}}{\textbf{distance of load from fulcrum}}$$

Problem 9.   The load on a first-order lever, similar to that shown in Fig. 16.7(a), is 1.2 kN. Determine the effort, the force ratio and the movement ratio when the distance between the fulcrum and the load is 0.5 m and the distance between the fulcrum and effort is 1.5 m. Assume the lever is 100% efficient.

Applying the principle of moments, for equilibrium:

anticlockwise moment = clockwise moment

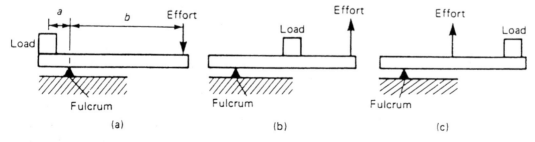

Figure 16.7

i.e. $1200 \text{ N} \times 0.5 \text{ m} = \text{effort} \times 1.5 \text{ m}$

Hence,

$$\text{effort} = \frac{1200 \times 0.5}{1.5} = \textbf{400 N}$$

$$\text{force ratio} = \frac{F_1}{F_e} = \frac{1200}{400} = \textbf{3}$$

Alternatively,

$$\text{force ratio} = \frac{b}{a} = \frac{1.5}{0.5} = \textbf{3}$$

This result shows that to lift a load of, say, 300 N, an effort of 100 N is required. Since, from equation (16.3),

$$\text{efficiency} = \frac{\text{force ratio}}{\text{movement ratio}} \times 100\%$$

then

$$\text{movement ratio} = \frac{\text{force ratio}}{\text{efficiency}} \times 100$$

$$= \frac{3}{100} \times 100 = \textbf{3}$$

This result shows that to raise the load by, say, 100 mm, the effort has to move 300 mm.

Problem 10.   A second-order lever, AB, is in a horizontal position. The fulcrum is at point C. An effort of 60 N applied at B just moves a load at point D, when BD is 0.5 m and BC is 1.25 m. Calculate the load and the force ratio of the lever.

A second-order lever system is shown in Fig. 16.7(b). Taking moments about the fulcrum as the load is just moving, gives:

anticlockwise moment = clockwise moment

i.e.

$$60 \text{ N} \times 1.25 \text{ m} = \text{load} \times 0.75 \text{ m}$$

Thus,

$$\text{load} = \frac{60 \times 1.25}{0.75} = 100 \text{ N}$$

From equation (16.1),

$$\text{force ratio} = \frac{\text{load}}{\text{effort}} = \frac{100}{60} = 1\frac{2}{3}$$

Alternatively,

force ratio

$$= \frac{\text{distance of effort from fulcrum}}{\text{distance of load from fulcrum}}$$

$$= \frac{1.25}{0.75} = 1\frac{2}{3}$$

16.7   Multi-choice questions on simple machines

(Answers on page 355.)

A simple machine requires an effort of 250 N moving through 10 m to raise a load of 1000 N through 2 m. Use this data to find the correct answers to questions 1 to 3, selecting these answers from:

(a) $\frac{1}{4}$ (b) 4   (c) 80%   (d) 20%   (e) 100
(f) 5   (g) 100%   (h) $\frac{1}{5}$ (i) 25%

1. Find the force ratio.
2. Find the movement ratio.
3. Find the efficiency.

The law of a machine is of the form $F_e = aF_1 + b$. An effort of 12 N is required to raise a load of 40 N and an effort of 6 N is required to raise a load of 16 N. The movement ratio of the machine is 5. Use this data to find the correct answers to questions 4 to 6, selecting these answers from:

(a) 80%   (b) 4   (c) 2.8   (d) $\frac{1}{4}$ (e) 1/2.8
(f) 25%   (g) 100%   (h) 2   (i) 25%

4. Determine the constant 'a'.
5. Find the limiting force ratio.
6. Find the limiting efficiency.
7. Which of the following statements is false?
    (a) A single-pulley system changes the line of action of the force but does not change the magnitude of the force, when losses are neglected.
    (b) In a two-pulley system, the force ratio is $\frac{1}{2}$ when losses are neglected.
    (c) In a two-pulley system, the movement ratio is 2.
    (d) The efficiency of a two-pulley system is 100% when losses are neglected.
8. Which of the following statements concerning a screw-jack is false?
    (a) A screw-jack changes both the line of action and the magnitude of the force.
    (b) For a single-start thread, the distance moved in 5 revolutions of the table is $5l$, where l is the lead of the screw.

(c) The distance moved by the effort is $2\pi r$, where r is the effective radius of the effort.

(d) The movement ratio is given by $2\pi r/5l$.

9. In a simple gear train, a follower has 50 teeth and the driver has 30 teeth. The movement ratio is:
(a) $\frac{3}{5}$ (b) 20 (c) $\frac{5}{3}$ (d) 80

10. Which of the following statements is true?
(a) An idler wheel between a driver and a follower is used to make the direction of the follower opposite to that of the driver.
(b) An idler wheel is used to change the movement ratio.
(c) An idler wheel is used to change the force ratio.
(d) An idler wheel is used to make the direction of the follower the same as that of the driver.

11. Which of the following statements is false?
(a) In a first-order lever, the fulcrum is between the load and the effort.
(b) In a second-order lever, the load is between the effort and the fulcrum.
(c) In a third-order lever, the effort is applied between the load and the fulcrum.
(d) The force ratio for a first-order lever system is given by

$$\frac{\text{distance of load from fulcrum}}{\text{distance of effort from fulcrum}}$$

12. In a second-order lever system, the load is 200 mm from the fulcrum and the effort is 500 mm from the fulcrum. If losses are neglected, an effort of 100 N will raise a load of:
(a) 100 N (b) 250 N (c) 400 N (d) 40 N

16.8 Short answer questions on simple machines

1. State what is meant by a simple machine.
2. Define force ratio.
3. Define movement ratio.
4. Define the efficiency of a simple machine in terms of the force and movement ratios.
5. State briefly why the efficiency of a simple machine cannot reach 100%.
6. With reference to the law of a simple machine, state briefly what is meant by the term 'limiting force ratio'.

7. Define limiting efficiency.
8. Explain why a four-pulley system has a force ratio of 4 when losses are ignored.
9. Give the movement ratio for a screw-jack in terms of the effective radius of the effort and the screw lead.
10. Explain the action of an idler gear.
11. Define the movement ratio for a two-gear system in terms of the teeth on the wheels.
12. Show that the action of an idler wheel does not affect the movement ratio of a gear system.
13. State the relationship between the speed of the first gear and the speed of the last gear in a compound train of four gears, in terms of the teeth on the wheels.
14. Sketch a second-order lever system.
15. Define the force ratio of a first-order lever system in terms of the distances of the load and effort from the fulcrum.

16.9 Further questions on simple machines

1. A simple machine raises a load of 825 N through a distance of 0.3 m. The effort is 250 N and moves through a distance of 3.3 m. Determine: (a) the force ratio, (b) the movement ratio, (c) the efficiency of the machine at this load.
[(a) 3.3 (b) 11 (c) 30%]

2. The efficiency of a simple machine is 50%. If a load of 1.2 kN is raised by an effort of 300 N, determine the movement ratio. [8]

3. An effort of 10 N applied to a simple machine moves a load of 40 N through a distance of 100 mm, the efficiency at this load being 80%. Calculate: (a) the movement ratio, (b) the distance moved by the effort.
[(a) 5 (b) 500 mm]

4. The effort required to raise a load using a simple machine, for various values of load is as shown:

Load (N)   2050 4120 7410 8240 10 300
Effort (N)   260 340 435 505   580

If the movement ratio for the machine is 30, determine (a) the law of the machine, (b) the limiting force ratio, (c) the limiting efficiency.
[(a) $F_e = 0.4F_1 + 170$ (b) 25 (c) $83\frac{1}{3}\%$]

5. For the data given in question 4, determine the values of force ratio and efficiency for

each value of the load. Hence plot graphs of effort, force ratio and efficiency to a base of load. From the graphs, determine the effort required to raise a load of 6 kN and the efficiency at this load.   [410 N, 48.8%]

6. In an experiment involving a screw-jack, the following data was obtained.

Load (N)   0   250   500   750   1000 1250
Effort (N) 10   30   50   70   90   110

Determine the law of the machine. If the movement ratio is 45, find the limiting force ratio and the limiting efficiency.
        $[F_e = 0.08\,F_1 + 10,\ 12.5,\ 27.8\%]$

7. For the data given in question 6, calculate the force ratio and efficiency for each value of the loads. Plot graphs of effort, force ratio and efficiency to a base of load and hence determine the effort required to raise a load of 850 N. Determine the efficiency of the screw-jack at this load.   [78 N, 24.2%]

8. A pulley system consists of four pulleys in an upper block and three pulleys in a lower block. Make a sketch of this arrangement showing how a movement ratio of 7 may be obtained. If the force ratio is 4.2, what is the efficiency of the pulley.   [60%]

9. A three-pulley lifting system is used to raise a load of 4.5 kN. Determine the effort required to raise this load when losses are neglected. If the actual effort required is 1.6 kN, determine the efficiency of the pulley system at this load.   [1.5 kN, 93.75%]

10. Sketch a simple screw-jack. The single-start screw of such a jack has a lead of 6 mm and the effective length of the operating bar from the centre of the screw is 300 mm. Calculate the load which can be raised by an effort of 150 N if the efficiency at this load is 20%.
        [9.425 kN]

11. A load of 1.7 kN is lifted by a screw-jack having a single-start screw of lead 5 mm. The effort is applied at the end of an arm of effective length 300 mm from the centre of the screw. Calculate the effort required if the efficiency at this load is 20%.   [22.55 N]

12. The driver gear of a gear system has 28 teeth and meshes with a follower gear having 168 teeth. Determine the movement ratio and the speed of the follower when the driver gear rotates at 60 revolutions per second.
        [6, 10 rev/s]

13. A compound gear train has a 30-tooth driver gear A, meshing with a 90-tooth follower gear B. Mounted on the same shaft as B and attached to it is a gear C with 60 teeth, meshing with a gear D on the output shaft having 120 teeth. Calculate the movement and force ratios if the overall efficiency of the gears is 72%.   [6, 4.32]

14. A compound gear train is as shown in Fig. 16.6. The movement ratio is 6 and the numbers of teeth on gears A, C and D are 25, 100 and 60, respectively. Determine the number of teeth on gear B and the force ratio when the efficiency is 60%.
        [250, 3.6]

15. Use sketches to show what is meant by: (a) a first-order, (b) a second-order, (c) a third-order lever system. Give one practical use for each type of lever.

16. In a second-order lever system, the force ratio is 2.5. If the load is at a distance of 0.5 m from the fulcrum, find the distance that the effort acts from the fulcrum if losses are negligible.   [1.25 m]

17. A lever AB is 2 m long and the fulcrum is at a point 0.5 m from B. Find the effort to be applied at A to raise a load of 0.75 kN at B when losses are negligible.   [250 N]

18. The load on a third-order lever system is at a distance of 750 mm from the fulcrum and the effort required to just move the load is 1 kN when applied at a distance of 250 mm from the fulcrum. Determine the value of the load and the force ratio if losses are negligible. $[333\tfrac{1}{3}\ \text{N},\ \tfrac{1}{3}]$

17

The effects of forces on materials

At the end of this chapter you should be able to:

- define force and state its unit
- recognize a tensile force and state relevant practical examples
- recognize a compressive force and state relevant practical examples
- recognize a shear force and state relevant practical examples
- define stress and state its unit
- calculate stress σ from $\sigma = F/A$
- define strain
- calculate strain ϵ from $\epsilon = x/l$
- define elasticity, plasticity and elastic limit
- state Hooke's law
- define Young's modulus of elasticity E and stiffness
- appreciate typical values for E
- calculate E from $E = \sigma/\epsilon$
- perform calculations using Hooke's law
- plot a load/extension graph from given data
- define ductility, brittleness and malleability with examples of each

17.1 Introduction

A force exerted on a body can cause a change in either the shape or the motion of the body. The unit of force is the **newton, N.**

No solid body is perfectly rigid and when forces are applied to it, changes in dimensions occur. Such changes are not always perceptible to the human eye since they are so small. For example, the span of a bridge will sag under the weight of a vehicle and a spanner will bend slightly when tightening a nut. It is important for engineers and designers to appreciate the effects of forces on materials, together with their mechanical properties.

The three main types of mechanical force that can act on a body are (i) tensile, (ii) compressive, and (iii) shear.

17.2 Tensile force

Tension is a force which tends to stretch a material, as shown in Fig. 17.1(a). Examples include:

(i) the rope or cable of a crane carrying a load is in tension;
(ii) rubber bands, when stretched, are in tension;
(iii) a bolt; when a nut is tightened, a bolt is under tension.

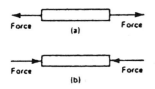

Figure 17.1

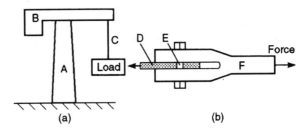

Figure 17.3

A tensile force, i.e one producing tension, increases the length of the material on which it acts.

17.3   Compressive force

Compression is a force which tends to squeeze or crush a material, as shown in Fig. 17.1(b). Examples include:

(i)   a pillar supporting a bridge is in compression;
(ii) the sole of a shoe is in compression;
(iii) the jib of a crane is in compression.

A compressive force, i.e. one producing compression, will decrease the length of the material on which it acts.

17.4   Shear force

Shear is a force which tends to slide one face of the material over an adjacent face. Examples include:

(i)   a rivet holding two plates together is in shear if a tensile force is applied between the plates (as shown in Fig. 17.2);
(ii) a guillotine cutting sheet metal, or garden shears, each provide a shear force;
(iii) a horizontal beam is subject to shear force;
(iv) transmission joints on cars are subject to shear forces.

A shear force can cause a material to bend, slide or twist.

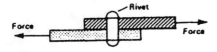

Figure 17.2

Problem 1.   Fig 17.3(a) represents a crane and Fig 17.3(b) a transmission joint. State the types of forces acting labelled A to F.

(a) For the crane, A, a supporting member, is in **compression**, B, a horizontal beam, is in **shear**, and C, a rope, is in **tension**.
(b) For the transmission joint, parts D and F are in **tension**, and E, the rivet or bolt, is in **shear**.

17.5   Stress

Forces acting on a material cause a change in dimensions and the material is said to be in a state of **stress**. Stress is the ratio of the applied force F to cross-sectional area A of the material. The symbol used for tensile and compressive stress is σ (Greek letter sigma). The unit of stress is the **Pascal, Pa**, where $1\ Pa = 1\ N/m^2$. Hence

$$\sigma = \frac{F}{A}\ \textbf{Pa}$$

where F is the force in newtons and A is the cross-sectional area in square metres. For tensile and compressive forces, the cross-sectional area is that which is at right angles to the direction of the force. For a shear force the shear stress is equal to F/A, where the cross-sectional area A is that which is parallel to the direction of the force. The symbol used for shear stress is the Greek letter tau, τ.

Problem 2.   A rectangular bar having a cross-sectional area of 75 mm$=^2$ has a tensile force of 15 kN applied to it. Determine the stress in the bar.

Cross-sectional area A = 75 mm^2 = 75 × 10^{-6} m^2; force F = 15 kN = 15 × 10^3 N. Stress in bar,

$$\sigma = \frac{F}{A} = \frac{15 \times 10^3 \, \text{N}}{75 \times 10^{-6} \, \text{m}^2} = 0.2 \times 10^9 \, \text{Pa}$$
$$= \mathbf{200 \, MPa}$$

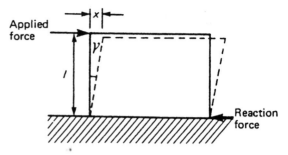

Figure 17.4

Problem 3. A circular wire has a tensile force of 60.0 N applied to it and this force produces a stress of 3.06 MPa in the wire. Determine the diameter of the wire.

Force F = 60.0 N; stress σ = 3.06 MPa = 3.06 × 10^6 Pa. Since $\sigma = F/A$, then

$$\text{area } A = \frac{F}{\sigma} = \frac{60.0 \, \text{N}}{3.06 \times 10^6 \, \text{Pa}}$$
$$= 19.61 \times 10^{-6} \, \text{m}^2$$
$$= 19.61 \, \text{mm}^2$$

Cross-sectional area $A = \pi d^2/4$. Hence 19.61 = $\pi d^2/4$, from which,

$$d^2 = \frac{4 \times 19.61}{\pi} \quad \text{and} \quad d = \sqrt{\left(\frac{4 \times 19.61}{\pi}\right)}$$

i.e. **diameter of wire = 5.0 mm**.

17.6   Strain

The fractional change in a dimension of a material produced by a force is called the **strain**. For a tensile or compressive force, strain is the ratio of the change of length to the original length. The symbol used for strain is ϵ (Greek epsilon). For a material of length l metres which changes in length by an amount x metres when subjected to stress

$$\boxed{\epsilon = \frac{x}{l}}$$

Strain is dimensionless and is often expressed as a percentage, i.e.

$$\text{percentage strain} = \frac{x}{l} \times 100$$

For a shear force, strain is denoted by the symbol γ (Greek letter gamma) and, with reference to Fig. 17.4, is given by:

$$\gamma = \frac{x}{l}$$

Problem 4. A bar 1.60 m long contracts by 0.1 mm when a compressive load is applied to it. Determine the strain and the percentage strain.

$$\text{Strain } \epsilon = \frac{\text{contraction}}{\text{original length}}$$

$$= \frac{0.1 \, \text{mm}}{1.60 \times 10^3 \, \text{mm}} = \frac{0.1}{1600} = \mathbf{0.000\,062\,5}$$

Percentage strain = 0.000 062 5 × 100 = **0.006 25%**

Problem 5. A wire of length 2.50 m has a percentage strain of 0.012% when loaded with a tensile force. Determine the extension of the wire.

Original length of wire = 2.50 m = 2500 mm.

$$\text{Strain} = \frac{0.012}{100} = 0.000\,12$$

$$\text{Strain } \epsilon = \frac{\text{extension } x}{\text{original length } l}$$

hence

$$\text{extension } x = \epsilon l = (0.000\,12)(2500)$$
$$= \mathbf{0.3 \, mm}$$

Problem 6. (a) A rectangular metal bar has a width of 10 mm and can support a maximum compressive stress of 20 MPa. Determine the minimum breadth of the bar when loaded with a force of 3 kN. (b) If the bar in (a) is 2 m long and decreases in length by 0.25 mm when the force is applied, determine the strain and the percentage strain.

(a) Since

$$\text{stress } \sigma = \frac{\text{force } F}{\text{area } A}$$

then

$$\text{area } A = \frac{F}{\sigma} = \frac{3000 \text{ N}}{20 \times 10^6 \text{ Pa}} = 150 \times 10^{-6} \text{ m}^2$$

$$= 150 \text{ mm}^2$$

Cross-sectional area = width × breadth, hence

$$\text{breadth} = \frac{\text{area}}{\text{width}} = \frac{150}{10} = \mathbf{15 \text{ mm}}$$

(b)

$$\text{Strain } \epsilon = \frac{\text{contraction}}{\text{original length}} = \frac{0.25}{2000}$$

$$= 0.000\,125$$

Percentage strain = 0.000 125 × 100
$$= \mathbf{0.0125\%}$$

Problem 7. A pipe has an outside diameter of 25 mm, an inside diameter of 15 mm and length 0.40 m and it supports a compressive load of 40 kN. The pipe shortens by 0.5 mm when the load is applied. Determine (a) the compressive stress, (b) the compressive strain in the pipe when supporting this load.

Compressive force F = 40 kN = 40 000 N.
Cross-sectional area of pipe, $A = (\pi D^2/4) - (\pi d^2/4)$, where D = outside diameter = 25 mm and d = inside diameter = 15 mm. Hence

$$A = \frac{\pi}{4}(25^2 - 15^2) \text{ mm}^2$$

$$= \frac{\pi}{4}(25^2 - 15^2) \times 10^{-6} \text{ m}^2$$

$$= 3.142 \times 10^{-4} \text{ m}^2$$

(a)

$$\text{Compressive stress } \sigma = \frac{F}{A}$$

$$= \frac{40\,000 \text{ N}}{3.142 \times 10^{-4} \text{ m}^2}$$

$$= 12.73 \times 10^7 \text{ Pa} = \mathbf{127.3 \text{ MPa}}$$

(b) Contraction of pipe when loaded, x = 0.5 mm = 0.0005 m. Original length of pipe, l = 0.4 m. Hence

$$\text{compressive strain } \epsilon = \frac{x}{l} = \frac{0.0005}{0.4}$$

$$= \mathbf{0.00125} \text{ (or 0.125\%)}$$

Problem 8. A circular hole of diameter 50 mm is to be punched out of a 2 mm thick metal plate. The shear stress needed to cause fracture is 500 MPa. Determine (a) the minimum force to be applied to the punch, (b) the compressive stress in the punch at this value.

(a) The area of metal to be sheared, A = perimeter of hole × thickness of plate. Perimeter of hole = πd = $\pi(0.050)$ = 0.1571 m. Hence shear area A = 0.1571 × 0.002 = 314.2 × 10^{-6} m². Since shear stress = force/area, shear force = shear stress × area = $(500 \times 10^6 \times 314.2 \times 10^{-6})$ N = **157.1 kN**, which is the minimum force to be applied to the punch.

(b)

$$\text{Area of punch} = \frac{\pi d^2}{4} = \frac{\pi(0.050)^2}{4}$$
$$= 0.001\,963 \text{ m}^2$$

$$\text{Compressive stress} = \frac{\text{force}}{\text{area}}$$

$$= \frac{157.1 \times 10^3 \text{ N}}{0.001\,963 \text{ m}^2}$$

$$= 8.003 \times 10^7 \text{ Pa}$$

$$= \mathbf{80.03 \text{ MPa}}$$

which is the compressive stress in the punch.

Problem 9. A rectangular block of plastic material 500 mm long by 20 mm wide by 300 mm high has its lower face glued to a bench and a force of 200 N is applied to the upper face and in line with it. The upper face moves 15 mm relative to the lower face. Determine (a) the shear stress, and (b) the shear strain in the upper face, assuming the deformation is uniform.

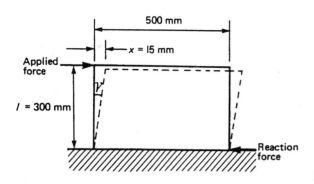

Figure 17.5

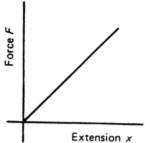

Figure 17.6

(a)
$$\text{Shear stress } \tau = \frac{\text{force}}{\text{area parallel to the force}}$$

Area of any face parallel to the force

$$= 500 \text{ mm} \times 20 \text{ mm}$$

$$= (0.5 \times 0.02) \text{ m}^2$$

$$= 0.01 \text{ m}^2$$

Hence

$$\text{shear stress } \tau = \frac{200 \text{ N}}{0.01 \text{ m}^2} = \textbf{20 000 Pa}$$
$$\textbf{or 20 kPa}$$

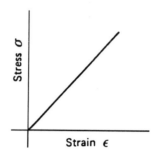

Figure 17.7

(b)
$$\text{Shear strain } \gamma = \frac{x}{l} \quad \text{(see side view in Fig. 17.5)}$$
$$= \frac{15}{300}$$
$$= \textbf{0.05 (or 5\%)}$$

17.7   Elasticity and elastic limit

Elasticity is the ability of a material to return to its original shape and size on the removal of external forces.

   Plasticity is the property of a material of being permanently deformed by a force without breaking. Thus if a material does not return to the original shape, it is said to be plastic.

   Within certain load limits, mild steel, copper, polythene and rubber are examples of elastic materials; lead and plasticine are examples of plastic materials.

   If a tensile force applied to a uniform bar of mild steel is gradually increased and the corresponding extension of the bar is measured, then provided the applied force is not too large, a graph depicting these results is likely to be as shown in Fig. 17.6. Since the graph is a straight line, **extension is directly proportional to the applied force.**

   If the applied force is large, it is found that the material no longer returns to its original length when the force is removed. The material is then said to have passed its elastic limit and the resulting graph of force/extension is no longer a straight line (see Chapter 18). Stress $= \sigma$ F/A, from section 17.5, and since, for a particular bar, A can be considered as a constant, then $F \propto \sigma$. Strain $\epsilon = x/l$, from section 17.6, and since for a particular bar l is constant, then $x \propto \epsilon$. Hence for stress applied to a material below the elastic limit a graph of stress/strain will be as shown in Fig. 17.7, and is a similar shape to the force/extension graph of Fig. 17.6.

17.8   Hooke's law

Hooke's law states:

Within the elastic limit, the extension of a material is proportional to the applied force.

It follows, from section 17.7, that:

Within the elastic limit of a material, the strain produced is directly proportional to the stress producing it.

Young's modulus of elasticity

Within the elastic limit, stress $\propto$ strain, hence stress = (a constant) $\times$ strain.

This constant of proportionality is called **Young's modulus of elasticity** and is given the symbol E. The value of E may be determined from the gradient of the straight line portion of the stress/strain graph. The dimensions of E are pascals (the same as for stress, since strain is dimensionless).

$$E = \frac{\sigma}{\epsilon} \text{ Pa}$$

Some typical values for Young's modulus of elasticity, E, include:

Aluminium 70 GPa (i.e. 70×10^9 Pa), brass 100 GPa, copper 96 GPa, diamond 1200 GPa, mild steel 210 GPa, lead 18 GPa, tungsten 410 GPa, cast iron 110 GPa, zinc 85 GPa.

Stiffness

A material having a large value of Young's modulus is said to have a high value of stiffness, where stiffness is defined as:

$$\text{Stiffness} = \frac{\text{force } F}{\text{extension } x}$$

For example, mild steel is much stiffer than lead. Since $E = \sigma/\epsilon$ and $\sigma = F/A$ and $\epsilon = x/l$, then

$$E = \frac{F/A}{x/l}$$

i.e.

$$E = \frac{Fl}{Ax} = \left(\frac{F}{x}\right)\left(\frac{l}{A}\right)$$

i.e.

$$E = (\text{stiffness}) \times \left(\frac{l}{A}\right)$$

Stiffness $(= F/x)$ is also the gradient of the force/extension graph, hence

$$E = (\text{gradient of force/extension graph}) \left(\frac{l}{A}\right)$$

Since l and A for a particular specimen are constant, the greater Young's modulus the greater the stiffness.

> **Problem 10.** A wire is stretched 2 mm by a force of 250 N. Determine the force that would stretch the wire 5 mm, assuming that the elastic limit is not exceeded.

Hooke's law states that extension x is proportional to force F, provided that the elastic limit is not exceeded, i.e. $x \propto F$ or $x = kF$ where k is a constant.

When $x = 2$ mm, $F = 250$ N, thus $2 = k(250)$, from which, constant $k = (2/250) = (1/125)$.

When $x = 5$ mm, then $5 = kF$, i.e. $5 = (1/125)F$, from which, force $F = 5(125) = 625$ N.

Thus to stretch the wire 5 mm a force of 625 N is required.

> **Problem 11.** A force of 10 kN applied to a component produces an extension of 0.1 mm. Determine (a) the force needed to produce an extension of 0.12 mm, and (b) the extension when the applied force is 6 kN, assuming in each case that the elastic limit is not exceeded.

From Hooke's law, extension x is proportional to force F within the elastic limit, i.e. $x \propto F$ or $x = kF$, where k is a constant. If a force of 10 kN produces an extension of 0.1 mm, then $0.1 = k(10)$ from which, constant $k = 0.1/10 = 0.01$.

(a) When extension $x = 0.12$ mm, then $0.12 = k(F)$, i.e. $0.12 = 0.01 F$, from which,

$$\text{force } F = \frac{0.12}{0.01} = \textbf{12 kN}$$

(b) When force $F = 6$ kN, then

$$\textbf{extension } x = k(6) = (0.01)(6)$$

$$= \textbf{0.06 mm}$$

> **Problem 12.** A copper rod of diameter 20 mm and length 2.0 m has a tensile force of 5 kN applied to it. Determine (a) the stress in the rod, (b) by how much the rod extends when the load is applied. Take the modulus of elasticity for copper as 96 GPa.

(a)   Force F = 5 kN = 5000 N.

Cross-sectional area $A = \dfrac{\pi d^2}{4}$

$= \dfrac{\pi(0.020)^2}{4} = 0.000\ 314\ \text{m}^2$

Stress $\sigma = \dfrac{F}{A} = \dfrac{5000\ \text{N}}{0.000\ 314\ \text{m}^2}$

$= 15.92 \times 10^6\ \text{Pa} = \mathbf{15.92\ MPa}$

(b)   Since $E = \sigma/\epsilon$ then

strain $\epsilon = \dfrac{\sigma}{E} = \dfrac{15.92 \times 10^6\ \text{Pa}}{96 \times 10^9\ \text{Pa}}$

$= 0.000\ 166$

Strain $\epsilon = x/l$,

hence extension, $x = \epsilon l = (0.000\ 166)(2.0)$

$= 0.000\ 332\ \text{m}$

i.e. **extension of rod is 0.332 mm·**

Problem 13.   A bar of thickness 15 mm and having a rectangular cross-section carries a load of 120 kN. Determine the minimum width of the bar to limit the maximum stress to 200 MPa. The bar, which is 1.0 m long, extends by 2.5 mm when carrying a load of 120 kN. Determine the modulus of elasticity of the material of the bar.

Force, F = 120 kN = 120 000 N. Cross-sectional area $A = (15x)10^{-6}\ \text{m}^2$, where x is the width of the rectangular bar in millimetres. Stress $\sigma = F/A$, from which

$A = \dfrac{F}{A} = \dfrac{120\ 000\ \text{N}}{200 \times 10^6\ \text{Pa}} = 6 \times 10^{-4}\ \text{m}^2$

$= 6 \times 10^2\ \text{mm} = 600\ \text{mm}^2$

Hence $600 = 15x$, from which, width of bar $x = 600/15 = \mathbf{40\ mm}$. Extension of bar = 2.5 mm = 0.0025 m. Strain $\epsilon = x/l = (0.0025/1.0) = 0.0025$.

Modulus of elasticity $E = \dfrac{\text{stress}}{\text{strain}}$

$= \dfrac{200 \times 10^6}{0.0025} = 80 \times 10^9 = \mathbf{80\ GPa}$

Problem 14.   An aluminium rod has a length of 200 mm and a diameter of 10 mm. When subjected to a compressive force the length of the rod is 199.6 mm. Determine (a) the stress in the rod when loaded, and (b) the magnitude of the force. Take the modulus of elasticity for aluminium as 70 GPa.

(a)   Original length of rod, l = 200 mm; final length of rod = 199.6 mm, hence contraction, x = 0.4 mm. Thus

strain $\epsilon = \dfrac{x}{l} = \dfrac{0.4}{200} = 0.002$

Modulus of elasticity, $E = (\text{stress}\ \sigma)/(\text{strain}\ \epsilon)$, hence

stress $\sigma = E\epsilon = 70 \times 10^9 \times 0.002$

$= 140 \times 10^6\ \text{Pa} = \mathbf{140\ MPa}$

(b)   Since stress $\sigma = (\text{force}\ F)/(\text{area}\ A)$, then force $F = \sigma A$.

Cross-sectional area, $A = (\pi d^2/4) = (\pi(0.010)^2/4) = 7.854 \times 10^{-5}\ \text{m}^2$. Hence

compressive force $F = \sigma A$

$= 140 \times 10^6 \times 7.854 \times 10^{-5}$

$= \mathbf{11.0\ kN}$

Problem 15.   A brass tube has an internal diameter of 120 mm and an outside diameter of 150 mm and is used to support a load of 5 kN. The tube is 500 mm long before the load is applied. Determine by how much the tube contracts when loaded, taking the modulus of elasticity for brass as 90 GPa.

Force in tube, F = 5 kN = 5000 N.
Cross-sectional area of tube, A is given by

$\dfrac{\pi}{4}(D^2 - d^2) = \dfrac{\pi}{4}(0.150^2 - 0.120^2)$

$= 0.006\ 362\ \text{m}^2$

Stress in tube, $\sigma = (F/A)$
$= (5000\ \text{N})/(0.006\ 362\ \text{m}^2) = 0.7859 \times 10^6\ \text{Pa}.$
Since the modulus of elasticity,
$E = (\text{stress}\ \sigma)/(\text{strain}\ \epsilon)$, then strain $\epsilon = \sigma/E$, i.e.

$$\frac{0.7859 \times 10^6 \text{ Pa}}{90 \times 10^9 \text{ Pa}} = 8.732 \times 10^{-6}$$

Strain ϵ = (contraction x)/(original length l), thus

$$\text{contraction } x = \epsilon l$$

$$= 8.732 \times 10^{-6} \times 0.500$$

$$= 4.37 \times 10^{-6} \text{ m}$$

Thus, when loaded, the tube contracts by 4.37 μm.

Problem 16. In an experiment to determine the modulus of elasticity of a sample of mild steel, a wire is loaded and the corresponding extension noted. The results of the experiment are as shown.

Load (N)	0	40	110	160
Extension (mm)	0	1.2	3.3	4.8
Load (N)	200	250	290	340
Extension (mm)	6.0	7.5	10.0	16.2

Draw the load/extension graph. The mean diameter of the wire is 1.3 mm and its length is 8.0 m. Determine the modulus of elasticity E of the sample, and the stress at the elastic limit.

A graph of load/extension is shown in Fig. 17.8

$$E = \frac{\sigma}{\epsilon} = \frac{F/A}{x/l} = \left(\frac{F}{x}\right)\left(\frac{l}{A}\right)$$

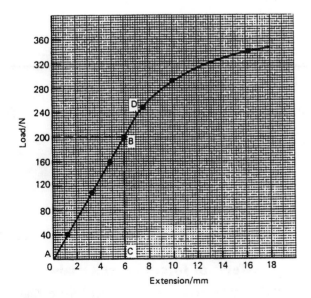

Figure 17.8

(F/x) is the gradient of the straight line part of the load/extension graph.

$$\text{Gradient } \frac{F}{x} = \frac{BC}{AC} = \frac{200 \text{ N}}{6 \times 10^{-3} \text{ m}}$$

$$= 33.33 \times 10^3 \text{ N/m}$$

Modulus of elasticity = (gradient of graph) (l/A). Length of specimen, l = 8.0 m.

$$\text{Cross-sectional area } A = \frac{\pi d^2}{4}$$

$$= \frac{\pi (0.0013)^2}{4} = 1.327 \times 10^{-6}$$

Hence modulus of elasticity =

$$(33.33 \times 10^3) \left(\frac{8.0}{1.327 \times 10^{-6}}\right) = 201 \text{ GPa}$$

The elastic limit is at point D in Fig. 17.8 where the graph no longer follows a straight line. This point corresponds to a load of 250 N as shown.

$$\text{Stress at elastic limit} = \frac{\text{force}}{\text{area}}$$

$$= \frac{250}{1.327 \times 10^{-6}} = 188.4 \times 10^6 \text{ Pa}$$

$$= \textbf{188.4 MPa}$$

17.9   Ductility, brittleness and malleability

Ductility is the ability of a material to be plastically deformed by elongation, without fracture. This is a property which enables a material to be drawn out into wires. For ductile materials such as mild steel, copper and gold, large extensions can result before fracture occurs with increasing tensile force. Ductile materials usually have a percentage elongation value of about 15% or more (see Chapter 18).

Brittleness is the property of a material manifested by fracture without appreciable prior plastic deformation. Brittleness is a lack of ductility, and brittle materials such as cast iron, glass, concrete, brick and ceramics, have virtually no plastic stage, the elastic stage being followed by immediate fracture. Little or no 'waist' occurs before fracture in a brittle material undergoing a tensile test.

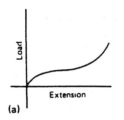
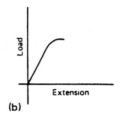
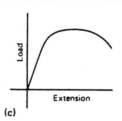

(a)   (b)   (c)

Figure 17.9

Malleability is the property of a material whereby it can be shaped when cold by hammering or rolling. A malleable material is capable of undergoing plastic deformation without fracture. Examples of malleable materials include lead, gold, putty and mild steel.

Problem 17.   Sketch typical load/extension curves for (a) an elastic non-metallic material, (b) a brittle material and (c) a ductile material. Give a typical example of each type of material.

(a) A typical load/extension curve for an elastic non-metallic material is shown in Fig. 17.9(a), and an example of such a material is **polythene**.
(b) A typical load/extension curve for a brittle material is shown in Fig. 17.9(b), and an example of such a material is **cast iron**.
(c) A typical load/extension curve for a ductile material is shown in Fig. 17.9(c), and an example of such a material is **mild steel**.

17.10 Multi-choice questions on the effects of forces on materials

(Answers on page 355.)

1. A wire is stretched 3 mm by a force of 150 N. Assuming the elastic limit is not exceeded, the force that will stretch the wire 5 mm is:
   (a) 150 N   (b) 250 N   (c) 90 N
2. For the wire in question 1, the extension when the applied force is 450 N is:
   (a) 1 mm   (b) 3 mm   (c) 9 mm
3. Due to the forces acting, a horizontal beam is in:
   (a) tension   (b) compression   (c) shear

4. Due to forces acting, a pillar supporting a bridge is in:
   (a) tension   (b) compression   (c) shear
5. Which of the following statements is false?
   (a) Elasticity is the ability of a material to return to its original dimensions after deformation by a load.
   (b) Plasticity is the ability of a material to retain any deformation produced in it by a load.
   (c) Ductility is the ability to be permanently stretched without fracturing.
   (d) Brittleness is the lack of ductility and a brittle material has a long plastic stage.
6. A circular rod of cross-sectional area 100 mm^2 has a tensile force of 100 kN applied to it. The stress in the rod is:
   (a) 1 MPa   (b) 1 GPa   (c) 1 kPa
   (d) 100 MPa
7. A metal bar 5.0 m long extends by 0.05 mm when a tensile load is applied to it. The percentage strain is:
   (a) 0.1   (b) 0.01   (c) 0.001   (d) 0.0001

An aluminium rod of length 1.0 m and cross-sectional area 500 mm^2 is used to support a load of 5 kN which causes the rod to contract by 100 μm. For questions 8 to 10, select the correct answer from the following list:
(a) 100 MPa   (b) 0.001   (c) 10 kPa
(d) 100 GPa   (e) 0.01   (f) 10 MPa
(g) 10 GPa   (h) 0.0001   (i) 10 Pa

8. The stress in the rod
9. The strain in the rod
10. Young's modulus of elasticity

17.12 Short answer questions on the effects of forces on materials

1. Name three types of mechanical force that can act on a body.

2. What is a tensile force? Name two practical examples of such a force.
3. What is a compressive force? Name two practical examples of such a force.
4. Define a shear force and name two practical examples of such a force.
5. Define elasticity and state two examples of elastic materials.
6. Define plasticity and state two examples of plastic materials.
7. State Hooke's law.
8. What is the difference between a ductile and a brittle material?
9. Define stress. What is the symbol used for (a) a tensile stress (b) a shear stress?
10. Strain is the ratio ——————
11. The ratio $\dfrac{\text{stress}}{\text{strain}}$ is called ....
12. State the units of (a) stress (b) strain (c) Young's modulus of elasticity.
13. Stiffness is the ratio ——————
14. Sketch on the same axes a typical load/extension graph for a ductile and a brittle material.
15. Define (a) ductility (b) brittleness (c) malleability.

17.12 Further questions on the effects of forces on materials

1. A rectangular bar having a cross-sectional area of 80 mm² has a tensile force of 20 kN applied to it. Determine the stress in the bar.
   [250 MPa]
2. A circular cable has a tensile force of 1 kN applied to it and the force produces a stress of 7.8 MPa in the cable. Calculate the diameter of the cable. [12.78 mm]
3. A square-sectioned support of side 12 mm is loaded with a compressive force of 10 kN. Determine the compressive stress in the support. [69.44 MPa]
4. A bolt having a diameter of 5 mm is loaded so that the shear stress in it is 120 MPa. Determine the value of the shear force on the bolt. [2.356 kN]
5. A split pin requires a force of 400 N to shear it. The maximum shear stress before shear occurs is 120 MPa. Determine the minimum diameter of the pin. [2.06 mm]
6. A wire of length 4.5 m has a percentage strain of 0.050% when loaded with a tensile force. Determine the extension in the wire.
   [2.25 mm]

7. A tube of outside diameter 60 mm and inside diameter 40 mm is subjected to a load of 60 kN. Determine the stress in the tube.
   [38.2 MPa]
8. A metal bar 2.5 m long extends by 0.05 mm when a tensile load is applied to it. Determine (a) the strain, (b) the percentage strain. [(a) 0.000 02 (b) 0.002%]
9. Explain, using appropriate practical examples, the difference between tensile, compressive and shear forces.
10. (a) State Hooke's law.
    (b) A wire is stretched 1.5 mm by a force of 300 N. Determine the force that would stretch the wire 4 mm, assuming the elastic limit of the wire is not exceeded.
    [(b) 800 N]
11. A rubber band extends 50 mm when a force of 300 N is applied to it. Assuming the band is within the elastic limit, determine the extension produced by a force of 60 N.
    [10 mm]
12. A force of 25 kN applied to a piece of steel produces an extension of 2 mm. Assuming the elastic limit is not exceeded, determine (a) the force required to produce an extension of 3.5 mm, (b) the extension when the applied force is 15 kN.
    [(a) 43.75 N (b) 1.22 mm]
13. A coil spring 300 mm long when unloaded extends to a length of 500 mm when a load of 40 N is applied. Determine the length of the spring when a load of 15 kN is applied.
    [375 mm]
14. A test to determine the load/extension graph for a specimen of copper gave the following results:

Load (kN)	8.5	15.0	23.5	30.0
Extension (mm)	0.04	0.07	0.11	0.14

Plot the load/extension graph, and from the graph determine (a) the load at an extension of 0.09 mm, and (b) the extension corresponding to a load of 12.0 N.
[(a) 19.1 kN (b) 0.057 mm]
15. Sketch on the same axes typical load/extension graphs for (a) a strong, ductile material (b) a brittle material.
16. Define (a) tensile stress (b) shear stress (c) strain (d) shear strain (e) Young's modulus of elasticity.
17. A circular bar is 2.5 m long and has a diameter of 60 mm. When subjected to a compressive load of 30 kN it shortens by 0.20 mm.

Determine Young's modulus of elasticity for the material of the bar. [132.6 GPa]

18. A bar of thickness 20 mm and having a rectangular cross-section carries a load of 82.5 kN. Determine (a) the minimum width of the bar to limit the maximum stress to 150 MPa, (b) the modulus of elasticity of the material of the bar if the 150 mm long bar extends by 0.8 mm when carrying a load of 200 kN. [(a) 27.5 mm (b) 68.2 GPa]

19. A metal rod of cross-sectional area 100 mm^2 carries a maximum tensile load of 20 kN. The modulus of elasticity for the material of the rod is 200 GPa. Determine the percentage strain when the rod is carrying its maximum load. [0.1%]

20. A metal tube 1.75 m long carries a tensile load and the maximum stress in the tube must not exceed 50 MPa. Determine the extension of the tube when loaded if the modulus of elasticity for the material is 70 GPa. [1.25 mm]

21. A piece of aluminium wire is 5 m long and has a cross-sectional area of 100 mm^2. It is subjected to increasing loads, the extension being recorded for each load applied. The results are:

Load (kN)	0	1.12	2.94	4.76	7.00	9.10
Extension (mm)	0	0.8	2.1	3.4	5.0	6.5

Draw the load/extension graph and hence determine the modulus of elasticity for the material of the wire. [70 GPa]

22. In an experiment to determine the modulus of elasticity of a sample of copper, a wire is loaded and the corresponding extension noted. The results are:

Load (kN)	0	20	34	72	94	120
Extension (mm)	0	0.7	1.2	2.5	3.3	4.2

Draw the load/extension graph and determine the modulus of elasticity of the sample if the mean diameter of the wire is 1.151 mm and its length is 4.0 m. [110 GPa]

18

Tensile testing

At the end of this chapter you should be able to:

- describe a tensile test
- recognize from a tensile test the limit of proportionality, the elastic limit and the yield point
- plot a load/extension graph from given data
- calculate from a load/extension graph, the modulus of elasticity, the yield stress, the ultimate tensile strength, percentage elongation and the percentage reduction in area

18.1   The tensile test

A **tensile test** is one in which a force is applied to a specimen of a material in increments and the corresponding extension of the specimen noted. The process may be continued until the specimen breaks into two parts and this is called **testing to destruction**. The testing is usually carried out using a universal testing machine which can apply either tensile or compressive forces to a specimen in small, accurately measured steps. **British Standard 18** gives the standard procedure for such a test. Test specimens of a material are made to standard shapes and sizes and two typical test pieces are shown in Fig. 18.1. The results of a tensile test may be plotted on a load/extension graph and a typical graph for a mild steel specimen is shown in Fig. 18.2.

(i)   Between A and B is the region in which Hooke's law applies and stress is directly proportional to strain. The gradient of AB is used when determining Young's modulus of elasticity (see Chapter 17).

(ii) Point B is the **limit of proportionality** and is the point at which stress is no longer proportional to strain when a further load is applied.

(iii) Point C is the **elastic limit** and a specimen loaded to this point will effectively return to

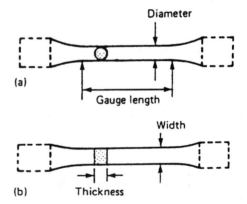

Figure 18.1

its original length when the load is removed, i.e. there is negligible permanent extension.

(iv) Point D is called the **yield point** and at this point there is a sudden extension with no increase in load. The yield stress of the material is given by:

Yield stress =

$$\frac{\textbf{load where yield begins to take place}}{\textbf{original cross-sectional area}}$$

The yield stress gives an indication of the ductility of the material (see Chapter 17).

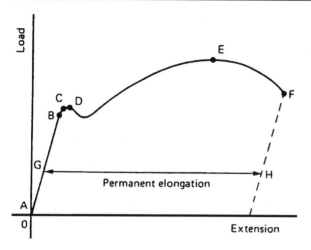

Figure 18.2

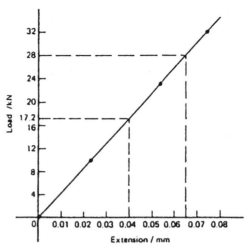

Figure 18.3

(v) Between points D and E extension takes place over the whole gauge length of the specimen.

(vi) Point E gives the maximum load which can be applied to the specimen and is used to determine the ultimate tensile strength (UTS) of the specimen (often just called the tensile strength)

$$\text{UTS} = \frac{\textbf{maximum load}}{\textbf{original cross-sectional area}}$$

(vii) Between points E and F the cross-sectional area of the specimen decreases, usually about half way between the ends, and a **waist** or **neck** is formed before fracture.

Percentage reduction in area =

$$\frac{\textbf{(original cross-sectional area)} - \textbf{(final cross-sectional area)}}{\textbf{original cross-sectional area}} \times \textbf{100\%}$$

The percentage reduction in area provides information about the malleability of the material (see Chapter 17). The value of stress at point F is greater than at point E since although the load on the specimen is decreasing as the extension increases, the cross-sectional area is also reducing.

(viii) At point F the specimen fractures.

(ix) Distance GH is called the **permanent elongation** and

Permanent elongation =

$$\frac{\textbf{increase in length during}}{\textbf{test to destruction}} \times \textbf{100\%}$$

Problem 1. A tensile test is carried out on a mild steel specimen. The results are shown in the following table of values.

Load (kN)	0	10	23	32
Extension (mm)	0	0.023	0.053	0.074

Plot a graph of load against extension, and from the graph determine (a) the load at an extension of 0.04 mm, and (b) the extension corresponding to a load of 28 kN.

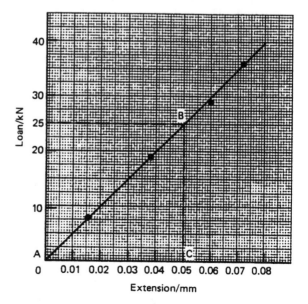

Figure 18.4

The load/extension graph is shown in Fig. 18.3. From the graph:

(a) when the extension is 0.04 mm, the load is **17.2 kN**;
(b) when the load is 28 kN, the extension is **0.065 mm**.

Problem 2. A tensile test is carried out on a mild steel specimen of gauge length 40 mm and cross-sectional area 100 mm². The results obtained for the specimen up to its yield point are given below.

Load (kN)	0	8	19	29	36
Extension (mm)	0	0.015	0.038	0.060	0.072

The maximum load carried by the specimen is 50 kN and its length after fracture is 52 mm. Determine (a) the modulus of elasticity, (b) the ultimate tensile strength, (c) the percentage elongation of the mild steel.

The load/extension graph is shown in Fig. 18.4.

(a) Gradient of straight line is given by

$$\frac{BC}{AB} = \frac{25\,000}{0.05 \times 10^{-3}} = 500 \times 10^{6} \text{ N/m}$$

Young's modulus of elasticity = (gradient of graph) (l/A), l = 40 mm (gauge length) = 0.040 m. Area, A = 100 mm² = 100×10^{-6} m².

Young's modulus of elasticity

$$= (500 \times 10^{6}) \left(\frac{0.040}{100 \times 10^{-6}} \right)$$

$$= 200 \times 10^{9} \text{ Pa}$$

$$= \textbf{200 GPa}$$

(b) Ultimate tensile strength =

$$\frac{\text{maximum load}}{\text{original cross-sectional area}}$$

$$= \frac{50\,000 \text{ N}}{100 \times 10^{-6} \text{ m}^2} = 500 \times 10^{6} \text{ Pa} = \textbf{500 MPa}$$

(c) Percentage elongation

$$= \frac{\text{increase in length}}{\text{original length}} \times 100$$

$$= \frac{52 - 40}{40} \times 100 = \frac{12}{40} \times 100$$

$$= 30\%$$

Problem 3. The results of a tensile test are: Diameter of specimen 15 mm; gauge length 40 mm; load at limit of proportionality 85 kN; extension at limit of proportionality 0.075 mm; maximum load 120 kN; final length at point of fracture 55 mm.

Determine (a) Young's modulus of elasticity, (b) the ultimate tensile strength, (c) the stress at the limit of proportionality, (d) the percentage elongation.

(a) Young's modulus of elasticity is given by

$$E = \frac{\text{stress}}{\text{strain}} = \frac{F/A}{x/l} = \frac{Fl}{Ax}$$

where the load at the limit of proportionality, F = 85 kN = 85 000 N.
l = gauge length = 40 mm = 0.040 m. A = cross-sectional area = $(\pi d^2/4)$ = $(\pi(0.015)^2/(4))$ = 0.000 176 7 m². x = extension = 0.075 mm = 0.000 075 m. Hence

Young's modulus of elasticity E

$$= \frac{Fl}{Ax}$$

$$= \frac{(85\,000)(0.040)}{(0.000\,176\,7)(0.000\,075)}$$

$$= 256.6 \times 10^{9} \text{ Pa}$$

$$= \textbf{256.6 GPa}$$

(b) Ultimate tensile strength

$$= \frac{\text{maximum load}}{\text{original cross-sectional area}}$$

$$= \frac{120\,000}{0.000\,176\,7} = 679 \times 10^{6} \text{ Pa}$$

$$= \textbf{679 MPa}$$

(c) Stress at limit of proportionality

$$= \frac{\text{load at limit of proportionality}}{\text{cross-sectional area}}$$

$$= \frac{85\,000}{0.000\,176\,7} = 481.0 \times 10^{6} \text{ Pa} = \textbf{481.0 MPa}$$

(d) Percentage elongation

$$= \frac{\text{increase in length}}{\text{original length}} \times 100$$

$$= \frac{(55 - 40) \text{ mm}}{40 \text{ mm}} \times 100 = \textbf{37.5\%}$$

Problem 4.   A rectangular zinc specimen is subjected to a tensile test and the data from the test is shown below. Width of specimen 40 mm; breadth of specimen 2.5 mm; gauge length 120 mm.

Load (kN)	10	17	25	30	35
Extension (mm)	0.15	0.25	0.35	0.55	1.00

Load (kN)	37.5	38.5	37	34	32
Extension (mm)	1.50	2.50	3.50	4.50	5.00

Fracture occurs when the extension is 5.0 mm and the maximum load recorded is 38.5 kN.

Plot the load/extension graph and hence determine (a) the stress at the limit of proportionality, (b) Young's modulus of elasticity, (c) the ultimate tensile strength, (d) the percentage elongation, (e) the stress at a strain of 0.01, (f) the extension at a stress of 200 MPa.

A load/extension graph is shown in Fig. 18.5.

(a) The limit of proportionality occurs at point P on the graph, where the initial gradient of the graph starts to change. This point has a load value of 26.5 kN.
Cross-sectional area of specimen = 40 mm × 2.5 mm = 100 mm^2 = 100 × 10^{-6} m^2.
Stress at the limit of proportionality is given by

$$\sigma = \frac{\text{force}}{\text{area}}$$

$$= \frac{26.5 \times 10^3 \text{ N}}{100 \times 10^{-6} \text{ m}^2}$$

$$= 265 \times 10^6 \text{ Pa} = \textbf{265 MPa}$$

(b) Gradient of straight line portion of graph is given by

$$\frac{BC}{AC} = \frac{25\,000 \text{ N}}{0.35 \times 10^{-3} \text{ m}}$$

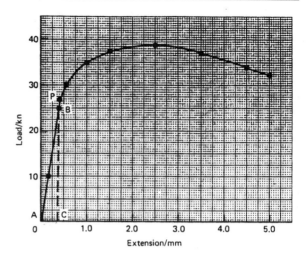

Figure 18.5

$$= 71.43 \times 10^6 \text{ N/m}$$

Young's modulus of elasticity

$$= (\text{gradient of graph}) \left(\frac{l}{A} \right)$$

$$= (71.43 \times 10^6) \left(\frac{120 \times 10^{-3}}{100 \times 10^{-6}} \right)$$

$$= 85.72 \times 10^9 \text{ Pa} = \textbf{85.72 GPa}$$

(c) Ultimate tensile strength

$$= \frac{\text{maximum load}}{\text{original cross-sectional area}}$$

$$= \frac{38.5 \times 10^3 \text{ N}}{100 \times 10^{-6} \text{ m}^2}$$

$$= 385 \times 10^6 \text{ Pa} = \textbf{385 MPa}$$

(d) Percentage elongation

$$= \frac{\text{extension at fracture point}}{\text{original length}} \times 100$$

$$= \frac{5.0 \text{ mm}}{120 \text{ mm}} \times 100 = \textbf{4.17\%}$$

(e) Strain ϵ = (extension x)/(original length l), from which, extension $x = \epsilon l = 0.01 \times 120 = 1.20$ mm.
From the graph, the load corresponding to an extension of 1.20 mm is 36 kN.
Stress at a strain of 0.01 is given by

$$\sigma = \frac{\text{force}}{\text{area}} = \frac{36\,000\,\text{N}}{100 \times 10^{-6}\,\text{m}^2}$$

$$= 360 \times 10^6\,\text{Pa} = \textbf{360 MPa}$$

(f) When the stress is 200 MPa, then force = area × stress, i.e.

$$(100 \times 10^{-6})(200 \times 10^6) = 20\,\text{kN}$$

From the graph, the corresponding extension is **0.30 mm**.

Problem 5. A mild steel specimen of cross-sectional area 250 mm² and gauge length 100 mm is subjected to a tensile test and the following data is obtained: within the limit of proportionality, a load of 75 kN produced an extension of 0.143 mm; load at yield point = 80 kN; maximum load on specimen = 120 kN; final cross-sectional area of waist at fracture = 90 mm²; the gauge length had increased to 135 mm at fracture.

Determine for the specimen: (a) Young's modulus of elasticity, (b) the yield stress, (c) the tensile strength, (d) the percentage elongation, (e) the percentage reduction in area.

(a) Force F = 75 kN = 75 000 N; gauge length l = 100 mm = 0.1 m; cross-sectional area A = 250 mm² = 250 × 10⁻⁶ m²; extension x = 0.143 mm = 0.143 × 10⁻³ m.

Young's modulus of elasticity, E

$$= \frac{\text{stress}}{\text{strain}} = \frac{F/A}{x/l} = \frac{Fl}{Ax}$$

$$= \frac{(75\,000)(0.1)}{(250 \times 10^{-6})(0.143 \times 10^{-3})}$$

$$= 210 \times 10^9\,\text{Pa} = \textbf{210 GPa}$$

(b) Yield stress =

$$\frac{\text{load when yield begins to take place}}{\text{original cross-sectional area}}$$

$$= \frac{80\,000\,\text{N}}{250 \times 10^{-6}\,\text{m}^2}$$

$$= 320 \times 10^6\,\text{Pa} = \textbf{320 MPa}$$

(c) Tensile strength

$$= \frac{\text{maximum load}}{\text{original cross-sectional area}}$$

$$= \frac{120\,000\,\text{N}}{250 \times 10^{-6}\,\text{m}^2} = 480 \times 10^6\,\text{Pa} = \textbf{480 MPa}$$

(d) Percentage elongation =

$$\left(\frac{\text{increase in length during test to destruction}}{\text{original length}} \right) \times 100$$

$$= \left(\frac{135 - 100}{100} \right) \times 100 = \textbf{35\%}$$

(e) Percentage reduction in area

$$= \frac{(\text{original cross-sectional area}) - (\text{final cross-sectional area})}{\text{original cross-sectional area}} \times 100$$

$$= \left(\frac{250 - 90}{250} \right) \times 100 = \left(\frac{160}{250} \right) 100 = \textbf{64\%}$$

18.2 Multi-choice questions on tensile testing

(Answers on page 355.)

A brass specimen having a cross-sectional area of 100 mm² and gauge length 100 mm is subjected to a tensile test from which the following information is obtained:

Load at yield point = 45 kN; Maximum load = 52.5 kN; final cross-sectional area of waist at fracture = 75 mm²; gauge length at fracture = 110 mm.

For questions 1 to 4, select the correct answer from the following list:

(a) 600 MPa   (b) 525 MPa   (c) 33⅓%   (d) 10%
(e) 9.09%   (f) 450 MPa   (g) 25%   (h) 700 MPa

1. The yield stress
2. The percentage elongation
3. The percentage reduction in area
4. The ultimate tensile strength

18.3   Short answer questions on tensile testing

1. What is a tensile test?
2. Which British Standard gives the standard procedure for a tensile test?
3. With reference to a load/extension graph for mild steel state the meaning of (a) the limit of proportionality (b) the elastic limit (c) the yield point (d) the percentage elongation.
4. Ultimate tensile strength is the ratio ――――
5. Yield stress is the ratio $\frac{\cdots\cdots\cdots}{\cdots\cdots\cdots}$
6. Define 'percentage reduction in area'.

18.4   Further questions on tensile testing

1. A tensile test is carried out on a specimen of mild steel of gauge length 40 mm and diameter 7.35 mm. The results are:

Load (kN)	0	10	17	25	30
Extension (mm)	0	0.05	0.08	0.11	0.14

Load (kN)	34	37.5	38.5	36
Extension (mm)	0.20	0.40	0.60	0.90

At fracture the final length of the specimen is 40.90 mm. Plot the load/extension graph and determine (a) the modulus of elasticity for mild steel, (b) the stress at the limit of proportionality, (c) the ultimate tensile strength, (d) the percentage elongation.
[(a) 202 GPa (b) 707 MPa (c) 907 MPa (d) 2.25%]

2. In a tensile test on a zinc specimen of gauge length 100 mm and diameter 15 mm a load of 100 kN produced an extension of 0.666 mm. Determine (a) the stress induced, (b) the strain, (c) Young's modulus of elasticity.
[(a) 566 MPa (b) 0.00666 (c) 85 GPa]

3. The results of a tensile test are:
Diameter of specimen 20 mm; gauge length 50 mm; load at limit of proportionality 80 kN; Extension at limit of proportionality 0.075 mm; maximum load 100 kN; final length at point of fracture 60 mm.
Determine (a) Young's modulus of elasticity, (b) the ultimate tensile strength, (c) the stress at the limit of proportionality, (d) the percentage elongation.
[(a) 169.8 GPa (b) 318.3 MPa (c) 254.6 MPa (d) 20%]

4. What is a tensile test? Make a sketch of a typical load/extension graph for a mild steel specimen to the point of fracture and mark on the sketch the following:
(a) the limit of proportionality, (b) the elastic limit, (c) the yield point.

5. An aluminium alloy specimen of gauge length 75 mm and of diameter 11.28 mm was subjected to a tensile test, with these results:

Load (kN)	0	2.0	6.5	11.5	13.6	16.0
Extension (mm)	0	0.012	0.039	0.069	0.080	0.107

Load (kN)	18.0	19.0	20.5	19.0
Extension (mm)	0.133	0.158	0.225	0.310

The specimen fractured at a load of 19.0 kN. Determine (a) the modulus of elasticity of the alloy, (b) the percentage elongation.
[(a) 125 GPa (b) 0.413%]

6. An aluminium test piece 10 mm in diameter and gauge length 50 mm gave the following results when tested to destruction:
Load at yield point 4.0 kN; maximum load 6.3 kN; extension at yield point 0.036 mm; diameter at fracture 7.7 mm.
Determine (a) the yield stress, (b) Young's modulus of elasticity, (c) the ultimate tensile strength, (d) the percentage reduction in area.
[(a) 50.93 MPa (b) 70.7 GPa (c) 80.2 MPa (d) 40.7%]

19

Linear momentum and impulse

At the end of this chapter you should be able to:

- define momentum and state its unit
- state Newton's first law of motion
- state the principle of conservation of momentum
- calculate momentum given mass and velocity
- state Newton's second law of motion
- define impulse
- appreciate when impulsive forces occur
- state Newton's third law of motion
- calculate impulse and impulsive force
- use the equation of motion $v^2 = u^2 + 2as$ in calculations

19.1 Linear momentum

The **momentum** of a body is defined as the product of its mass and its velocity, i.e. **momentum = mu**, where m = mass (in kg) and u = velocity (in m/s). The unit of momentum is kg m/s.

Since velocity is a vector quantity, **momentum is a vector quantity**, i.e. it has both magnitude and direction.

Newton's first law of motion states:

a body continues in a state of rest or in a state of uniform motion in a straight line unless acted on by some external force.

Hence the momentum of a body remains the same provided no external forces act on it.

The principle of conservation of momentum for a closed system (i.e. one on which no external forces act) may be stated as:

the total linear momentum of a system is a constant.

The total momentum of a system before collision in a given direction is equal to the total momen-

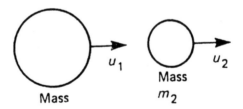

Figure 19.1

tum of the system after collision in the same direction. In Fig. 19.1, masses m_1 and m_2 are travelling in the same direction with velocity $u_1 > u_2$. A collision will occur, and applying the principle of conservation of momentum:

total momentum before impact
= total momentum after impact

i.e.

$$m_1u_1 + m_2u_2 = m_1v_1 + m_2v_2$$

where v_1 and v_2 are the velocities of m_1 and m_2 after impact.

Problem 1 . Determine the momentum of a pile driver of mass 400 kg when it is moving downwards with a speed of 12 m/s.

Momentum = mass × velocity

$$= 400 \text{ kg} \times 12 \text{ m/s}$$

$$= \textbf{4800 kg m/s downwards}$$

Problem 2.   A cricket ball of mass 150 g has a momentum of 4.5 kg m/s. Determine the velocity of the ball in km/h.

Momentum = mass × velocity. Hence

$$\text{velocity} = \frac{\text{momentum}}{\text{mass}} = \frac{4.5 \text{ kg m/s}}{150 \times 10^{-3} \text{ kg}}$$

$$= 30 \text{ m/s}$$

30 m/s = 30 × 3.6 km/h = **108 km/h = velocity of cricket ball**.

Problem 3.   Determine the momentum of a railway wagon of mass 50 tonnes moving at a velocity of 72 km/h.

Momentum = mass × velocity. Mass = 50 t = 50 000 kg (since 1 t = 1000 kg); velocity = 72 km/h = (72/3.6) m/s = 20 m/s. Hence

momentum = 50 000 kg × 20 m/s

$$= 1\,000\,000 \text{ kg m/s}$$

$$= \textbf{10}^{\textbf{6}} \textbf{ kg m/s}$$

Problem 4.   A wagon of mass 10 t is moving at a speed of 6 m/s and collides with another wagon of mass 15 t, which is stationary. After impact, the wagons are coupled together. Determine the common velocity of the wagons after impact.

Mass $m_1 = 10 \text{ t} = 10\,000$ kg, $m_2 = 15\,000$ kg. Velocity $u_1 = 6$ m/s, $u_2 = 0$.

Total momentum before impact

$$= m_1 u_1 + m_2 u_2$$

$$= (10\,000 \times 6) + (15\,000 \times 0)$$

$$= 60\,000 \text{ kg m/s}$$

Let the common velocity of the wagons after impact be v m/s. Since total momentum before impact = total momentum after impact:

$$60\,000 = m_1 v + m_2 v$$

$$= v(m_1 + m_2) = v(25\,000)$$

Hence

$$v = \frac{60\,000}{25\,000} = 2.4 \text{ m/s}$$

i.e. **the common velocity after impact is 2.4 m/s in the direction in which the 10 t wagon is initially travelling**.

Problem 5.   A body has a mass of 30 g and is moving with a velocity of 20 m/s. It collides with a second body which has a mass of 20 g and which is moving with a velocity of 15 m/s. Assuming that the bodies both have the same velocity after impact, determine this common velocity, (a) when the initial velocities have the same line of action and the same sense, and (b) when the initial velocities have the same line of action but are opposite in sense.

Mass $m_1 = 30 \text{ g} = 0.030$ kg, $m_2 = 20 \text{ g} = 0.020$ kg, velocity $u_1 = 20$ m/s, $u_2 = 15$ m/s.

(a)   When the velocities have the same line of action and the same sense, both u_1 and u_2 are considered as positive values.

Total momentum before impact

$$= m_1 u_1 + m_2 u_2$$

$$= (0.030 \times 20) + (0.020 \times 15)$$

$$= 0.60 + 0.30$$

$$= 0.90 \text{ kg m/s}$$

Let the common velocity after impact be v m/s. Total momentum before impact = total momentum after impact, i.e.

$$0.90 = m_1 v + m_2 v = v(m_1 + m_2)$$

$$0.90 = v(0.030 + 0.020)$$

from which, **common velocity**, v = (0.90/0.050) = **18 m/s in the direction in which the bodies are initially travelling**.

(b)   When the velocities have the same line of action but are opposite in sense, one is considered as positive and the other negative. Taking the direction of mass m_1 as positive gives: velocity $u_1 = +20$ m/s and $u_2 = -15$ m/s.

Total momentum before impact

$$= m_1 u_1 + m_2 u_2$$

$$= (0.030 \times 20) + (0.020 \times -15)$$

$$= 0.60 - 0.30$$

$$= +0.30 \text{ kg m/s}$$

and since it is positive this indicates a momentum in the same direction as that of mass m_1. If the common velocity after impact is v m/s then

$$0.30 = v(m_1 + m_2) = v(0.050)$$

from which, **common velocity**, $v = (0.30/0.050) = $ **6 m/s in the direction that the 30 g mass is initially travelling**.

Problem 6.   A ball of mass 50 g is moving with a velocity of 4 m/s when it strikes a stationary ball of mass 25 g. The velocity of the 50 g ball after impact is 2.5 m/s in the same direction as before impact. Determine the velocity of the 25 g ball after impact.

Mass $m_1 = 50$ g $= 0.050$ kg, $m_2 = 25$ g $= 0.025$ kg. Initial velocity $u_1 = 4$ m/s, $u_2 = 0$; final velocity $v_1 = 2.5$ m/s, v_2 is unknown.

Total momentum before impact

$$= m_1 u_1 + m_2 u_2$$

$$= (0.050 \times 4) + (0.025 \times 0)$$

$$= 0.20 \text{ kg m/s}$$

Total momentum after impact

$$= m_1 v_1 + m_2 v_2$$

$$= (0.050 \times 2.5) + (0.025 v_2)$$

$$= 0.125 + 0.025 v_2$$

Total momentum before impact = total momentum after impact, hence $0.20 = 0.125 + 0.025 v_2$, from which,

velocity of 25 g ball after impact, v_2

$$= \frac{0.20 - 0.125}{0.025}$$

$$= \textbf{3 m/s}$$

Problem 7.   Three masses, P, Q and R lie in a straight line. P has a mass of 5 kg and is moving towards Q at 8 m/s. Q has a mass of 7 kg and a velocity of 4 m/s, and is moving towards R. Mass R is stationary. P collides with Q, and P and Q then collide with R. Determine the mass of R assuming all three masses have a common velocity of 2 m/s after the collision of P and Q with R.

Mass $m_P = 5$ kg, $m_Q = 7$ kg; velocity $u_P = 8$ m/s, $u_Q = 4$ m/s.

Total momentum before P collides with Q

$$= m_P u_P + m_Q u_Q$$

$$= (5 \times 8) + (7 \times 4)$$

$$= 68 \text{ kg m/s}$$

Let P and Q have a common velocity of v_1 m/s after impact.

Total momentum after P and Q collide

$$= m_P v_1 + m_Q v_1$$

$$= v_1(m_P + m_Q) = 12 v_1$$

Total momentum before impact = total momentum after impact, i.e. $68 = 12v_1$, from which, common velocity of P and Q, $v_1 = (68/12) = 5\frac{2}{3}$ m/s. Total momentum after P and Q collide with R $= (m_{P+Q} \times 2) + (m_R \times 2)$ (since the common velocity after impact $= 2$ m/s) $= (12 \times 2) + (12 m_R)$. Total momentum before P and Q collide with R = total momentum after P and Q collide with R, i.e.

$$\left(m_{P+Q} \times 5\frac{2}{3}\right) = (12 \times 2) + 2 m_R$$

i.e.

$$12 \times 5\frac{2}{3} = 24 + 2 m_R$$

$$68 - 24 = 2 m_R$$

from which, **mass of R**, $m_R = 44/2 = $ **22 kg**.

19.2   Impulse and impulsive forces

Newton's second law of motion states:

the rate of change of momentum is directly proportional to the applied force producing the change, and takes place in the direction of this force.

In the SI system, the units are such that:

the applied force

= rate of change of momentum

$$= \frac{\text{change of momentum}}{\text{time taken}} \qquad (19.1)$$

When a force is suddenly applied to a body due to either a collision with another body or being hit by an object such as a hammer, the time taken in equation (19.1) is very small and difficult to measure. In such cases, the total effect of the force is measured by the change of momentum it produces.

Forces which act for very short periods of time are called **impulsive forces**. The product of the impulsive force and the time during which it acts is called the **impulse** of the force and is equal to the change of momentum produced by the impulsive force, i.e.

> **impulse = applied force × time**
> **= change in linear momentum**

Examples where impulsive forces occur include when a gun recoils and when a free-falling mass hits the ground. Solving problems associated with such occurrences often requires the use of the equation of motion: $v^2 = u^2 + 2as$, from Chapter 13.

When a pile is being hammered into the ground, the ground resists the movement of the pile and this resistance is called a **resistive force**. **Newton's third law of motion may be stated as:**

for every force there is an equal and opposite force.

The force applied to the pile is the resistive force. The pile exerts an equal and opposite force on the ground.

In practice, when impulsive forces occur, energy is not entirely conserved and some energy is changed into heat, noise, and so on.

Problem 8. The average force exerted on the workpiece of a press-tool operation is 150 kN, and the tool is in contact with the workpiece for 50 ms. Determine the change in momentum.

From above, change of linear momentum = applied force × time (= impulse). Hence

change in momentum of workpiece

$$= 150 \times 10^3 \text{ N} \times 50 \times 10^{-3} \text{ s}$$

$$= \textbf{7500 kg m/s} \text{ (since } 1 \text{ N} = 1 \text{ kg m/s}^2)$$

Problem 9. A force of 15 N acts on a body of mass 4 kg for 0.2 s. Determine the change in velocity.

$$\text{Impulse} = \text{applied force} \times \text{time}$$

$$= \text{change in linear momentum}$$

i.e. 15 N × 0.2 s = mass × change in velocity

$$= 4 \text{ kg} \times \text{change in velocity}$$

from which,

$$\textbf{change in velocity} = \frac{15 \text{ N} \times 0.2 \text{ s}}{4 \text{ kg}}$$

$$= \textbf{0.75 m/s} \quad (\text{since } 1 \text{ N} = 1 \text{ kg m/s}^2)$$

Problem 10. A mass of 8 kg is dropped vertically on to a fixed horizontal plane and has an impact velocity of 10 m/s. The mass rebounds with a velocity of 6 m/s. If the mass–plane contact time is 40 ms, calculate (a) the impulse, and (b) the average value of the impulsive force on the plane.

(a) Impulse = change in momentum

$$= m(u_1 - v_1)$$

where u_1 = impact velocity = 10 m/s and v_1 = rebound velocity = –6 m/s (v_1 is negative since it acts in the opposite direction to u_1). Thus

$$m(u_1 - v_1) = 8 \text{ kg } (10 - -6) \text{ m/s}$$

$$= 8 \times 16 = \textbf{128 kg m/s}$$

(b) Impulsive force $= \dfrac{\text{impulse}}{\text{time}}$

$$= \frac{128 \text{ kg m/s}}{40 \times 10^{-3} \text{ s}} = \textbf{3200 N} \text{ or } \textbf{3.2 kN}$$

Problem 11. The hammer of a pile-driver of mass 1 t falls a distance of 1.5 m on to a pile. The blow takes place in 25 ms and the hammer does not rebound. Determine the average applied force exerted on the pile by the hammer.

Initial velocity $u = 0$; acceleration due to gravity, $g = 9.81$ m/s^2, and distance, $s = 1.5$ m. Using the equation of motion $v^2 = u^2 + 2gs$ then $v^2 = 0^2 + 2(9.81)(1.5)$ from which impact velocity, $v = \sqrt{[2(9.81)(1.5)]} = 5.425$ m/s. Neglecting the small distance moved by the pile and hammer after impact,

momentum lost by hammer

= the change of momentum

= mv = 1000 kg × 5.425 m/s

Rate of change of momentum

$$= \frac{\text{change of momentum}}{\text{change of time}} = \frac{1000 \times 5.425}{25 \times 10^{-3}}$$

= 217 000 N

Since the impulsive force is the rate of change of momentum, the **average force exerted on the pile is 217 kN**.

Problem 12.   A mass of 40 g having a velocity of 15 m/s collides with a rigid surface and rebounds with a velocity of 5 m/s. The duration of the impact is 0.3 ms. Determine (a) the impulse, and (b) the impulsive force at the surface.

Mass $m = 40$ g $= 0.040$ kg; initial velocity, $u = 15$ m/s; final velocity $v = -5$ m/s (negative since the rebound is in the opposite direction to velocity u).

(a)   Momentum before impact = mu = 0.040 × 15 = 0.6 kg m/s.
Momentum after impact = mv = 0.040 × –5 = –0.2 kg m/s.
Impulse = change of momentum = 0.6 – (–0.2) = 0.8 kg m/s.

(b)   Impulsive force = $\dfrac{\text{change of momentum}}{\text{change of time}}$

$$= \frac{0.8 \text{ kg m/s}}{0.20 \times 10^{-3} \text{ s}}$$

= **4000 N** or **4 kN**

Problem 13.   A gun of mass 1.5 t fires a shell of mass 15 kg horizontally with a velocity of 500 m/s. Determine (a) the initial velocity of recoil, and (b) the uniform force necessary to stop the recoil of the gun in 200 mm.

Mass of gun, $m_g = 1.5$ t $= 1500$ kg; mass of shell, $m_S = 15$ kg; initial velocity of shell, $u_S = 500$ m/s.

(a)   Momentum of shell = $m_S u_S$ = 15 × 500 = 7500 kg m/s.
Momentum of gun = $m_g v$ = 1500v, where v = initial velocity of recoil of the gun.
By the principle of conservation of momentum, initial momentum = final momentum, i.e. 0 = 7500 + 1500v, from which,

velocity $v = \dfrac{-7500}{1500} = -5$ m/s (the

negative sign indicating recoil velocity)

i.e. **the initial velocity of recoil = 5 m/s**.

(b)   The retardation of the recoil, a, may be determined using $v^2 = u^2 + 2as$, where v, the final velocity, is zero, u, the initial velocity, is 5 m/s and s, the distance, is 200 mm, i.e. 0.2 m.
Rearranging $v^2 = u^2 + 2as$ for a gives:

$$a = \frac{v^2 - u^2}{2s} = \frac{0^2 - 5^2}{2(0.2)} = \frac{-25}{0.4} = -62.5 \text{ m/s}^2$$

Force necessary to stop recoil in 200 mm

= mass × acceleration

= 1500 kg × 62.5 m/s^2

= **93 750 N** or **93.75 kN**

Problem 14.   A vertical pile of mass 100 kg is driven 200 mm into the ground by the blow of a 1 t hammer which falls through 750 mm. Determine (a) the velocity of the hammer just before impact, (b) the velocity immediately after impact (assuming the hammer does not bounce), and (c) the resistive force of the ground assuming it to be uniform.

(a)   For the hammer, $v^2 = u^2 + 2gs$, where v = final velocity, u = initial velocity = 0, g = 9.81 m/s^2 and s = distance = 750 mm = 0.75 m. Hence $v^2 = 0^2 + 2(9.81)(0.75)$, from which,

velocity of hammer, just before impact,

$v = \sqrt{[2(9.81)(0.75)]}$

= **3.84 m/s**

(b)   Momentum of hammer just before impact

= mass × velocity

= 1000 kg × 3.84 m/s

= 3840 kg m/s

Momentum of hammer and pile after impact = momentum of hammer before impact. Hence 3840 kg m/s = (mass of hammer and pile) × (velocity immediately after impact), i.e. 3840 = (1000 + 100)(v), from which,

velocity immediately after impact, v

$$= \frac{3840}{1100} = \textbf{3.49 m/s}$$

(c)   Resistive force of ground = mass × acceleration. The acceleration is determined using $v^2 = u^2 + 2as$ where v = final velocity = 0, u = initial velocity = 3.49 m/s and s = distance driven in ground = 200 mm = 0.2 m. Hence $0^2 = (3.49)^2 + 2(a)(0.2)$, from which,

$$\text{acceleration } a = \frac{-(3.49)^2}{2(0.2)} = -30.45 \text{ m/s}^2$$

(the minus sign indicates retardation)

Thus resistive force of ground

= mass × acceleration

= 1100 kg × 30.45 m/s²

= **33.5 kN**

19.3   Multi-choice questions on linear momentum and impulse

(Answers on page 355.)

1. A mass of 100 g has a momentum of 100 kg m/s. The velocity of the mass is:
   (a) 10 m/s   (b) 10^2 m/s   (c) 10^{-3} m/s
   (d) 10^3 m/s

2. A rifle bullet has a mass of 50 g. The momentum when the muzzle velocity is 108 km/h is:
   (a) 54 kg m/s   (b) 1.5 kg m/s
   (c) 15 000 kg m/s   (d) 21.6 kg m/s

A body P of mass 10 kg has a velocity of 5 m/s and the same line of action as a body Q of mass 2 kg and having a velocity of 25 m/s. The bodies collide, and their velocities are the same after impact. In questions 3 to 6, select the correct answer from the following:
(a) $\frac{25}{3}$ m/s   (b) 360 kg m/s   (c) 0   (d) 30 m/s
(e) 160 kg m/s   (f) 100 kg m/s   (g) 20 m/s

3. Determine the total momentum of the system before impact when P and Q have the same sense.

4. Determine the total momentum of the system before impact when P and Q have the opposite sense.

5. Determine the velocity of P and Q after impact if their sense is the same before impact.

6. Determine the velocity of P and Q after impact if their sense is opposite before impact.

7. A force of 100 N acts on a body of mass 10 kg for 0.1 s. The change in velocity of the body is:
   (a) 1 m/s   (b) 100 m/s   (c) 0.1 m/s
   (d) 0.01 m/s

A vertical pile of mass 200 kg is driven 100 mm into the ground by the blow of a 1 t hammer which falls through 1.25 m. In questions 8 to 12, take g as 10 m/s² and select the correct answer from the following:
(a) 25 m/s   (b) $\frac{25}{6}$ m/s   (c) 5 kg m/s
(d) 0   (e) $\frac{625}{6}$ kN   (f) 5000 kg m/s
(g) 5 m/s   (h) 12 kN

8. Calculate the velocity of the hammer immediately before impact.

9. Calculate the momentum of the hammer just before impact.

10. Calculate the momentum of the hammer and pile immediately after impact assuming they have the same velocity.

11. Calculate the velocity of the hammer and pile immediately after impact assuming they have the same velocity.

12. Calculate the resistive force of the ground, assuming it to be uniform.

19.4   Short answer questions on linear momentum and impulse

1. Define momentum.
2. State Newton's first law of motion.
3. State the principle of the conservation of momentum.
4. State Newton's second law of motion.
5. Define impulse.
6. What is meant by an impulsive force?
7. State Newton's third law of motion.

19.5 Further questions on linear momentum and impulse

(Where necessary, take g as 9.81 m/s^2.)

1. Determine the momentum in a mass of 50 kg having a velocity of 5 m/s.   [250 kg m/s]
2. A milling machine and its component have a combined mass of 400 kg. Determine the momentum of the table and component when the feed rate is 360 mm/min.   [2.4 kg m/s]
3. The momentum of a body is 160 kg m/s when the velocity is 2.5 m/s. Determine the mass of the body.   [64 kg]
4. Calculate the momentum of a car of mass 750 kg moving at a constant velocity of 108 km/h.   [22 500 kg m/s]
5. A football of mass 200 g has a momentum of 5 kg m/s. What is the velocity of the ball in km/h.   [90 km/h]
6. A wagon of mass 8 t is moving at a speed of 5 m/s and collides with another wagon of mass 12 t, which is stationary. After impact, the wagons are coupled together. Determine the common velocity of the wagons after impact.   [2 m/s]
7. A ball of mass 40 g is moving with a velocity of 5 m/s when it strikes a stationary ball of mass 30 g. The velocity of the 40 g ball after impact is 4 m/s in the same direction as before impact. Determine the velocity of the 30 g ball after impact.   [$1\frac{1}{3}$ m/s]
8. A car of mass 800 kg was stationary when hit head-on by a lorry of mass 2000 kg travelling at 15 m/s. Assuming no brakes are applied and the car and lorry move as one, determine the speed of the wreckage immediately after collision.   [10.71 m/s]
9. A body has a mass of 25 g and is moving with a velocity of 30 m/s. It collides with a second body which has a mass of 15 g and which is moving with a velocity of 20 m/s. Assuming that the bodies both have the same speed after impact, determine their common velocity (a) when the speeds have the same line of action and the same sense, and (b) when the speeds have the same line of action but are opposite in sense.
    (a) $26\frac{1}{4}$ m/s (b) $11\frac{1}{4}$ m/s]
10. Three masses, X, Y and Z, lie in a straight line. X has a mass of 15 kg and is moving towards Y at 20 m/s. Y has a mass of 10 kg and a velocity of 5 m/s and is moving towards Z. Mass Z is stationary. X collides with Y, and X and Y then collide with Z. Determine the mass of Z assuming all three masses have a common velocity of 4 m/s after the collision of X and Y with Z.   [62.5 kg]
11. The sliding member of a machine tool has a mass of 200 kg. Determine the change in momentum when the sliding speed is increased from 10 mm/s to 50 mm/s.   [8 kg m/s]
12. A force of 48 N acts on a body of mass 8 kg for 0.25 s. Determine the change in velocity.   [1.5 m/s]
13. In a press-tool operation, the tool is in contact with the workpiece for 40 ms. If the average force exerted on the workpiece is 90 kN, determine the change in momentum.   [3600 kg m/s]
14. The speed of a car of mass 800 kg is increased from 54 km/h to 63 km/h in 2 s. Determine the average force in the direction of motion necessary to produce the change in speed.   [1000 N]
15. A 10 kg mass is dropped vertically on to a fixed horizontal plane and has an impact velocity of 15 m/s. The mass rebounds with a velocity of 5 m/s. If the contact time of mass and plane is 0.025 s, calculate (a) the impulse, and (b) the average value of the impulsive force on the plane.   [(a) 200 kg m/s (b) 8 kN]
16. The hammer of a pile driver of mass 1.2 t falls 1.4 m on to a pile. The blow takes place in 20 ms and the hammer does not rebound. Determine the average applied force exerted on the pile by the hammer.   [314.5 kN]
17. A tennis ball of mass 60 g is struck from rest with a racket. The contact time of ball on racket is 10 ms and the ball leaves the racket with a velocity of 25 m/s. Calculate (a) the impulse, and (b) the average force exerted by a racket on the ball.   [(a) 1.5 kg m/s (b) 150 N]
18. A gun of mass 1.2 t fires a shell of mass 12 kg with a velocity of 400 m/s. Determine (a) the initial velocity of recoil, and (b) the uniform force necessary to stop the recoil of the gun in 150 mm.   [(a) 4 m/s (b) 64 kN]
19. In making a steel stamping, a mass of 100 kg falls on to the steel through a distance of 1.5 m and is brought to rest after moving through a further distance of 15 mm. Determine the magnitude of the resisting force, assuming a uniform resistive force is exerted by the steel.   [98.1 kN]

20. A vertical pile of mass 150 kg is driven 120 mm into the ground by the blow of a 1.1 t hammer which falls through 800 mm. Assuming the hammer and pile remain in contact, determine (a) the velocity of the hammer just before impact, (b) the velocity immediately after impact, and (c) the resistive force of the ground, assuming it to be uniform.

[(a) 3.96 m/s (b) 3.48 m/s (c) 63.08 kN]

20

Torque

At the end of this chapter you should be able to:

- define a couple
- define a torque and state its unit
- calculate torque given force and radius
- calculate work done given torque and angle turned through
- calculate power given torque and angular velocity
- appreciate kinetic energy = $I\omega^2/2$ where I is the moment of inertia
- appreciate that torque $T = I\alpha$ where α is the angular acceleration
- calculate torque given I and α
- calculate kinetic energy given I and ω
- understand power transmission by means of belt and pulley
- perform calculations involving torque, power and efficiency of belt drives

20.1 Couple and torque

When two equal forces act on a body as shown in Fig. 20.1, they cause the body to rotate, and the system of forces is called a **couple**.

The turning moment of a couple is called a **torque**, T. In Fig. 20.1, torque = magnitude of either force × perpendicular distance between the forces, i.e.

$$\boxed{T = Fd}$$

The unit of torque is the **newton metre**, N m.

When a force F newtons is applied at a radius r metres from the axis of, say, a nut to be turned by a spanner (as shown in Fig. 20.2), the torque T applied to the nut is given by $T = Fr\,\textbf{N\,m}$.

Problem 1. Determine the torque when a pulley wheel of diameter 300 mm has a force of 80 N applied at the rim.

Torque $T = Fr$, where force $F = 80$ N and radius $r = (300/2) = 150$ mm $= 0.15$ m.

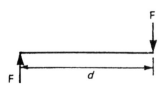

Figure 20.1

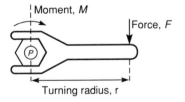

Figure 20.2

Hence **torque** $T = (80)(0.15) = \mathbf{12\ N\ m}$.

Problem 2.   Determine the force applied tangentially to a bar of a screwjack at a radius of 800 mm, if the torque required is 600 N m.

Torque T = force × radius, from which

$$\mathbf{force} = \frac{\text{torque}}{\text{radius}}$$

$$= \frac{600\ \text{N m}}{800 \times 10^{-3}\ \text{m}} = \mathbf{750\ N}$$

Problem 3.   The circular handwheel of a valve of diameter 500 mm has a couple applied to it composed of two forces, each of 250 N. Calculate the torque produced by the couple.

Torque produced by couple, $T = Fd$, where force $F = 250$ N and distance between the forces, $d = 500$ mm = 0.5 m.
Hence **torque** $T = (250)(0.5) = \mathbf{125\ N\ m}$.

20.2   Work done and power transmitted by a constant torque

Figure 20.3(a) shows a pulley wheel of radius r metres attached to a shaft and a force F newtons applied to the rim at point P.

Figure 20.3(b) shows the pulley wheel having turned through an angle θ radians as a result of the force F being applied. The force moves through a distance s, where arc length $s = r\theta$. Work done = force × distance moved by force = $F \times r\theta$ N m = $Fr\theta$ J. However, Fr is the torque T, hence **work done = $T\theta$ joules**.

$$\text{Average power} = \frac{\text{work done}}{\text{time taken}}$$

$$= \frac{T\theta}{\text{time taken}},$$

for a constant torque T

However, (angle θ)(time taken) = angular velocity, ω rad/s. Hence

power, $P = T\omega$ watts

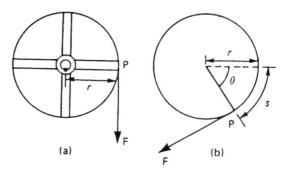

Figure 20.3

Angular velocity, $\omega = 2\pi n$ rad/s where n is the speed in rev/s. Hence

---

power $T = 2\pi nT$ watts

Problem 4.   A constant force of 150 N is applied tangentially to a wheel of diameter 140 mm. Determine the work done, in joules, in 12 revolutions of the wheel.

Torque $T = Fr$, where $F = 150$ N and radius $r = (140/2) = 70$ mm = 0.070 m. Hence torque $T = (150)(0.070) = 10.5$ N m. Work done = $T\theta$ joules, where torque $T = 10.5$ N m and angular displacement $\theta = 12$ revolutions = $12 \times 2\pi$ rad = 24π rad. Hence

work done $= (10.5)(24\pi) = \mathbf{792\ J}$

Problem 5.   Calculate the torque developed by a motor whose spindle is rotating at 1000 rev/min and developing a power of 2.50 kW.

Power $P = 2\pi nT$ (from above), from which

$$\text{torque}\ T = \frac{P}{2\pi n}\ \text{N m}$$

where power $P = 2.50$ kW = 2500 W and speed $n = 1000/60$ rev/s. Thus

$$\text{torque}\ T = \frac{2500}{2\pi(1000/60)} = \frac{2500 \times 60}{2\pi \times 1000}$$

$$= \mathbf{23.87\ N\ m}$$

Problem 6. An electric motor develops a power of 3.75 kW and a torque of 12.5 N m. Determine the speed of rotation of the motor in rev/min.

Power $P = 2\pi nT$, from which

$$\text{speed } n = \frac{P}{2\pi T} \text{ rev/s}$$

where power $P = 3.75$ kW $= 3750$ W and torque $T = 12.5$ N m. Hence

$$\text{speed } n = \frac{3750}{2\pi(12.5)} = 47.75 \text{ rev/s}$$

The speed of rotation of the motor $= 47.75 \times 60$ **= 2865 rev/min.**

Problem 7. In a turning-tool test, the tangential cutting force is 50 N. If the mean diameter of the workpiece is 40 mm, calculate (a) the work done per revolution of the spindle, (b) the power required when the spindle speed is 300 rev/min.

(a) Work done $= T\theta$, where $T = Fr$; force $F = 50$ N; radius $r = (40/2) = 20$ mm $= 0.02$ m; and angular displacement $\theta = 1$ rev $= 2\pi$ rad. Hence

work done per revolution of spindle

$$= Fr\theta$$

$$= (50)(0.02)(2\pi)$$

$$= \textbf{6.28 J}$$

(b) Power $P = 2\pi nT$, where torque $T = Fr = (50)(0.02) = 1$ N m and speed $n = (300/60) = 5$ rev/s. Hence

power required, $P = 2\pi(5)(1) = \textbf{31.42 W}$

Problem 8. A pulley is 600 mm in diameter and the difference in tension on the two sides of the driving belt is 1.5 kN. If the speed of the pulley is 500 rev/min, determine (a) the torque developed, (b) the work done in 3 minutes.

(a) Torque $T = Fr$ where force $F = 1.5$ kN $= 1500$ N and radius $r = (600/2) = 300$ mm $= 0.3$ m. Hence

torque developed $= (1500)(0.3) = \textbf{450 N m}$

(b) Work done $= T\theta$, where torque $T = 450$ N m and angular displacement in 3 minutes, $\theta = (3 \times 500)$ revs $(3 \times 500 \times 2\pi)$ rad. Hence

work done $= (450)(3 \times 50 \times 2\pi)$
$= 4.24 \times 10^6$ J $= \textbf{4.24 MJ}$

Problem 9. A motor connected to a shaft develops a torque of 5 kN m. Determine the number of revolutions made by the shaft if the work done is 9 MJ.

Work done $= T\theta$, from which angular displacement $\theta = $ work done/torque. The work done $= 9$ MJ $= 9 \times 10^6$ J and torque $= 5$ kN $= 5000$ N m. Hence

$$\text{angular displacement } \theta = \frac{9 \times 10^6}{5000}$$

$$= 1800 \text{ rad}$$

2π rad $= 1$ rev, hence

the number of revolutions made by the shaft $= 1800/2\pi$

$$= \textbf{286.5 revs}$$

20.3 Kinetic energy and moment of inertia

The tangential velocity v of a particle of mass m moving at an angular velocity ω rad/s at a radius r metres (see Fig. 20.4) is given by $v = \omega r$ m/s.

The kinetic energy of a particle of mass m is given by:

$$\textbf{Kinetic energy} = \frac{1}{2} mv^2 \text{ (from Chapter 15)}$$

$$= \frac{1}{2} m(\omega r)^2$$

$$= \frac{1}{2} m\omega^2 r^2 \textbf{ joules}$$

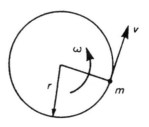

Figure 20.4

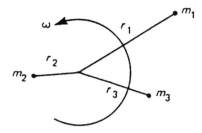

Figure 20.5

The total kinetic energy of a system of masses rotating at different radii about a fixed axis but with the same angular velocity, as shown in Fig. 20.5, is given by:

Total kinetic energy

$$= \frac{1}{2} m_1 \omega^2 r_1^2 + \frac{1}{2} m_2 \omega^2 r_2^2 + \frac{1}{2} m_3 \omega^2 r_3^2$$

$$= (m_1 r_1^2 + m_2 r_2^2 + m_3 r_3^2) \frac{\omega^2}{2}$$

In general, this may be written as:

Total kinetic energy $= (\Sigma m r^2) \dfrac{\omega^2}{2} = I \dfrac{\omega^2}{2}$

where $I(= \Sigma m r^2)$ is called the **moment of inertia** of the system about the axis of rotation and has units of kg m^2.

The moment of inertia of a system is a measure of the amount of work done to give the system an angular velocity of ω rad/s, or the amount of work which can be done by a system turning at ω rad/s.

From section 20.2, work done $= T\theta$, and if this work is available to increase the kinetic energy of a rotating body of moment of inertia I, then:

$$T\theta = I \left(\frac{\omega_2^2 - \omega_1^2}{2} \right)$$

where ω_1 and ω_2 are the initial and final angular velocities, i.e.

$$T\theta = I \left(\frac{\omega_2 + \omega_1}{2} \right) (\omega_2 - \omega_1)$$

However, $(\omega_2 + \omega_1)/2$ is the mean angular velocity, i.e. θ/t, where t is the time, and $(\omega_2 - \omega_1)$ is the change in angular velocity, i.e. αt, where α is the angular acceleration. Hence

$$T\theta = I \left(\frac{\theta}{t} \right) (\alpha t)$$

from which,

$$\boxed{\text{torque } T = I\alpha}$$

where I is the moment of inertia in kg m^2, α is the angular acceleration in rad/s^2, and T is the torque in N m.

Problem 10. A shaft system has a moment of inertia of 37.5 kg m^2 Determine the torque required to give it an angular acceleration of 5.0 rad/s^2.

Torque $T = I\alpha$, where moment of inertia $I = 37.5$ kg m^2 and angular acceleration $\alpha = 5.0$ rad/s^2. Hence

torque $T = (37.5)(5.0) = $ **187.5 N m.**

Problem 11.   A shaft has a moment of inertia of 31.4 kg m^2. What angular acceleration of the shaft would be produced by an accelerating torque of 495 N m?

Torque $T = I\alpha$, from which angular acceleration $\alpha = T/I$, where torque $T = 495$ N m and moment of inertia $I = 31.4$ kg m^2. Hence

angular acceleration $\alpha = \dfrac{495}{31.4} = $ **15.76 rad/s^2**

Problem 12.   A body of mass 100 g is fastened to a wheel and rotates in a circular path of 500 mm in diameter. Determine the increase in kinetic energy of the body when the speed of the wheel increases from 450 rev/min to 750 rev/min.

From above,

$$\text{kinetic energy} = I \frac{\omega^2}{2}$$

Thus,

$$\text{increase in kinetic energy} = I \left(\frac{\omega_2^2 - \omega_1^2}{2} \right)$$

where moment of inertia $I = mr^2$, mass $m = 100$ g $= 0.1$ kg and radius $r = (500/2) = 250$ mm $= 0.25$ m. Initial angular velocity, $\omega_1 = 450$ rev/min $= (450 \times 2\pi)/60$ rad/s $= 47.12$ rad/s, and final angular velocity, $\omega_2 = 750$ rev/min $= (750 \times 2\pi)/60$ rad/s $= 78.54$ rad/s.

Thus

    increase in kinetic energy

$$= I\left(\frac{\omega_2^2 - \omega_1^2}{2}\right) = (mr^2)\left(\frac{\omega_2^2 - \omega_1^2}{2}\right)$$

$$= (0.1)(0.25^2)\left(\frac{78.54^2 - 47.12^2}{2}\right) = \mathbf{12.34\ J}$$

Problem 13. A system consists of three small masses rotating at the same speed about the same fixed axis. The masses and their radii of rotation are: 15 g at 250 mm, 20 g at 180 mm and 30 g at 200 mm. Determine (a) the moment of inertia of the system about the given axis (b) the kinetic energy in the system if the speed of rotation is 1200 rev/min.

(a) Moment of inertia of the system, $I = \Sigma mr^2$, i.e.

$$I = [(15 \times 10^{-3}\ \text{kg})(0.25\ \text{m})^2]$$
$$+ [(20 \times 10^{-3}\ \text{kg})(0.18\ \text{m})^2]$$
$$+ [(30 \times 10^{-3}\ \text{kg})(0.20\ \text{m})^2]$$
$$= (9.375 \times 10^{-4}) + (6.48 \times 10^{-4})$$
$$+ (12 \times 10^{-4})$$
$$= 27.855 \times 10^{-4}\ \text{kg m}^2 = \mathbf{2.7855 \times 10^{-3}\ kg\ m^2}$$

(b) Kinetic energy $= I\dfrac{\omega^2}{2}$, where

moment of inertia $I = 2.7855 \times 10^{-3}$ kg m^2 and angular velocity $\omega = 2\pi n = 2\pi(1200/60)$ rad/s $= 40\pi$ rad/s. Hence

kinetic energy in the system

$$= (2.7855 \times 10^{-3})\frac{(40\pi)^2}{2}$$

$$= \mathbf{21.99\ J}$$

Problem 14. A shaft with its rotating parts has a moment of inertia of 20 kg m^2. It is accelerated from rest by an accelerating torque of 45 N m. Determine the speed of the shaft in rev/min (a) after 15 s, (b) after the first 5 revolutions.

(a) Since torque $T = I\alpha$, then angular acceleration, $\alpha = (T/I) = (45/20) = 2.25$ rad/s^2.

The angular velocity of the shaft is initially zero, i.e. $\omega_1 = 0$. The angular velocity after 15 s, $\omega_2 = \omega_1 + \alpha t = 0 + (2.25)(15) = 33.75$ rad/s, i.e.

speed of shaft after 15 s

$$= (33.75)\left(\frac{60}{2\pi}\right)\ \text{rev/min}$$

$$= \mathbf{322.3\ rev/min}$$

(b) Work done $= T\theta$ where torque $T = 45$ N m, and angular displacement $\theta = 5$ revolutions $= 5 \times 2\pi$ rad $= 10\pi$ rad. Hence work done $= (45)(10\pi) = 1414$ J. This work done results in an increase in kinetic energy, given by $I(\omega^2/2)$, where moment of inertia $I = 20$ kg m^2 and $\omega =$ angular velocity. Hence $1414 = (20)(\omega^2/2)$, from which

$$\omega = \sqrt{\left(\frac{1414 \times 2}{20}\right)} = 11.89\ \text{rad/s}$$

Hence

speed of shaft after the first 5 revolutions

$$= 11.89 \times \frac{60}{2\pi} = \mathbf{113.5\ rev/min}$$

Problem 15. The accelerating torque on a turbine rotor is 250 N m.

(a) Determine the gain in kinetic energy of the rotor while it turns through 100 revolutions (neglecting any frictional and other resisting torques).
(b) If the moment of inertia of the rotor is 25 kg m^2 and the speed at the beginning of the 100 revolutions is 450 rev/min, determine its speed at the end.

(a) The kinetic energy gained is equal to the work done by the accelerating torque of 250 N m over 100 revolutions; i.e.

gain in kinetic energy = work done =

$$T\theta = (250)(100 \times 2\pi)$$

$$= \mathbf{157.08\ kJ}$$

(b) Initial kinetic energy of rotation is given by

$$I\frac{\omega_1^2}{2} = \frac{(25)\left(\dfrac{450 \times 2\pi}{60}\right)^2}{2}$$

$$= 27.76\ \text{kJ}$$

The final kinetic energy is the sum of the initial kinetic energy and the kinetic energy gained, i.e.

$$\frac{I\omega_2^2}{2} = 27.76 \text{ kJ} + 157.08 \text{ kJ} = 184.84 \text{ kJ}$$

Hence

$$\frac{(25)\omega_2^2}{2} = 184\,840$$

from which

$$\omega_2 = \sqrt{\left(\frac{184\,840 \times 2}{25}\right)}$$

$$= 121.6 \text{ rad/s}$$

Thus

speed at end of 100 revolutions

$$= \frac{121.6 \times 60}{2\pi} \text{ rev/min}$$

$$= \mathbf{1161 \text{ rev/min}}$$

Problem 16. A shaft with its associated rotating parts has a moment of inertia of 55.4 kg m². Determine the uniform torque required to accelerate the shaft from rest to a speed of 1650 rev/min while it turns through 12 revolutions.

From above,

$$T\theta = I\left(\frac{\omega_2^2 - \omega_1^2}{2}\right)$$

where angular displacement $\theta = 12 \text{ rev} = 12 \times 2\pi \text{ rad} = 24\pi \text{ rad}$, final speed, $\omega_2 = 1650 \text{ rev/min} = (1650/60) \times 2\pi = 172.79 \text{ rad/s}$, initial speed, $\omega_1 = 0$, and moment of inertia, $I = 55.4 \text{ kg m}^2$. Hence

torque required,

$$T = \frac{I}{\theta}\left(\frac{\omega_2^2 - \omega_1^2}{2}\right)$$

$$= \frac{55.4}{24\pi}\left(\frac{(172.79)^2 - (0)^2}{2}\right)$$

$$= \mathbf{10.97 \text{ kN m}.}$$

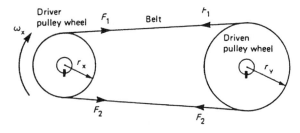

Figure 20.6

20.4 Power transmission and efficiency

A common and simple method of transmitting power from one shaft to another is by means of a **belt** passing over pulley wheels which are keyed to the shafts, as shown in Fig. 20.6. Typical applications include an electric motor driving a lathe or a drill, and an engine driving a pump or generator.

For a belt to transmit power between two pulleys there must be a difference in tensions in the belt on either side of the driving and driven pulleys. For the direction of rotation shown in Fig. 20.6, $F_2 > F_1$.

The torque T available at the driving wheel to do work is given by:

$$\boxed{T = (F_2 - F_1)r_x \text{ N m}}$$

and the available power P is given by:

$$\boxed{P = T\omega = (F_2 - F_1)r_x\,\omega_x \text{ watts}}$$

From section 20.3, the linear velocity of a point on the driver wheel, $v_x = r_x\omega_x$. Similarly, the linear velocity of a point on the driven wheel, $v_y = r_y\omega_y$. Assuming no slipping, $v_x = v_y$, i.e. $r_x\omega_x = r_y\omega_y$. Hence $r_x(2\pi n_x) = r_y(2\pi n_y)$ from which

$$\frac{r_x}{r_y} = \frac{n_y}{n_x}$$

Percentage efficiency

$$= \frac{\text{useful work output}}{\text{energy output}} \times 100$$

or

$$\mathbf{efficiency = \frac{power\ output}{power\ input} \times 100\%}$$

Problem 17. An electric motor has an efficiency of 75% when running at 1450 rev/min. Determine the output torque when the power input is 3.0 kW.

$$\text{Efficiency} = \frac{\text{power output}}{\text{power input}} \times 100$$

hence

$$75 = \frac{\text{power output}}{3000} \times 100$$

from which,

$$\text{power output} = \frac{75}{100} \times 3000 = 2250 \text{ W}$$

From section 20.2, power output, $P = 2\pi nT$, from which

$$\text{torque } T = \frac{P}{2\pi n}$$

where $n = (1450/60)$ rev/s. Hence

$$\text{output torque} = \frac{2250}{2\pi\left(\dfrac{1450}{60}\right)} = \textbf{14.82 N m}$$

Problem 18. A 15 kW motor is driving a shaft at 1150 rev/min by means of pulley wheels and a belt. The tensions in the belt on each side of the driver pulley wheel are 400 N and 50 N. The diameters of the driver and driven pulley wheels are 500 mm and 750 mm respectively. Determine (a) the efficiency of the motor, (b) the speed of the driven pulley wheel.

(a) From above, power output from motor = $(F_2 - F_1)r_x\omega_x$. Force $F_2 = 400$ and $F_1 = 50$ N, hence $(F_2 - F_1) = 350$ N. Radius $r_x = (500/2) = 250$ mm $= 0.25$ m and angular velocity, $\omega_x = (1150 \times 2\pi)/60$ rad/s. Hence power output from motor

$$(F_2 - F_1)r_x\omega_x = (350)(0.25)\left(\frac{1150 \times 2\pi}{60}\right)$$

$$= 10.54 \text{ kW}$$

Power input = 15 kW. Hence

efficiency of the motor $= \dfrac{\text{power output}}{\text{power input}}$

$$= \frac{10.54}{15} \times 100$$

$$= \textbf{70.27\%}$$

(b) From above,

$$\frac{r_x}{r_y} = \frac{n_y}{n_x}$$

from which,

speed of driven pulley wheels, n_y

$$= \frac{n_x r_x}{r_y} = \frac{(1150)(0.25)}{(0.75/2)}$$

$$= \textbf{767 rev/min}$$

Problem 19. A crane lifts a load of mass 5 tonne to a height of 25 m. If the overall efficiency of the crane is 65% and the input power to the hauling motor is 100 kW, determine how long the lifting operation takes.

The increase in potential energy is the work done and is given by mgh, where mass $m = 5 t = 5000$ kg, $g = 9.81$ m/s^2 and height $h = 25$ m. Hence work done $= mgh = (5000)(9.81)(25) = 1.226$ MJ. Input power = 100 kW = 100 000 W.

$$\text{Efficiency} = \frac{\text{output power}}{\text{input power}} \times 100$$

hence

$$65 = \frac{\text{output power}}{100\,000} \times 100$$

from which,

$$\text{output power} = \frac{65}{100} \times 100\,000$$

$$= 65\,000 \text{ W} = \frac{\text{work done}}{\text{time taken}}$$

Thus

time taken for lifting operation

$$= \frac{\text{work done}}{\text{output power}}$$

$$= \frac{1.226 \times 10^6 \text{ J}}{65\,000 \text{ W}} = \textbf{18.86 s}$$

Problem 20.   The tool of a shaping machine has a mean cutting speed of 250 mm/s and the average cutting force on the tool in a certain shaping operation is 1.2 kN. If the power input to the motor driving the machine is 0.75 kW, determine the overall efficiency of the machine.

Velocity v = 250 mm/s = 0.25 m/s, force F = 1.2 kN = 1200 N. Power output required at the cutting tool (i.e. power output), P = force × velocity = 1200 N × 0.25 m/s = 300 W. Power input = 0.75 kW = 750 W. Hence

efficiency of the machine

$$= \frac{\text{output power}}{\text{input power}} \times 100 = \frac{300}{750} \times 100$$

$$= \textbf{40\%}$$

Problem 21.   Calculate the input power of the motor driving a train at a constant speed of 72 km/h on a level track, if the efficiency of the motor is 80% and the resistance due to friction is 20 kN.

Force resisting motion = 20 kN = 20 000 N; velocity = 72 km/h = (72/3.6) = 20 m/s.
Output power from motor = resistive force × velocity of train = 20 000 × 20 = 400 kW.

$$\text{Efficiency} = \frac{\text{power output}}{\text{power input}} \times 100$$

hence

$$80 = \frac{400}{\text{power input}} \times 100$$

from which,

$$\textbf{power input} = 400 \times \frac{100}{80} = \textbf{500 kW}$$

20.5   Multi-choice questions on torque

(Answers on page 355.)

1. A force of 100 N is applied to the rim of a pulley wheel of diameter 200 mm. The torque is:
(a) 2 N m   (b) 20 kN m   (c) 10 N m
(d) 20 N m

2. The work done on a shaft to turn it through 5π radians is 25π N m. The torque applied to the shaft is:
(a) 0.2 N m   (b) 125 π^2 N m   (c) 30π N m
(d) 5 N m

3. A 5 kW electric motor is turning at 50 rad/s. The torque developed at this speed is:
(a) 100 N m   (b) 250 N m   (c) 0.01 N m
(d) 0.1 N m

4. The force applied tangentially to a bar of a screwjack at a radius of 500 mm if the torque required is 1 kN m is:
(a) 2 N   (b) 2 kN   (c) 500 N   (d) 0.5 N

5. A 10 kW motor developing a torque of $(200/\pi)$ N m is running at a speed of:
(a) $(\pi/20)$ rev/s   (b) 50π rev/s   (c) 25 rev/s
(d) $(20/\pi)$ rev/s

6. A shaft and its associated rotating parts has a moment of inertia of 50 kg m². The angular acceleration of the shaft to produce an accelerating torque of 5 kN m is:
(a) 10 rad/s²   (b) 250 rad/s²   (c) 0.01 rad/s²
(d) 100 rad/s²

7. A motor has an efficiency of 25% when running at 3000 rev/min. If the output torque is 10 N m, the power output is:
(a) 4π kW   (b) 0.25π kW   (c) 15π kW
(d) 75π kW

8. In a belt-pulley wheel system, the effective tension in the belt is 500 N and the diameter of the driver wheel is 200 mm. If the power output from the driving motor is 5 kW, the driver pulley wheel turns at:
(a) 50 rad/s   (b) 2500 rad/s   (c) 100 rad/s
(d) 0.1 rad/s

20.6   Short answer questions on torque

1. What is meant by a couple?
2. Define torque.
3. State the unit of torque.
4. State the relationship between work, torque T and angular displacement θ.
5. State the relationship between power P, torque T and angular velocity ω.
6. Define moment of inertia and state the symbol used.
7. State the unit of moment of inertia.
8. State the relationship between torque, moment of inertia and angular acceleration.
9. State one method of power transmission commonly used.
10. Define efficiency.

20.7 Further questions on torque

1. Determine the torque developed when a force of 200 N is applied tangentially to a spanner at a distance of 350 mm from the centre of the nut. [70 N m]
2. During a machining test on a lathe, the tangential force on the tool is 150 N. If the torque on the lathe spindle is 12 N m, determine the diameter of the workpiece. [160 mm]
3. A constant force of 4 kN is applied tangentially to the rim of a pulley wheel of diameter 1.8 m attached to a shaft. Determine the work done, in joules, in 15 revolutions of the pulley wheel. [339.3 kJ]
4. A motor connected to a shaft develops a torque of 3.5 kN m. Determine the number of revolutions made by the shaft if the work done is 11.52 MJ. [523.8 rev]
5. A wheel is turning with an angular velocity of 18 rad/s and develops a power of 810 W at this speed. Determine the torque developed by the wheel. [45 N m]
6. Calculate the torque provided at the shaft of an electric motor which develops an output power of 2.4 kW at 1800 rev/min. [12.73 N m]
7. Determine the angular velocity of a shaft when the power available is 2.75 kW and the torque is 200 N m. [13.75 rad/s]
8. The drive shaft of a ship supplies a torque of 400 kN m to its propeller at 400 rev/min. Determine the power delivered by the shaft. [16.76 MW]
9. A motor is running at 1460 rev/min and produces a torque of 180 N m. Determine the average power developed by the motor. [27.52 kW]
10. A wheel is rotating at 1720 rev/min and develops a power of 600 W at this speed. Calculate (a) the torque, (b) the work done, in joules, in a quarter of an hour. [(a) 3.33 N m (b) 540 kJ]
11. A force of 60 N is applied to a lever of a screwjack at a radius of 220 mm. If the lever makes 25 revolutions, determine (a) the work done on the jack (b) the power, if the time taken to complete 25 revolutions is 40 s. [(a) 2.073 kJ (b) 51.84 W]
12. A shaft system has a moment of inertia of 51.4 kg m^2. Determine the torque required to give it an angular acceleration of 5.3 rad/s^2. [272.4 N m]
13. A shaft has an angular acceleration of 20 rad/s^2 and produces an accelerating torque of 600 N m. Determine the moment of inertia of the shaft. [30 kg m^2]
14. A uniform torque of 3.2 kN m is applied to a shaft while it turns through 25 revolutions. Assuming no frictional or other resistances, calculate the increase in kinetic energy of the shaft (i.e. the work done). If the shaft is initially at rest and its moment of inertia is 24.5 kg m^2, determine its rotational speed, in rev/min, at the end of the 25 revolutions. [502.65 kJ, 1934 rev/min]
15. An accelerating torque of 30 N m is applied to a motor, while it turns through 10 revolutions. Determine the increase in kinetic energy. If the moment of inertia of the rotor is 15 kg m^2 and its speed at the beginning of the 10 revolutions is 1200 rev/min, determine its speed at the end. [1.885 kJ, 1209.5 rev/min]
16. A shaft with its associated rotating parts has a moment of inertia of 48 kg m^2. Determine the uniform torque required to accelerate the shaft from rest to a speed of 1500 rev/min while it turns through 15 revolutions. [6.283 kN m]
17. A small body, of mass 82 g, is fastened to a wheel and rotates in a circular path of 456 mm diameter. Calculate the increase in kinetic energy of the body when the speed of the wheel increases from 450 rev/min to 950 rev/min. [16.36 J]
18. A system consists of three small masses rotating at the same speed about the same fixed axis. The masses and their radii of rotation are: 16 g at 256 mm, 23 g at 192 mm and 31 g at 176 mm. (a) Find the moment of inertia of the system about the given axis, (b) if the speed of rotation is 1250 rev/min, find the kinetic energy in the system. [(a) 2.857 × 10^{-3} kg m^2 (b) 24.48 J]
19. A shaft with its rotating parts has a moment of inertia of 16.42 kg m^2. It is accelerated from rest by an accelerating torque of 43.6 N m. Find the speed of the shaft (a) after 15 s, (b) after the first four revolutions. [(a) 380.3 rev/min (b) 110.3 rev/min]
20. The driving torque on a turbine rotor is 203 N m, neglecting frictional and other resisting torques. (a) What is the gain in kinetic energy of the rotor while it turns through 100 revolutions? (b) If the moment of inertia of the rotor is 23.2 kg m^2 and the

speed at the beginning of the 100 revolutions is 600 rev/min, what will be its speed at the end?   [(a) 127.55 kJ (b) 1167 rev/min]

21. A motor has an efficiency of 72% when running at 2600 rev/min. If the output torque is 16 N m at this speed, determine the power supplied to the motor.   [6.05 kW]

22. The difference in tensions between the two sides of a belt round a driver pulley of radius 240 mm is 200 N. If the driver pulley wheel is on the shaft of an electric motor running at 700 rev/min and the power input to the motor is 5 kW, determine the efficiency of the motor. Determine also the diameter of the driven pulley wheel if its speed is to be 1200 rev/min.   [70.37%, 280 mm]

23. A winch is driven by a 4 kW electric motor and is lifting a load of 400 kg to a height of 5.0 m. If the lifting operation takes 8.6 s, calculate the overall efficiency of the winch and motor.   [57.03%]

24. A belt and pulley system transmits a power of 5 kW from a driver to a driven shaft. The driver pulley wheel has a diameter of 200 mm and rotates at 600 rev/min. The diameter of the driven pulley wheel is 400 mm. Determine the tension in the slack side of the belt and the speed of the driven pulley when the tension in the tight side of the belt is 1.2 kN.   [404.2 N, 300 rev/min]

25. The average force on the cutting tool of a lathe is 750 N and the cutting speed is 400 mm/s. Determine the power input to the motor driving the lathe if the overall efficiency is 55%.   [545.5 W]

26. A ship's anchor has a mass of 5 tonne. Determine the work done in raising the anchor from a depth of 100 m. If the hauling gear is driven by a motor whose output is 80 kW and the efficiency of the haulage is 75%, determine how long the lifting operation takes.   [4.905 MJ, 1 min 22 s]

21

Pressure in fluids

At the end of this chapter you should be able to:

- define pressure and state its unit
- calculate pressure given force and cross-sectional area
- understand the factors governing pressure within fluids
- calculate pressure p in a fluid from $p = \rho g h$
- distinguish between atmospheric pressure, gauge pressure and absolute pressure
- state Archimedes' principle
- use Archimedes' principle to calculate apparent loss of weight, $W = V \rho g$
- recognize various ways of measuring pressure

21.1 Pressure

The pressure acting on a surface is defined as the perpendicular force per unit area of surface. The unit of pressure is the **pascal**, Pa, where 1 pascal is equal to 1 newton per square metre. Thus pressure,

$$p = \frac{F}{A} \text{ pascals}$$

where F is the force in newtons acting at right angles to a surface of area A square metres.

When a force of 20 N acts uniformly over, and perpendicular to, an area of 4 m², then the pressure on the area, p, is given by

$$p = \frac{20 \text{ N}}{4 \text{ m}^2} = 5 \text{ Pa}$$

Problem 1.   A table loaded with books has a force of 250 N acting in each of its legs. If the contact area between each leg and the floor is 50 mm², find the pressure each leg exerts on the floor.

From above, pressure p = (force)/(area). Hence

$$p = \frac{250 \text{ N}}{50 \text{ mm}^2} \times \frac{10^6 \text{ mm}^2}{1 \text{ m}^2}$$

$$= 5 \times 10^6 \text{ N/m}^2 = \textbf{5 MPa}$$

That is, **the pressure exerted by each leg on the floor is 5 MPa**.

Problem 2.   Calculate the force exerted by the atmosphere on a pool of water which is 30 m long by 10 m wide, when the atmospheric pressure is 100 kPa.

From above, pressure = (force)/(area), hence, force = pressure × area. The area of the pool is 30 m × 10 m, i.e. 300 m². Thus, force on pool, F = 100 kPa × 300 m² and since 1 Pa = 1 N/m²,

$$F = (100 \times 10^3) \frac{\text{N}}{\text{m}^2} \times 300 \text{ m}^2$$

$$= 3 \times 10^7 \text{ N} = 30 \times 10^6 \text{ N} = 30 \text{ MN}$$

That is, **the force on the pool of water is 30 MN**.

Problem 3.   A circular piston exerts a pressure of 80 kPa on a fluid, when the force applied to the piston is 0.2 kN. Find the diameter of the piston.

From above, pressure = (force)/(area), Hence, area = (force)/(pressure). Force in newtons is

$$0.2 \text{ kN} \times \frac{1000 \text{ N}}{1 \text{ kN}} = 200 \text{ N}$$

Pressure in pascals is $80 \text{ kPa} = 80\,000 \text{ Pa} = 80\,000 \text{ N/m}^2$. Hence

$$\text{area} = \frac{200 \text{ N}}{80\,000 \text{ N/m}^2} = 0.0025 \text{ m}^2$$

Since the piston is circular, its area is given by $\pi d^2/4$, where d is the diameter of the piston. Hence,

$$\text{area} = \frac{\pi d^2}{4} = 0.0025$$

$$d^2 = 0.0025 \times \frac{4}{\pi} = 0.003\,183$$

i.e. $d = 0.0564$ m, i.e. 56.4 mm

Hence, **the diameter of the piston is 56.4 mm**.

21.2   Fluid pressure

A fluid is either a liquid or a gas and there are four basic factors governing the pressure within fluids.

(a) The pressure at a given depth in a fluid is equal in all directions, see Fig. 21.1(a).
(b) The pressure at a given depth in a fluid is independent of the shape of the container in which the fluid is held. In Fig. 21.1(b), the pressure at X is the same as the pressure at Y.
(c) Pressure acts at right angles to the surface containing the fluid. In Fig. 21.1(c), the pressures at points A to F all act at right angles to the container.
(d) When a pressure is applied to a fluid, this pressure is transmitted equally in all directions. In Fig. 21.1(d), if the mass of the fluid is neglected, the pressures at points A to D are all the same.

The pressure, p, at any point in a fluid depends on three factors:

(a) the density of the fluid, ρ, in kg/m³;
(b) the gravitational acceleration, g, taken as approximately 9.8 m/s² (or the gravitational field force in N/kg); and
(c) the height of fluid vertically above the point, h metres. The relationship connecting these quantities is:

$$p = \rho g h \text{ pascals}$$

When the container shown in Fig. 21.2 is filled with water of density 1000 kg/m³, the pressure due to the water at a depth of 0.03 m below the surface is given by:

$$p = \rho g h$$
$$= (1000 \times 9.8 \times 0.03) \text{ Pa}$$
$$= 294 \text{ Pa}$$

Problem 4.   A tank contains water to a depth of 600 mm. Calculate the water pressure (a) at a depth of 350 mm and (b) at the base of the tank. Take the density of water as 1000 kg/m³ and the gravitational acceleration as 9.8 m/s².

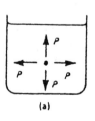
(a)

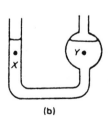
(b)

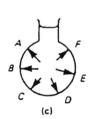
(c)

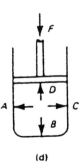
(d)

Figure 21.1

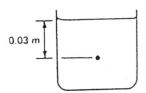

0.03 m

Figure 21.2

From above, pressure p at any point in a fluid is given by $p = \rho g h$ pascals, where ρ is the density in kg/m^3, g is the gravitational acceleration in m/s^2 and h is the height of fluid vertically above the point.

(a) At a depth of 350 mm, i.e. 0.35 m

$p = 1000 \times 9.8 \times 0.35$

$= \textbf{3430 Pa} = \textbf{3.43 kPa}$

(b) At the base of the tank, the vertical height of the water is 600 mm, that is, 0.6 m. Hence

$p = 1000 \times 9.8 \times 0.6$

$= \textbf{5880 Pa} = \textbf{5.88 kPa}$

Problem 5. A storage tank contains petrol to a height of 4.7 m. If the pressure at the base of the tank is 32.3 kPa, determine the density of the petrol. Take the gravitational field force as 9.8 m/s^2.

From above, pressure $p = \rho g h$ pascals, where ρ is the density in kg/m^3, g is the gravitational acceleration in m/s^2 and h is the vertical height of the petrol. Transposing gives:

$$\rho = \frac{p}{gh}$$

The pressure p is 32.2 kPa, that is, 32 200 Pa, hence,

$$\text{density } \rho = \frac{32\ 200}{9.8 \times 4.7} = 699 \text{ kg/m}^3$$

That is, the **density of the petrol is 699 kg/m^3**.

Problem 6. A vertical tube is partly filled with mercury of density 13 600 kg/m^3. Find the height, in millimetres, of the column of mercury, when the pressure at the base of the tube is 101 kPa. Take the gravitational field force as 9.8 m/s^2.

From above, pressure $p = \rho g h$, hence vertical height h is given by

$$h = \frac{p}{\rho g}$$

Pressure $p = 101$ kPa $= 101\ 000$ Pa, thus

$$h = \frac{101\ 000}{13\ 600 \times 9.8} = 0.758 \text{ m}$$

That is, the **height of the column of mercury is 758 mm**.

21.3 Atmospheric pressure

The air above the earth's surface is a fluid, having a density, ρ, which varies from approximately 1.225 kg/m^3 at sea level to zero in outer space. Since $p = \rho g h$, where height h is several thousands of metres, the air exerts a pressure on all points on the earth's surface. This pressure, called **atmospheric pressure**, has a value of approximately 100 kilopascals. Two terms are commonly used when measuring pressures:

(a) **absolute pressure**, meaning the pressure above that of an absolute vacuum (i.e. zero pressure); and

(b) **gauge pressure**, meaning the pressure above that normally present due to the atmosphere. Thus:

absolute pressure = atmospheric pressure + gauge pressure

Thus, a gauge pressure of 50 kPa is equivalent to an absolute pressure of (100 + 50) kPa, i.e. 150 kPa, since the atmospheric pressure is approximately 100 kPa.

For more detail on atmospheric, absolute and gauge pressure, see Chapter 39 on the measurement of pressure.

Problem 7. Calculate the absolute pressure at a point on a submarine, at a depth of 30 m below the surface of the sea, when the atmospheric pressure is 101 kPa. Take the density of sea water as 1030 kg/m^3 and the gravitational acceleration as 9.8 m/s^2.

From section 21.2, the pressure due to the sea, that is, the gauge pressure (p_g) is given by $p_g = \rho g h$ pascals, i.e.

$p_g = 1030 \times 9.8 \times 30 = 302\,820$ Pa

$\quad = 302.82$ kPa

From above,

$\qquad$ absolute pressure = atmospheric pressure + gauge pressure

$\qquad = (101 + 302.82)$ kPa $= 403.82$ kPa

That is, the **absolute pressure at a depth of 30 m is 403.82 kPa**.

21.4 Archimedes' principle

Archimedes' principle states that:

If a solid body floats, or is submerged, in a liquid, the liquid exerts an upthrust on the body equal to the gravitational force on the liquid displaced by the body.

In other words, if a solid body is immersed in a liquid, the apparent loss of weight is equal to the weight of liquid displaced.

If V is the volume of the body below the surface of the liquid, then the apparent loss of weight W is given by:

$$\boxed{W = V\omega = V\rho g}$$

where ω is the specific weight (i.e. weight per unit volume) and ρ is the density.

If a body floats on the surface of a liquid all of its weight appears to have been lost. The weight of liquid displaced is equal to the weight of the floating body.

Problem 8. A body weighs 2.760 N in air and 1.925 N when completely immersed in water of density 1000 kg/m³. Calculate (a) the volume of the body, (b) the density of the body and (c) the relative density of the body. Take the gravitational acceleration as 9.81 m/s².

(a) The apparent loss of weight is 2.760 N − 1.925 N = 0.835 N. This is the weight of water displaced, i.e. $V\rho g$, where V is the volume of the body and ρ is the density of water, i.e.

0.835 N $= V \times 1000$ kg/m³ $\times 9.81$ m/s²

$\qquad = V \times 9.81$ kN/m³

Hence

$V = \dfrac{0.835}{9.81 \times 10^3}$ m³ $= 8.512 \times 10^{-5}$ m³

$\quad = \mathbf{8.512 \times 10^4\ mm^3}$

(b) The density of the body

$= \dfrac{\text{mass}}{\text{volume}} = \dfrac{\text{weight}}{g \times V}$

$= \dfrac{2.760\ \text{N}}{9.81\ \text{m/s}^2 \times 8.512 \times 10^{-5}\ \text{m}^3}$

$= \dfrac{\dfrac{2.760}{9.81}\ \text{kg} \times 10^5}{8.512\ \text{m}^3}$

$= \mathbf{3305\ kg/m^3} = \mathbf{3.305\ tonne/m^3}$

(c) Relative density $= \dfrac{\text{density}}{\text{density of water}}$

(from Chapter 7)

The relative density of the body is therefore

$\dfrac{3305\ \text{kg/m}^3}{1000\ \text{kg/m}^3} = \mathbf{3.305}$

Problem 9. A rectangular watertight box is 560 mm long, 420 mm wide and 210 mm deep. It weighs 223 N.

(a) If it floats with its sides and ends vertical in water of density 1030 kg/m³, what depth of the box will be submerged?

(b) If the box is held completely submerged in water of density 1030 kg/m³, by a vertical chain attached to the underside of the box, what is the force in the chain?

(a) The apparent weight of a floating body is zero. That is, the weight of the body is equal to the weight of liquid displaced. This is given by: $V\rho g$ where V is the volume of liquid displaced, and ρ is the density of the liquid. Here,

223 N $= V \times 1030$ kg/m³ $\times 9.81$ m/s²

$\qquad = V \times 10.104$ kN/m³

Hence,

$V = \dfrac{223\ \text{N}}{10.104\ \text{kN/m}^3} = 22.07 \times 10^{-3}$ m³

This volume is also given by *lbd*, where *l* = length of box, *b* = breadth of box, and *d* = depth of box submerged, i.e.

22.07×10^{-3} m^3 = 0.56 m $\times$ 0.42 m $\times$ *d*

Hence,

depth submerged, $d = \dfrac{22.07 \times 10^{-3}}{0.56 \times 0.42}$

= 0.093 84 m,

= **93.84 mm**

(b) The volume of water displaced is the total volume of the box. The upthrust or buoyancy of the water, i.e. the 'apparent loss of weight', is greater than the weight of the box. The force in the chain accounts for the difference.

Volume of water displaced,

V = 0.56 m $\times$ 0.42 m $\times$ 0.21 m

= 4.9392 $\times 10^{-2}$ m^3

Weight of water displaced = $V\rho g$ = 4.9392 $\times 10^{-2}$ m^3 $\times$ 1030 kg/m^3 $\times$ 9.81 m/s^2 = 499.1 N. The force in the chain = weight of water displaced – weight of box = 499.1 N – 223 N = **276.1 N**.

21.5 Measurement of pressure

There are various ways of measuring pressure, and these include by:

(a) barometers,
(b) manometers,
(c) pressure gauges, and
(d) vacuum gauges.

The construction and principle of operation of each of these devices are described fully in Chapter 39.

Let us look briefly at just one of these instruments – the manometer. A manometer is a device used for measuring relatively small pressures, either above or below atmospheric pressure. A simple U-tube manometer is shown in Fig. 21.3. Pressure *p* acting in, say, a gas main, pushes the liquid in the U-tube until equilibrium is obtained. At equilibrium: pressure in gas main, *p* = (atmospheric pressure, p_a) + (pressure due to the column of liquid, ρgh), i.e.

$$p = p_a + \rho gh$$

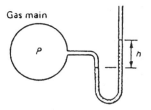

Figure 21.3

Thus, for example, if the atmospheric pressure, p_a is 101 kPa, the liquid in the U-tube is water of density 1000 kg/m^3 and height, *h* is 300 mm, then

absolute gas pressure

= (101 000 + 1000 $\times$ 9.8 $\times$ 0.3) Pa

= (101 000 + 2940) Pa = 103 940 Pa

= 103.94 kPa

The gauge pressure of the gas is 2.94 kPa.

By filling the U-tube with a more dense liquid, say mercury having a density of 13 600 kg/m^3, for a given height of U-tube, the pressure which can be measured is increased by a factor of 13.6.

21.6 Multi-choice questions on pressure in fluids

(Answers on page 355.)

1. A force of 50 N acts uniformly over and at right angles to a surface. When the area of the surface is 5 m^2, the pressure on the area is:
   (a) 250 Pa (b) 10 Pa (c) 45 Pa (d) 55 Pa
2. Which of the following statements is false? The pressure at a given depth in a fluid
   (a) is equal in all directions
   (b) is independent of the shape of the container
   (c) acts at right angles to the surface containing the fluid
   (d) depends on the area of the surface
3. A container holds water of density 1000 kg/m^3. Taking the gravitational acceleration as 10 m/s^2, the pressure at a depth of 100 mm is:
   (a) 1 kPa (b) 1 MPa (c) 100 Pa (d) 1 Pa
4. If the water in question 3 is now replaced by a fluid having a density of 2000 kg/m^3, the pressure at a depth of 100 mm is:

(a) 2 kPa   (b) 500 kPa   (c) 200 Pa
(d) 0.5 Pa

5. The gauge pressure of fluid in a pipe is 70 kPa and the atmospheric pressure is 100 kPa. The absolute pressure of the fluid in the pipe is:
(a) 7 MPa   (b) 30 kPa   (c) 170 kPa
(d) $\frac{10}{7}$ kPa

6. A U-tube manometer contains mercury of density 13 600 kg/m³. When the difference in the height of the mercury levels is 100 mm and taking the gravitational acceleration as 10 m/s², the gauge pressure is:
(a) 13.6 Pa   (b) 13.6 MPa   (c) 13 710 Pa
(d) 13.6 kPa

7. The mercury in the U-tube of question 6 is to be replaced by water of density 1000 kg/m³. The height of the tube to contain the water for the same gauge pressure is:
(a) (1/13.6) of the original height
(b) 13.6 times the original height
(c) 13.6 m more than the original height
(d) 13.6 m less than the original height

21.7   Short answer questions on pressure in fluids

1. Define pressure.
2. State the unit of pressure.
3. Define a fluid.
4. State the four basic factors governing the pressure in fluids.
5. Write down a formula for determining the pressure at any point in a fluid in symbols, defining each of the symbols and giving their units.
6. What is meant by atmospheric pressure?
7. State the approximate value of atmospheric pressure.
8. State what is meant by gauge pressure.
9. State what is meant by absolute pressure.
10. State the relationship between absolute, gauge and atmospheric pressures.
11. State Archimedes' principle.
12 Name four pressure measuring devices.

21.8   Further questions on pressure in fluids

Take the gravitational acceleration as 9.8 m/s², the density of water as 1000 kg/m³ and the density of mercury as 13 600 kg/m³.

1. A force of 280 N is applied to a piston of a hydraulic system of cross-sectional area 0.010 m². Determine the pressure produced by the piston in the hydraulic fluid.   [28 kPa]

2. Find the force on the piston of question 1 to produce a pressure of 450 kPa.   [4.5 kN]

3. If the area of the piston in question 1 is halved and the force applied is 280 N, determine the new pressure in the hydraulic fluid.   [56 kPa]

4. Determine the pressure acting at the base of a dam, when the surface of the water is 35 m above base level.   [343 kPa]

5. An uncorked bottle is full of sea water of density 1030 kg/m³. Calculate, correct to 3 significant figures, the pressures on the side wall of the bottle at depths of (a) 30 mm and (b) 70 mm below the top of the bottle.
[(a) 303 Pa (b) 707 Pa]

6. A U-tube manometer is used to determine the pressure at a depth of 500 mm below the free surface of a fluid. If the pressure at this depth is 6.86 kPa, calculate the density of the liquid used in the manometer.   [1400 kg/m³]

7. The height of a column of mercury in a barometer is 750 mm. Determine the atmospheric pressure, correct to 3 significant figures.   [100 kPa]

8. A U-tube manometer containing mercury gives a height reading of 250 mm of mercury when connected to a gas cylinder. If the barometer reading at the same time is 756 mm of mercury, calculate the absolute pressure of the gas in the cylinder, correct to 3 significant figures.   [134 kPa]

9. A water manometer connected to a condenser shows that the pressure in the condenser is 350 mm below atmospheric pressure. If the barometer is reading 760 mm of mercury, determine the absolute pressure in the condenser, correct to 3 significant figures.   [97.9 kPa]

10. A Bourdon pressure gauge shows a pressure of 1.151 MPa. If the absolute pressure is 1.25 MPa, find the atmospheric pressure in millimetres of mercury.   [743 mm]

11. A body of volume 0.124 m³ is completely immersed in water of density 1000 kg/m³. What is the apparent loss of weight of the body?   [1.215 kN]

12. A body of weight 27.4 N and volume 1240 cm³ is completely immersed in water of specific weight 9.81 kN/m³. What is its apparent weight?   [15.24 N]

13. A body weighs 512.6 N in air and 256.8 N when completely immersed in oil of density 810 kg/m^3. What is the volume of the body?   [32.22 dm^3 = 0.032 22 m^3]

14. A body weighs 243 N in air and 125 N when completely immersed in water. What will it weigh when completely immersed in oil of relative density 0.8?   [148.6 N]

15. A watertight rectangular box, 1.2 m long and 0.75 m wide, floats with its sides and ends vertical in water of density 1000 kg/m^3. If the depth of the box in the water is 280 mm, what is its weight?   [2.575 kN]

16. A body weighs 18 N in air and 13.7 N when completely immersed in water of density 1000 kg/m^3. What is the density and relative density of the body?   [4.186 tonne/m^3, 4.186]

17. A watertight rectangular box is 660 mm long and 320 mm wide. Its weight is 336 N. If it floats with its sides and ends vertical in water of density 1020 kg/m^3, what will be its depth in the water?   [159 mm]

18. A watertight drum has a volume of 0.165 m^3 and a weight of 115 N. It is completely submerged in water of density 1030 kg/m^3, held in position by a single vertical chain attached to the underside of the drum. What is the force in the chain?   [1.552 kN]

22

Heat energy

At the end of this chapter you should be able to:

- distinguish between heat and temperature
- appreciate that temperature is measured on the Celsius or the thermo-dynamic scale
- convert temperatures from Celsius into kelvin and vice versa
- recognize several temperature measuring devices
- define specific heat capacity, c
- recognize typical values of specific heat capacity
- calculate the quantity of heat energy Q using $Q = mc(t_2 - t_1)$
- understand change of state from solid to liquid to gas, and vice versa
- distinguish between sensible and latent heat
- define specific latent heat of fusion
- define specific latent heat of vaporization
- recognize typical values of latent heats of fusion and vaporization
- calculate quantity of heat Q using $Q = mL$
- describe the principle of operation of a simple refrigerator

22.1 Introduction

Heat is a form of energy and is measured in joules.

Temperature is the degree of hotness or coldness of a substance. Heat and temperature are thus **not** the same thing. For example, twice the heat energy is needed to boil a full container of water than half a container – that is, different amounts of heat energy are needed to cause an equal rise in the temperature of different amounts of the same substance.

Temperature is measured either (i) on the **Celsius (°C) scale** (formerly Centigrade), where the temperature at which ice melts, i.e. the freez-ing point of water, is taken as 0°C and the point at which water boils under normal atmospheric pressure is taken as 100°C, or (ii) on the **thermo-dynamic scale**, in which the unit of temperature is the kelvin (K). The kelvin scale uses the same temperature interval as the Celsius scale but its zero takes the 'absolute zero of temperature' which is at about –273°C. Hence,

kelvin temperature = degree Celsius + 273

i.e.

$$K = (°C) + 273$$

Thus, for example, 0°C = 273 K, 25°C = 298 K and 100° = 373 K.

Problem 1.   Convert the following temperatures into the kelvin scale:
(a) 37°C (b) –28°C.

From above, kelvin temperature = degree Celsius + 273.

(a) 37°C corresponds to a kelvin temperature of 37 + 273, i.e. 310 K.
(b) –28°C corresponds to a kelvin temperature of –28 + 273, i.e. 245 K.

Problem 2.   Convert the following temperatures into the Celsius scale:
(a) 365 K (b) 213 K.

From above, K = (°C) + 273.
Hence, degree Celsius = kelvin temperature –273.

(a) 365 K corresponds to 365 – 273, i.e. 92°C.
(b) 213 K corresponds to 213 – 273, i.e. –60°C.

22.2   The measurement of temperature

A **thermometer** is an instrument which measures temperature. Any substance which possesses one or more properties which vary with temperature can be used to measure temperature. These properties include changes in length, area or volume, electrical resistance or in colour. Examples of temperature measuring devices include:

(i) **liquid-in-glass thermometer**, which uses the expansion of a liquid with increase in temperature as its principle of operation;
(ii) **thermocouples**, which use the e.m.f. set up when the junction of two dissimilar metals is heated;
(iii) **resistance thermometer**, which uses the change in electrical resistance caused by temperature change; and
(iv) **pyrometers**, which are devices for measuring very high temperatures, using the principle that all substances emit radiant energy when hot, the rate of emission depending on their temperature.

Each of these temperature measuring devices, together with others, are described fully in Chapter 37.

22.3   Specific heat capacity

The **specific heat capacity** of a substance is the quantity of heat energy required to raise the temperature of 1 kg of the substance by 1°C. The symbol used for specific heat capacity is c and the units are J/(kg °C) or J/(kg K). (Note that these units may also be written as $J\,kg^{-1}\,°C^{-1}$ or $J\,kg^{-1}\,K^{-1}$.)

Some typical values of specific heat capacity for the range of temperature 0°C to 100°C include:

Water, 4190 J/(kg °C),   Ice, 2100 J/(kg °C)
Aluminium, 950 J/(kg °C), Copper, 390 J/(kg °C),
Iron, 500 J/(kg °C),   Lead, 130 J/(kg °C)

Hence to raise the temperature of 1 kg of iron by 1°C requires 500 J of energy, to raise the temperature of 5 kg of iron by 1°C requires (500×5) J of energy, and to raise the temperature of 5 kg of iron by 40°C requires $(500 \times 5 \times 40)$ J of energy, i.e. 100 kJ.

In general, the quantity of heat energy, Q, required to raise a mass m kg of a substance with a specific heat capacity c J/(kg °C) from temperature t_1 °C to t_2 °C is given by:

$$Q = mc(t_2 - t_1) \text{ joules}$$

Problem 3.   Calculate the quantity of heat required to raise the temperature of 5 kg of water from 0°C to 100°C. Assume the specific heat capacity of water is 4200 J/(kg °C)

Quantity of heat energy, Q

$= mc(t_2 - t_1)$

$= 5 \text{ kg} \times 4200 \text{ J/(kg °C)} \times (100 - 0) \text{ °C}$

$= 5 \times 4200 \times 100$

$= \mathbf{2\ 100\ 000\ J}$ or **2100 kJ** or **2.1 MJ**

Problem 4.   A block of cast iron having a mass of 10 kg cools from a temperature of 150°C to 50°C. How much energy is lost by the cast iron? Assume the specific heat capacity of iron is 500 J/(kg °C).

Quantity of heat energy, Q

$= mc(t_2 - t_1)$

= 10 kg × 500 J/(kg °C) × (50 − 150) °C

= 10 × 500 × (−100)

= **−500 000 J or −500 kJ or −0.5 MJ**

(Note that the minus sign indicates that heat is given out or lost.)

Problem 5.   Some lead having a specific heat capacity of 130 J/(kg °C) is heated from 27°C to its melting point at 327°C. If the quantity of heat required is 780 kJ determine the mass of the lead.

Quantity of heat, $Q = mc(t_2 − t_1)$, hence, $780 × 10^3 J = m × 130 J/(kg °C) × (327 − 27) °C$, i.e.

$$780\,000 = m × 130 × 300$$

from which,

$$\text{mass } m = \frac{780\,000}{130 × 300} \text{ kg} = 20 \text{ kg}$$

Problem 6.   273 kJ of heat energy are required to raise the temperature of 10 kg of copper from 15°C to 85°C. Determine the specific heat capacity of copper.

Quantity of heat, $Q = mc(t_2 − t_1)$, hence, $273 × 10^3 J = 10 \text{ kg} × c × (85 − 15)°C$ where $c =$ specific heat capacity, i.e.

$$273\,000 = 10 × c × 70$$

from which

specific heat capacity of copper,

$$c = \frac{273\,000}{10 × 70} = \textbf{390 J/(kg °C)}$$

Problem 7.   5.7 MJ of heat energy are supplied to 30 kg of aluminium which is initially at a temperature of 20°C. If the specific heat capacity of aluminium is 950 J(kg °C), determine its final temperature.

Quantity of heat, $Q = mc(t_2 − t_1)$, hence, $5.7 × 10^6 J = 30 \text{ kg} × 950 J/(kg °C) × (t_2 − 20) °C$, from which

$$(t_2 − 20) = \frac{5.7 × 10^6}{30 × 950} = 200$$

Hence the final temperature, $t_2 = 200 + 20 = $ **220°C**.

Problem 8.   A copper container of mass 500 g contains 1 litre of water at 293 K. Calculate the quantity of heat required to raise the temperature of the water and container to boiling point assuming there are no heat losses. Assume that the specific heat capacity of copper is 390 J/(kg K), the specific heat capacity of water is 4.2 kJ/(kg K) and 1 litre of water has a mass of 1 kg.

Heat is required to raise the temperature of the water, and also to raise the temperature of the copper container.

For the water: $m = 1 \text{ kg}$; $t_1 = 293 \text{ K}$; $t_2 = 373 \text{ K}$ (i.e. boiling point); $c = 4.2 \text{ kJ/(kg K)}$.

Quantity of heat required for the water is given by

$$Q_W = mc(t_2 − t_1)$$
$$= (1 \text{ kg})\left(4.2 \frac{\text{kJ}}{\text{kg K}}\right)(373 − 293) \text{ K}$$
$$= 4.2 × 80 \text{ kJ}$$

i.e. $Q_W = 336 \text{ kJ}$

For the copper container: $m = 500 \text{ g} = 0.5 \text{ kg}$; $t_1 = 293 \text{ K}$; $t_2 = 373 \text{ K}$; $c = 390 \text{ J/(kg K)} = 0.39 \text{ kJ(kg K)}$.

Quantity of heat required for the copper container is given by

$$Q_C = mc(t_2 − t_1)$$
$$= (0.5 \text{ kg})(0.39 \text{ kJ/(kg K)})(80 \text{ K})$$

i.e. $Q_C = 15.6 \text{ kJ}$

Total quantity of heat required, $Q = Q_W + Q_C = 336 + 15.6 = $ **351.6 kJ**.

22.4   Change of state

A material may exist in any one of three states – solid, liquid or gas. If heat is supplied at a constant rate to some ice initially at, say, −30°C, its temperature rises as shown in Fig. 22.1. Initially the temperature increases from −30°C to 0°C as shown by the line AB. It then remains constant at 0°C for the time BC required for the ice to melt into water.

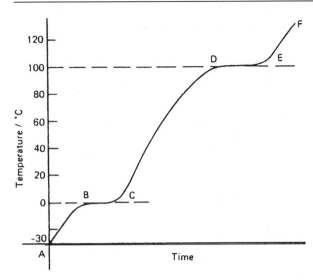

Figure 22.1

When melting commences the energy gained by continual heating is offset by the energy required for the change of state and the temperature remains constant even though heating is continued. When the ice is completely melted to water, continual heating raises the temperature to 100°C, as shown by CD in Fig. 22.1. The water then begins to boil and the temperature again remains constant at 100°C, shown as DE, until all the water has vaporized.

Continual heating raises the temperature of the steam as shown by EF in the region where the steam is termed superheated.

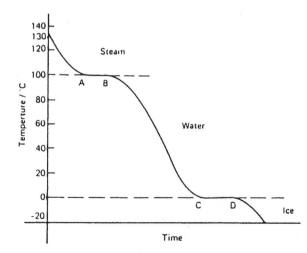

Figure 22.2

Changes of state from solid to liquid or liquid to gas occur without change of temperature and such changes are reversible processes. When heat energy flows to or from a substance and causes a change of temperature, such as between A and B, between C and D and between E and F in Fig. 22.1, it is called **sensible heat** (since it can be 'sensed' by a thermometer).

Heat energy which flows to or from a substance while the temperature remains constant, such as between B and C and between D and E in Fig. 22.1, is called **latent heat** (latent means concealed or hidden).

Problem 9. Steam initially at a temperature of 130°C is cooled to a temperature of 20°C below the freezing point of water, the loss of heat energy being at a constant rate. Make a sketch, and briefly explain, the expected temperature/time graph representing this change.

A temperature/time graph representing the change is shown in Fig. 22.2. Initially steam cools until it reaches the boiling point of water at 100°C. Temperature then remains constant, i.e. between A and B, even though it is still giving off heat (i.e. latent heat). When all the steam at 100°C has changed to water at 100°C it starts to cool again until it reaches the freezing point of water at 0°C. From C to D the temperature again remains constant until all the water is converted to ice. The temperature of the ice then decreases as shown.

22.5 Latent heats of fusion and vaporization

The **specific latent heat of fusion** is the heat required to change 1 kg of a substance from the solid state to the liquid state (or vice versa) at constant temperature.

The **specific latent heat of vaporization** is the heat required to change 1 kg of a substance from a liquid to a gaseous state (or vice versa) at constant temperature. The units of the specific latent heats of fusion and vaporization are J/kg, or more often kJ/kg, and some typical values are shown in Table 22.1.

The quantity of heat Q supplied or given out during a change of state is given by:

$$\boxed{Q = mL}$$

where m is the mass in kilograms and L is the specific latent heat. Thus, for example, the heat required to convert 10 kg of ice at 0°C to water at 0°C is given by 10 kg × 335 kJ/kg, i.e. 3350 kJ or 3.35 MJ.

Table 22.1

	Latent heat of fusion (kJ/kg)	Melting point (°C)
Mercury	11.8	−39
Lead	22	327
Silver	100	957
Ice	335	0
Aluminium	387	660

	Latent heat of vaporization (kJ/kg)	Boiling point (°C)
Oxygen	214	−183
Mercury	286	357
Ethyl alcohol	857	79
Water	2257	100

Besides changing temperature, the effects of supplying heat to a material can involve changes in dimensions, as well as in colour, state and electrical resistance. Most substances expand when heated and contract when cooled, and there are many practical applications and design implications of thermal movement (see Chapter 23).

Problem 10. How much heat is needed to melt completely 12 kg of ice at 0°C? Assume the latent heat of fusion of ice is 335 kJ/kg.

Quantity of heat required,

$Q = mL$,

= 12 kg × 335 kJ/kg

= **4020 kJ** or **4.02 MJ**

Problem 11. Calculate the heat required to convert 5 kg of water at 100°C to superheated steam at 100°C. Assume the latent heat of vaporization of water is 2260 kJ/kg.

Quantity of heat required, Q

= mL

= 5 kg × 2260 kJ/kg

= **11 300 kJ** or **11.3 MJ**

Problem 12. Determine the heat energy needed to convert 5 kg of ice initially at −20°C completely to water at 0°C. Assume the specific heat capacity of ice is 2100 J/(kg °C) and the specific latent heat of fusion of ice is 335 kJ/kg.

Quantity of heat energy needed, Q = sensible heat + latent heat. The quantity of heat needed to raise the temperature of ice from −20°C to 0°C, i.e. sensible heat, is given by

$$Q_1 = mc(t_2 - t_1)$$
$$= 5 \text{ kg} \times 2100 \text{ J/(kg °C)} \times (0 - -20) \text{ °C}$$
$$= (5 \times 2100 \times 20) \text{ J} = 210 \text{ kJ}$$

The quantity of heat needed to melt 5 kg of ice at 0°C, i.e. the latent heat, $Q_2 = mL = 5$ kg × 335 kJ/kg = 1675 kJ.
Total heat energy needed, $Q = Q_1 + Q_2 = 210 + 1675 =$ **1885 kJ.**

Problem 13. Calculate the heat energy required to convert completely 10 kg of water at 50°C into steam at 100°C, given that the specific heat capacity of water is 4200 J/(kg°C) and the specific latent heat of vaporization of water is 2260 kJ/kg.

Quantity of heat required = sensible heat + latent heat.

Sensible heat, $Q_1 = mc(t_2 - t_1)$
$$= 10 \text{ kg} \times 4200 \text{ J/(kg °C)}$$
$$\times (100 - 50) \text{ °C}$$
$$= 2100 \text{ kJ}$$

Latent heat, $Q_2 = mL = 10$ kg × 2260 kJ/kg = 22 6000 kJ.

Total heat energy required, $Q = Q_1 + Q_2$

= (2100 + 22 600) kJ

= **24 700 kJ** or **24.70 MJ**

Problem 14.   Determine the amount of heat energy needed to change 400 g of ice, initially at –20°C, into steam at 120°C. Assume the following: latent heat of fusion of ice = 335 kJ/kg; latent heat of vaporization of water = 2260 kJ/kg; specific heat capacity of ice = 2.14 kJ/(kg °C); specific heat capacity of water = 4.2 kJ(kg °C); specific heat capacity of steam = 2.01 kJ/(kg°C).

The energy needed is determined in five stages:

(i)   Heat energy needed to change the temperature of ice from –20°C to 0°C is given by:

$$Q_1 = mc(t_2 - t_1) = 0.4 \text{ kg} \times 2.14 \text{ kJ/(kg °C)} \times (0 - -20) \text{ °C}$$

$$= 17.12 \text{ kJ}$$

(ii)   Latent heat needed to change ice at 0°C into water at 0°C is given by:

$$Q_2 = mL_f = 0.4 \text{ kg} \times 335 \text{ kJ/kg} = 134 \text{ kJ}$$

(iii)   Heat energy needed to change the temperature of water from 0°C (i.e. melting point) to 100°C (i.e. boiling point) is given by:

$$Q_3 = mc(t_2 - t_1)$$

$$= 0.4 \text{ kg} \times 4.2 \text{ kJ/(kg °C)} \times 100°C$$

$$= 168 \text{ kJ}$$

(iv)   Latent heat needed to change water at 100°C into steam at 100°C is given by:

$$Q_4 = mL_V = 0.4 \text{ kg} \times 2260 \text{ kJ/kg} = 904 \text{ kJ}$$

(v)   Heat energy needed to change steam at 100°C into steam at 120°C is given by:

$$Q_5 = mc(t_2 - t_2)$$

$$= 0.4 \text{ kg} \times 2.01 \text{ kJ/(kg °C)} \times 20°C$$

$$= 16.08 \text{ kJ}$$

Total heat energy needed, Q

$$= Q_1 + Q_2 + Q_3 + Q_4 + Q_5$$

$$= 17.12 + 134 + 168 + 904 + 16.08$$

$$= \textbf{1239.2 kJ}$$

22.6   A simple refrigerator

The boiling point of most liquids may be lowered if the pressure is lowered. In a simple refrigera-
tor a working fluid, such as ammonia or freon, has the pressure acting on it reduced. The resulting lowering of the boiling point causes the liquid to vaporize. In vaporizing, the liquid takes in the necessary latent heat from its surroundings, i.e. the freezer, which thus becomes cooled. The vapour is immediately removed by a pump to a condenser which is outside of the cabinet, where it is compressed and changed back into a liquid, giving out latent heat. The cycle is repeated when the liquid is pumped back to the freezer to be vaporized.

22.7   Multi-choice questions on heat energy

(Answers on page 356.)

1. Heat energy is measured in:
   (a) kelvin   (b) watts   (c) kilograms
   (d) joules
2. A change of temperature of 20°C is equivalent to a change in thermodynamic temperature of (a) 293 K (b) 20 K
3. A temperature of 20°C is equivalent to (a) 293 K (b) 20 K
4. The unit of specific heat capacity is:
   (a) joules per kilogram
   (b) joules
   (c) joules per kilogram kelvin
   (d) cubic metres
5. The quantity of heat required to raise the temperature of 500 g of iron by 2°C, given that the specific heat capacity is 500 J/(kg °C), is:
   (a) 500 kJ   (b) 0.5 kJ   (c) 2 J   (d) 250 kJ
6. The heat energy required to change 1 kg of a substance from a liquid to a gaseous state at the same temperature is called:
   (a) specific heat capacity
   (b) specific latent heat of vaporization
   (c) sensible heat
   (d) specific latent heat of fusion

22.8   Short answer questions on heat energy

1. Differentiate between temperature and heat.
2. Name two scales on which temperature is measured.
3. Name any four temperature measuring devices.

4. Define specific heat capacity and name its unit.

5. Differentiate between sensible and latent heat.

6. The quantity of heat, Q, required to raise a mass m kg from temperature t_1 °C to t_2 °C, the specific heat capacity being c, is given by $Q =$ ....

7. What is meant by the specific latent heat of fusion?

8. Define the specific latent heat of vaporization.

9. Explain briefly the principle of operation of a simple refrigerator.

22.9   Further questions on heat energy

1. Convert the following temperatures into the kelvin scale:
   (a) 51°C   (b) –78°C   (c) 183°C.
   [(a) 324 K (b) 195 K (c) 456 K]

2. Convert the following temperatures into the Celsius scale:
   (a) 307 K   (b) 237 K   (c) 415 K.
   [(a) 34°C (b) –36°C (c) 142°C]

3. (a) What is the difference between heat and temperature?
   (b) State three temperature measuring devices and state the principle of operation of each.

4. Determine the quantity of heat energy (in megajoules) required to raise the temperature of 10 kg of water from 0°C to 50°C. Assume the specific heat capacity of water is 4200 J/(kg°C).   [2.1 MJ]

5. Some copper, having a mass of 20 kg, cools from a temperature of 120°C to 70°C. If the specific heat capacity of copper is 390 J/(kg °C), how much heat energy is lost by the copper?   [390 kJ]

6. A block of aluminium having a specific heat capacity of 950 J/(kg °C) is heated from 60°C to its melting point at 660°C. If the quantity of heat required is 2.85 MJ, determine the mass of the aluminium block.   [5 kg]

7. 20.8 kJ of heat energy is required to raise the temperature of 2 kg of lead from 16°C to 96°C. Determine the specific heat capacity of lead.   [130 J/(kg °C)]

8. 250 kJ of heat energy is supplied to 10 kg of iron which is initially at a temperature of 15°C. If the specific heat capacity of iron is 500 J/(kg °C), determine its final temperature.   [65°C]

9. Some ice, initially at –40°C, has heat supplied to it at a constant rate until it becomes superheated steam at 150°C. Sketch a typical temperature/time graph expected and use it to explain the difference between sensible and latent heat.

10. How much heat is needed to melt completely 25 kg of ice at 0°C. Assume the specific latent heat of fusion of ice is 335 kJ/kg.   [8.375 MJ]

11. Determine the heat energy required to change 8 kg of water at 100°C to superheated steam at 100°C. Assume the specific latent heat of vaporization of water is 2260 kJ/kg.   [18.08 MJ]

12. Calculate the heat energy required to convert 10 kg of ice initially at –30°C completely into water at 0°C. Assume the specific heat capacity of ice is 2.1 kJ/(kg°C) and the specific latent heat of fusion of ice is 335 kJ/kg.   [3.98 MJ]

13. Determine the heat energy needed to convert completely 5 kg of water at 60°C to steam at 100°C, given that the specific heat capacity of water is 4.2 kJ/(kg °C) and the specific latent heat of vaporization of water is 2260 kJ/kg.   [12.14 MJ]

23

Thermal expansion

At the end of this chapter you should be able to:

- appreciate that expansion and contraction occurs with change of temperature
- describe practical applications where expansion and contraction must be allowed for
- understand the expansion and contraction of water
- define the coefficient of linear expansion α
- recognize typical values for the coefficient of linear expansion
- calculate the new length l_2, after expansion or contraction, using $l_2 = l_1[1 + \alpha(t_2 - t_1)]$
- define the coefficient of superficial expansion β
- calculate the new surface area A_2, after expansion or contraction, using $A_2 = A_1[1 + \beta(t_2 - t_1)]$
- appreciate that $\beta = 2\alpha$
- define the coefficient of cubic expansion
- recognize typical values for the coefficient of cubic expansion
- appreciate that $\gamma \approx 3\alpha$
- calculate the new volume V_2, after expansion or contraction, using $V_2 = V_1[1 + \gamma(t_2 - t_1)]$

23.1 Introduction

When heat is applied to most materials, **expansion** occurs in all directions. Conversely, if heat energy is removed from a material (i.e. the material is cooled) **contraction** occurs in all directions. The effects of expansion and contraction each depend on the **change of temperature** of the material.

23.2 Practical applications of thermal expansion

Some practical applications where expansion and contraction of solid materials must be allowed for include:

(i) Overhead electrical transmission lines are hung so that they are slack in summer, otherwise their contraction in winter may snap the conductors or bring down pylons.

(ii) Gaps need to be left in lengths of railway lines to prevent buckling in hot weather (except where these are continuously welded).

(iii) Ends of large bridges are often supported on rollers to allow them to expand and contract freely.

(iv) Fitting a metal collar to a shaft or a steel tyre to a wheel is often achieved by first heating them so that they expand, fitting them in position, and then cooling them so that the contraction holds them firmly in place. This is known as a 'shrink-fit'. By a similar

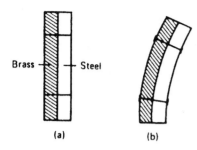

Figure 23.1

method hot rivets are used for joining metal sheets.

(v) The amount of expansion varies with different materials. Fig. 23.1(a) shows a bimetallic strip at room temperature (i.e. two different strips of metal riveted together). When heated, brass expands more than steel, and since the two metals are riveted together the bimetallic strip is forced into an arc as shown in Fig. 23.1(b). Such a movement can be arranged to make or break an electric circuit and bimetallic strips are used, in particular, in thermostats (which are temperature-operated switches) used to control central heating systems, cookers, refrigerators, toasters, irons, hot-water and alarm systems.

(vi) Motor engines use the rapid expansion of heated gases to force a piston to move.

(vii) Designers must predict, and allow for, the expansion of steel pipes in a steam-raising plant so as to avoid damage and consequent danger to health.

23.3 Expansion and contraction of water

Water is a liquid which at low temperature displays an unusual effect. If cooled, contraction occurs until, at about 4°C, the volume is at a minimum. As the temperature is further decreased from 4°C to 0°C expansion occurs, i.e. the volume increases. When ice is formed, considerable expansion occurs and it is this expansion which often causes frozen water pipes to burst.

A practical application of the expansion of a liquid is with thermometers, where the expansion of a liquid, such as mercury or alcohol, is used to measure temperature.

23.4 Coefficient of linear expansion

The amount by which unit length of a material expands when the temperature is raised one degree is called the **coefficient of linear expansion** of the material and is represented by α (Greek alpha).

The units of the coefficient of linear expansion are m/(mK), although it is usually quoted as just /K or K^{-1}. For example, copper has a coefficient of linear expansion value of $17 \times 10^{-6} K^{-1}$, which means that a 1 m long bar of copper expands by 0.000 017 m if its temperature is increased by 1 K (or 1°C). If a 6 m long bar of copper is subjected to a temperature rise of 25 K then the bar will expand by $(6 \times 0.000\,017 \times 25)$ m, i.e. 0.002 55 m or 2.55 mm. (Since the kelvin scale uses the same temperature interval as the Celsius scale, a **change** of temperature of, say, 50°C, is the same as a change of temperature of 50 K.)

If a material, initially of length l_1 and at a temperature of t_1 and having a coefficient of linear expansion α, has its temperature increased to t_2, then the new length l_2 of the material is given by:

New length = original length + expansion

i.e. $l_2 = l_1 + l_1\alpha(t_2 - t_1)$

i.e. $l_2 = l_1[1 + \alpha(t_2 - t_1)]$ (23.1)

Some typical values for the coefficient of linear expansion include:

Aluminium	$23 \times 10^{-6} K^{-1}$	Brass	$18 \times 10^{-6} K^{-1}$
Concrete	$12 \times 10^{-6} K^{-1}$	Copper	$17 \times 10^{-6} K^{-1}$
Gold	$14 \times 10^{-6} K^{-1}$	Invar (nickel-	
Iron	$11\text{–}12 \times 10^{-6} K^{-1}$	steel alloy)	$0.9 \times 10^{-6} K^{-1}$
Steel	$15\text{–}16 \times 10^{-6} K^{-1}$	Nylon	$100 \times 10^{-6} K^{-1}$
Zinc	$31 \times 10^{-6} K^{-1}$	Tungsten	$4.5 \times 10^{-6} K^{-1}$

> Problem 1. The length of an iron steam pipe is 20.0 m at a temperature of 18°C. Determine the length of the pipe under working conditions when the temperature is 300°C. Assume the coefficient of linear expansion of iron is $12 \times 10^{-6} K^{-1}$.

Length $l_1 = 20.0$ m; temperature $t_1 = 18$°C; $t_2 = 300$°C; $\alpha = 12 \times 10^{-6} K^{-1}$. Length of pipe at 300°C is given by

$$l_2 = l_1[1 + \alpha(t_2 - t_1)]$$
$$= 20.0[1 + (12 \times 10^{-6})(300 - 18)]$$

$$= 20.0[1 + 0.003\ 384] = 20.0[1.003\ 384]$$

$$= \textbf{20.067 68 m}$$

i.e. an increase in length of 0.067 68 m, i.e. 67.68 mm.

In practice, allowances are made for such expansions. U-shaped expansion joints are connected into pipelines carrying hot fluids to allow some 'give' to take up the expansion.

Problem 2.   An electrical overhead transmission line has a length of 80.0 m between its supports at 15°C. Its length increases by 92 mm at 65°C. Determine the coefficient of linear expansion of the material of the line.

Length $l_1 = 80.0$ m; $l_2 = 80.0 + 92$ mm $= 80.092$ m; temperature $t_1 = 15$°C; temperature $t_2 = 65$°C. Length $l_2 = l_1[1 + \alpha(t_2 - t_1)]$, i.e.

$$80.092 = 80.0[1 + \alpha(65 - 15)]$$
$$80.092 = 80.0 + (80.0)(\alpha)(50)$$

i.e.

$$80.092 - 80.0 = (80.0)(\alpha)(50)$$

Hence the coefficient of linear expansion, α

$$= \frac{0.092}{(80.0)(50)}$$

$$= 0.000\ 023$$

i.e. $\alpha = \textbf{23} \times \textbf{10}^{-6}\ \textbf{K}^{-1}$ (which is aluminium – see above).

Problem 3.   A measuring tape made of copper measures 5.0 m at a temperature of 288 K. Calculate the percentage error in measurement when the temperature has increased to 313 K. Take the coefficient of linear expansion of copper as 17×10^{-6} K^{-1}.

Length $l_1 = 5.0$ m; temperature $t_1 = 288$ K; $t_2 = 313$ K; $\alpha = 17 \times 10^{-6}$ K^{-1}. Length at 313 K is given by

$$l_2 = l_1[1 + \alpha(t_2 - t_1)]$$

$$= 5.0[1 + (17 \times 10^{-6})(313 - 288)$$

$$= 5.0[1 + (17 \times 10^{-6})(25)]$$

$$= 5.0[1 + 0.000\ 425]$$

$$= 5.0[1.000\ 425] = 5.002\ 125\ \text{m}$$

i.e. the length of the tape has increased by 0.002 125 m.

Percentage error in measurement at 313 K

$$= \frac{\text{increase in length}}{\text{original length}} \times 100\%$$

$$= \frac{0.002\ 125}{5.0} \times 100$$

$$= \textbf{0.0425\%}$$

Problem 4.   The copper tubes in a boiler are 4.20 m long at a temperature of 20°C. Determine the length of the tubes (a) when surrounded only by feed water at 10°C, (b) when the boiler is operating and the mean temperature of the tubes is 320°C. Assume the coefficient of linear expansion of copper to be 17×10^{-6} K^{-1}.

(a)   Initial length, $l_1 = 4.20$ m; initial temperature, $t_1 = 20$°C; final temperature, $t_2 = 10$°C; $\alpha = 17 \times 10^{-6}$ K^{-1}. Final length at 10°C is given by

$$l_2 = l_1[1 + \alpha(t_2 - t_1)]$$

$$= 4.20[1 + (17 \times 10^{-6})(10 - 20)]$$

$$= 4.20[1 - 0.000\ 17] = \textbf{4.1993 m}$$

i.e. the tube contracts by 0.7 mm when the temperature decreases from 20°C to 10°C.

(b) $l_1 = 4.20$ m; $t_1 = 20$°C; $t_2 = 320$°C; $\alpha = 17 \times 10^{-6}$ K^{-1}. Final length at 320°C is given by

$$l_2 = l_1[1 + \alpha(t_2 - t_1)]$$

$$= 4.20[1 + (17 \times 10^{-6})(320 - 20)]$$

$$= 4.20[1 + 0.0051] = \textbf{4.2214 m}$$

i.e. the tube extends by 21.4 mm when the temperature rises from 20°C to 320°C.

23.5   Coefficient of superficial expansion

The amount by which unit area of a material increases when the temperature is raised by one degree is called the **coefficient of superficial** (i.e. **area) expansion** and is represented by β (Greek beta).

If a material having an initial surface area A_1 at temperature t_1 and having a coefficient of superficial expansion β, has its temperature increased to t_2, then the new surface area A_2 of the material is given by:

New surface area = original surface area
+ increase in area

i.e.

$$A_2 = A_1 + A_1\beta(t_2 - t_1)$$

i.e.

$$A_2 = A_1[1 + \beta(t_2 - t_1)] \qquad (23.2)$$

It may be shown (in Problem 5 below) that the coefficient of superficial expansion is twice the coefficient of linear expansion, i.e. $\beta = 2\alpha$, to a very close approximation.

Problem 5.   Show that for a rectangular area of material having dimensions l by b the coefficient of superficial expansion $\beta \approx 2\alpha$, where α is the coefficient of linear expansion.

Initial area, $A_1 = lb$. For a temperature rise of 1 K, side l will expand to $(l + l\alpha)$ and side b will expand to $(b + b\alpha)$. Hence the new area of the rectangle A_2 is given by:

$$A_2 = (l + l\alpha)(b + b\alpha) = l(1 + \alpha)b(1 + \alpha)$$

$$= lb(1 + \alpha)^2$$

$$= lb(1 + 2\alpha + \alpha^2) \approx lb(1 + 2\alpha)$$

since α^2 is very small (see typical values in section 23.4). Hence $A_2 \approx A_1(1 + 2\alpha)$. For a temperature rise of $(t_2 - t_1)$ K

$$A_2 \approx A_1[1 + 2\alpha(t_2 - t_1)]$$

Thus from equation (23.2), $\beta \approx 2\alpha$.

23.6   Coefficient of cubic expansion

The amount by which unit volume of a material increases for a one degree rise of temperature is called the **coefficient of cubic** (or **volumetric**) **expansion** and is represented by γ (Greek gamma).

If a material having an initial volume V_1 at temperature t_1 and having a coefficient of cubic expansion γ, has its temperature raised to t_2, then the new volume V_2 of the material is given by:

New volume = initial volume
+ increase in volume

i.e.

$$V_2 = V_1 + V_1\gamma(t_2 - t_1)$$

i.e.

$$V_2 = V_1[1 + \gamma(t_2 - t_1)] \qquad (23.3)$$

It may be shown (in Problem 6 below) that the coefficient of cubic expansion is three times the coefficient of linear expansion, i.e. $\gamma = 3\alpha$, to a very close approximation. A liquid has no definite shape and only its cubic or volumetric expansion need be considered. Thus with expansions in liquids, equation (23.3) is used.

Problem 6.   Show that for a rectangular block of material having dimensions l, b and h, the coefficient of cubic expansion $\gamma \approx 3\alpha$, where α is the coefficient of linear expansion.

Initial volume, $V_1 = lbh$. For a temperature rise of 1 K, side l expands to $(l + l\alpha)$, side b expands to $(b + b\alpha)$ and side h expands to $(h + h\alpha)$. Hence the new volume of the block V_2 is given by:

$$V_2 = (l + l\alpha)(b + b\alpha)(h + h\alpha)$$

$$= l(1 + \alpha)b(1 + \alpha)h(1 + \alpha)$$

$$= lbh(1 + \alpha)^3 = lbh(1 + 3\alpha + 3\alpha^2 + \alpha^3)$$

$$\approx lbh(1 + 3\alpha)$$

since terms in α^2 and α^3 are very small. Hence $V_2 \approx V_1(1 + 3\alpha)$. For a temperature rise of $(t_2 - t_1)$ K,

$$V_2 \approx V_1[1 + 3\alpha(t_2 - t_1)]$$

Thus from equation (23.3), $\gamma \approx 3\alpha$.

Some typical values for the coefficient of cubic expansion measured at 20°C (i.e. 293 K) include:

Ethyl alcohol	$1.1 \times 10^{-3}\,K^{-1}$	Mercury	$1.82 \times 10^{-4}\,K^{-1}$
Paraffin oil	$9 \times 10^{-2}\,K^{-1}$	Water	$2.1 \times 10^{-4}\,K^{-1}$

The coefficient of cubic expansion γ is only constant over a limited range of temperature.

Problem 7.   A brass sphere has a diameter of 50 mm at a temperature of 289 K. If the temperature of the sphere is raised to 789 K, determine the increase in (a) the diameter, (b) the surface area, (c) the volume of the sphere. Assume the coefficient of linear expansion for brass is $18 \times 10^{-6}\,K^{-1}$.

(a) Initial diameter, $l_1 = 50$ mm; initial temperature, $t_1 = 289$ K; final temperature, $t_2 = 789$ K; $a = 18 \times 10^{-6}$ K^{-1}. New diameter at 789 K is given by

$$l_2 = l_1[1 + \alpha(t_2 - t_1)]$$

from equation (23.1), i.e.

$$l_2 = 50[1 + (18 \times 10^{-6})(789 - 289)]$$
$$= 50[1 + 0.009] = 50.45 \text{ mm}$$

Hence the increase in the diameter is **0.45 mm**.

(b) Initial surface area of sphere is given by

$$A_1 = 4\pi r^2 = 4\pi \left(\frac{50}{2}\right)^2$$

$$= 2500\pi \text{ mm}^2$$

New surface area at 789 K is given by

$$A_2 = A_1[1 + \beta(t_2 - t_1)]$$

from equation (23.2), i.e.

$$A_2 = A_1[1 + 2\alpha(t_2 - t_1)]$$

since $\beta = 2\alpha$, to a very close approximation. Thus

$$A_2 = 2500\pi[1 + 2(18 \times 10^{-6})(500)]$$
$$= 2500\pi[1 + 0.018]$$
$$= 2500\pi + 2500\pi(0.018)$$

Hence increase in surface area $= 2500\pi(0.018) = $ **141.4 mm^2**.

(c) Initial volume of sphere is given by

$$V_2 = \frac{4}{3}\pi r^3 = \frac{4}{3}\pi \left(\frac{50}{2}\right)^3 \text{ mm}^3$$

New volume at 789 K is given by

$$V_2 = V_1[1 + \gamma(t_2 - t_1)]$$

from equation (23.3), i.e.

$$V_2 = V_1[1 + 3\alpha(t_2 - t_1)]$$

since $\gamma = 3\alpha$, to a very close approximation. Thus

$$V_2 = \frac{4}{3}\pi(25)^3[1 + 3(18 \times 10^{-6})(500)]$$

$$= \frac{4}{3}\pi(25)^3[1 + 0.027]$$

$$= \frac{4}{3}\pi(25)^3 + \frac{4}{3}\pi(25)^3(0.027)$$

Hence the increase in volume $= \frac{4}{3}\pi(25)^3(0.027) = $ **1767 mm^3**.

Problem 8. Mercury contained in a thermometer has a volume of 476 mm^3 at 15°C. Determine the temperature at which the volume of mercury is 478 mm^3, assuming the coefficient of cubic expansion for mercury to be 1.8×10^{-4} K^{-1}.

Initial volume, $V_1 = 476$ mm^3; final volume $V_2 = 478$ mm^3; initial temperature $t_1 = 15$°C; $\gamma = 1.8 \times 10^{-4}$ K^{-1}.
Final volume, $V_2 = V_1[1 + \gamma(t_2 - t_1)]$, from equation (23.3), i.e. $V_2 = V_1 + V_1\gamma(t_2 - t_1)$, from which

$$(t_2 - t_1) = \frac{V_2 - V_1}{V_1\gamma} = \frac{478 - 476}{(476)(1.8 \times 10^{-4})}$$

$$= 23.34°C$$

Hence $t_2 = 23.34 + 15 = 38.34$°C. Hence the temperature at which the volume of mercury is 478 mm^3 is **38.34°C**.

Problem 9. A rectangular glass block has a length of 100 mm, width 50 mm and depth 20 mm at 293 K. When heated to 353 K its length increases by 0.054 mm. What is the coefficient of linear expansion of the glass?

Find also (a) the increase in surface area, (b) the change in volume resulting from the change of length.

Final length, $l_2 = l_1[1 + \alpha(t_2 - t_1)]$, from equation (23.1), hence increase in length is given by $l_2 - l_1 = l_1\alpha(t_2 - t_1)$.
Hence $0.054 = (100)(\alpha)(353 - 293)$ from which the coefficient of linear expansion is given by

$$\alpha = \frac{0.054}{(100)(60)} = 9 \times 10^{-6} \text{ K}^{-1}$$

(a) Initial surface area of glass,

$$A_1 = (2 \times 100 \times 50) + (2 \times 50 \times 20)$$
$$+ (2 \times 100 \times 20)$$
$$= 10\,000 + 2000 + 4000 = 16\,000 \text{ mm}^2.$$

Final surface area of glass,

$$A_2 = A_1[1 + \beta(t_2 - t_1)] = A_1[1 + 2\alpha(t_2 - t_1)],$$

since $\beta = 2\alpha$ to a very close approximation. Hence

increase in surface area $= A_1(2\alpha)(t_2 - t_1)$

$= (16\,000)(2 \times 9 \times 10^{-6})(60)$

$= \textbf{17.28 mm}^2$

(b) Initial volume of glass, $V_1 = 100 \times 50 \times 20$
$= 100\,000$ mm^3.
Final volume of glass, $V_2 = V_1[1 + \gamma(t_2 - t_1)]$
$= V_1[1 + 3\alpha(t_2 - t_1)]$, since $\gamma = 3\alpha$ to a very close approximation. Hence

increase in volume of glass

$= V_1(3\alpha)(t_2 - t_1)$

$= (100\,000)(3 \times 9 \times 10^{-6})(60)$

$= \textbf{162 mm}^3$

23.7 Multi-choice questions on thermal expansion

(Answers on page 356.)

1. When the temperature of a rod of copper is increased, its length:
   (a) stays the same (b) increases
   (c) decreases
2. The amount by which unit length of a material increases when the temperature is raised one degree is called the coefficient of:
   (a) cubic expansion (b) superficial expansion
   (c) linear expansion.
3. The symbol used for volumetric expansion is:
   (a) γ (b) β (c) l (d) α
4. A material of length l_1, at temperature θ_1 K is subjected to a temperature rise of θ K. The coefficient of linear expansion of the material is α K^{-1}. The material expands by:
   (a) $l_2(1 + \alpha\theta)$ (b) $l_1\alpha(\theta - \theta_1)$
   (c) $l_1[1 + \alpha(\theta - \theta_1)]$ (d) $l_1\alpha\theta$
5. Some iron has a coefficient of linear expansion of 12×10^{-6} K^{-1}. A 100 mm length of iron piping is heated through 20 K. The pipe extends by:
   (a) 0.24 mm (b) 0.024 mm (c) 2.4 mm
   (d) 0.0024 mm
6. If the coefficient of linear expansion is A, the coefficient of superficial expansion is B and the coefficient of cubic expansion is C, which of the following is false?
   (a) $C = 3A$ (b) $A = B/2$ (c) $B = \frac{3}{2}C$
   (d) $A = C/3$
7. The length of a 100 mm bar of metal increases by 0.3 mm when subjected to a temperature rise of 100 K. The coefficient of linear expansion of the metal is:

(a) 3×10^{-3} K^{-1} (b) 3×10^{-4} K^{-1}
(c) 3×10^{-5} K^{-1} (d) 3×10^{-6} K^{-1}

8. A liquid has a volume V_1 at temperature θ_1. The temperature is increased to θ_2. If γ is the coefficient of cubic expansion, the increase in volume is given by:
   (a) $V_1\gamma(\theta_2 - \theta_1)$ (b) $V_1\gamma\theta_2$ (c) $V_1 + V_1\gamma\theta_2$
   (d) $V_1[1 + \gamma(\theta_2 - \theta_1)]$
9. Which of the following statements is false?
   (a) Gaps need to be left in lengths of railway lines to prevent buckling in hot weather.
   (b) Bimetallic strips are used in thermostats, a thermostat being a temperature-operated switch.
   (c) As the temperature of water is decreased from 4°C to 0°C contraction occurs.
   (d) A change of temperature of 15°C is equivalent to a change of temperature of 15 K.
10. The volume of a rectangular block of iron at a temperature t_1 is V_1. The temperature is raised to t_2 and the volume increases to V_2. If the coefficient of linear expansion of iron is α, then volume V_1 is given by:
    (a) $V_2[1 + \alpha(t_2 - t_1)]$
    (b) $V_2/1 + 3\alpha(t_2 - t_1)$
    (c) $3V_2\alpha(t_2 - t_1)$
    (d) $1 + \alpha(t_2 - t_1)/V_2$

23.8 Short answer questions on thermal expansion

1. When heat is applied to most solids and liquids .... occurs.
2. When solids and liquids are cooled they usually ....
3. State three practical applications where the expansion of metals must be allowed for.
4. State a practical disadvantage where the expansion of metals occurs.
5. State one practical advantage of the expansion of liquids.
6. What is meant by the 'coefficient of expansion'.
7. Name the symbol and the unit used for the coefficient of linear expansion.
8. Define the 'coefficient of superficial expansion' and state its symbol.
9. Describe how water displays an unexpected effect between 0°C and 4°C.

10. Define the 'coefficient of cubic expansion' and state its symbol.

23.9   Further questions on thermal expansion

1. A length of lead piping is 50.0 m long at a temperature of 16°C. When hot water flows through it the temperature of the pipe rises to 80°C. Determine the length of the hot pipe if the coefficient of linear expansion of lead is 29×10^{-6} K^{-1}.   [50.0928 m]

2. A rod of metal is measured at 285 K and is 3.521 m long. At 373 K the rod is 3.523 m long. Determine the value of the coefficient of linear expansion for the metal.   [6.45×10^{-6} K^{-1}]

3. A copper overhead transmission line has a length of 40.0 m between its supports at 20°C. Determine the increase in length at 50°C if the coefficient of linear expansion of copper is 17×10^{-6} K^{-1}.   [20.4 mm]

4. A brass measuring tape measures 2.10 m at a temperature of 15°C. Determine (a) the increase in length when the temperature has increased to 40°C, (b) the percentage error in measurement at 40°C. Assume the coefficient of linear expansion of brass to be 18×10^{-6} K^{-1}.   [(a) 0.945 mm (b) 0.045%]

5. A silver plate has an area of 800 mm^2 at 15°C. Determine the increase in the area of the plate when the temperature is raised to 100°C. Assume the coefficient of linear expansion of silver to be 19×10^{-6} K^{-1}.   [2.584 mm^2]

6. A pendulum of a 'grandfather' clock is 2.0 m long and made of steel. Determine the change in length of the pendulum if the temperature rises by 15 K. Assume the coefficient of linear expansion of steel to be 15×10^{-6} K^{-1}.   [0.45 mm]

7. A brass shaft is 15.02 mm in diameter and has to be inserted in a hole of diameter 15.0 mm. Determine by how much the shaft must be cooled to make this possible, without using force. Take the coefficient of linear expansion of brass as 18×10^{-6} K^{-1}.   [74 K]

8. A temperature control system is operated by the expansion of a zinc rod which is 200 mm long at 15°C. If the system is set so that the source of heat supply is cut off when the rod has expanded by 0.20 mm, determine the temperature to which the system is limited. Assume the coefficient of linear expansion of zinc to be 31×10^{-6} K^{-1}.   [47.26°C]

9. A brass collar of bore diameter 49.8 mm is to be shrunk on to a shaft of 50.0 mm diameter, both of these dimensions being measured at 20°C. To what temperature must the collar be heated so that it just slides on to the shaft. Assume the coefficient of linear expansion to be 18×10^{-6} K^{-1}.   [243.1°C]

10. A length of steel railway line is 30.0 m long when the temperature is 288 K. Determine the increase in length of the line when the temperature is raised to 303 K. Assume the coefficient of linear expansion of steel to be 15×10^{-6} K^{-1}.   [6.75 mm]

11. A steel girder is 8.0 m long at a temperature of 20°C. Determine the length of the girder at a temperature of 50°C. Take the coefficient of linear expansion for steel to be 15×10^{-6} K^{-1}.   [8.0036 m]

12. An aluminium ball is heated to a temperature of 733 K when it has a diameter of 41.25 mm. It is then placed over a hole of diameter 41.0 mm. At what temperature will the ball just drop through the hole? Assume the coefficient of linear expansion of aluminium to be 23×10^{-6} K^{-1}.   [509.5 K]

13. At 283 K a thermometer contains 440 mm^3 of alcohol. Determine the temperature at which the volume is 480 mm^3 assuming that the coefficient of cubic expansion of the alcohol is 12×10^{-4} K^{-1}.   [358.8 K]

14. A zinc sphere has a radius of 30.0 mm at a temperature of 20°C. If the temperature of the sphere is raised to 420°C, determine the increase in: (a) the radius, (b) the surface area, (c) the volume of the sphere. Assume the coefficient of linear expansion for zinc to be 31×10^{-6} K^{-1}.
    [(a) 0.372 mm (b) 280.5 mm^2 (c) 4207 mm^3]

15. A block of cast iron has dimensions of 50 mm by 30 mm by 10 mm at 15 °C. Determine the increase in volume when the temperature of the block is raised to 75°C. Assume the coefficient of linear expansion of cast iron to be 11×10^{-6} K^{-1}.   [29.7 mm^3]

16. Two litres of water, initially at 20°C, is heated to 40°C. Determine the volume of water at 40°C if the coefficient of volumetric expansion of water within this range is 30×10^{-5} K^{-1}.   [2.012 l]

17. Determine the increase in volume, in litres, of 3 m^3 of water when heated from 293 K to

boiling point if the coefficient of cubic expansion is 2.1×10^{-4} K^{-1} (1 l $\approx 10^{-3}$ m^3).

[50.4 l]

18. Determine the reduction in volume when the temperature of 0.5 l of ethyl alcohol is reduced from 40°C to –15°C. Take the coefficient of cubic expansion for ethyl alcohol as 1.1×10^{-3} K^{-1}.

[0.030 25 l]

24

Ideal gas laws

At the end of this chapter you should be able to:

- state Boyle's law

- perform calculations using Boyle's law

- state Charles' law

- perform calculations using Charles' law

- state the Pressure law

- perform calculations using the Pressure law

- state Dalton's law of partial pressure

- understand the term 'ideal gas'

- perform calculations using Dalton's law of partial pressures

- recognize the combined gas law $\dfrac{p_1 V_1}{T_1} = \dfrac{p_2 V_2}{T_2}$

- perform calculations using the combined gas law

- recognize the characteristic gas equation $pV = mRT$

- understand STP

- perform calculations using the characteristic gas equation

24.1 Introduction

The relationships which exist between pressure, volume and temperature in a gas are given in a set of laws called **the gas laws**.

24.2 Boyle's law

Boyle's law states:

the volume V of a fixed mass of gas is inversely proportional to its absolute pressure p at constant temperature.

i.e. $p \propto 1/V$ or $p = k/V$ or

$$pV = k,$$

at constant temperature, where p = absolute pressure in pascals (Pa), V = volume in m³, and k = a constant.

Changes which occur at constant temperature are called **isothermal** changes. When a fixed mass of gas at constant temperature changes from pressure p_1 and volume V_1 to pressure p_2 and volume V_2 then:

$$p_1 V_1 = p_2 V_2$$

Problem 1.   A gas occupies a volume of 0.10 m³ at a pressure of 1.8 MPa. Determine (a) the pressure if the volume is changed to 0.06 m³ at constant temperature, and (b) the volume if the pressure is changed to 2.4 MPa at constant temperature.

(a)   Since the change occurs at constant temperature (i.e. an isothermal change), Boyle's law applies, i.e. $p_1 V_1 = p_2 V_2$, where $p_1 = 1.8$ MPa, $V_1 = 0.10$ m³ and $V_2 = 0.06$ m³. Hence $(1.8)(0.10) = p_2 (0.06)$ from which,

$$\textbf{pressure } \boldsymbol{p_2} = \frac{1.8 \times 0.10}{0.06} = \textbf{3 MPa}$$

(b) $p_1 V_1 = p_2 V_2$ where $p_1 = 1.8$ MPa, $V_1 = 0.10$ m³ and $p_2 = 2.4$ MPa. Hence $(1.8)(0.10) = (2.4)V_2$ from which

$$\textbf{volume } \boldsymbol{V_2} = \frac{(1.8)(0.10)}{2.4} = \textbf{0.075 m}^3$$

Problem 2.   In an isothermal process, a mass of gas has its volume reduced from 3200 mm³ to 2000 mm³. If the initial pressure of the gas is 110 kPa, determine the final pressure.

Since the process is isothermal, it takes place at constant temperature and hence Boyle's law applies, i.e. $p_1 V_1 = p_2 V_2$, where $p_1 = 110$ kPa, $V_1 = 3200$ mm³ and $V_2 = 2000$ mm³. Hence $(110)(3200) = p_2(2000)$, from which,

$$\textbf{final pressure, } \boldsymbol{p_2} = \frac{(110)(3200)}{2000} = \textbf{176 kPa}$$

Problem 3.   Some gas occupies a volume of 1.5 m³ in a cylinder at a pressure of 250 kPa. A piston, sliding in the cylinder, compresses the gas isothermally until the volume is 0.5 m³. If the area of the piston is 300 cm², calculate the force on the piston when the gas is compressed.

An isothermal process means constant temperature and thus Boyle's law applies, i.e. $p_1 V_1 = p_2 V_2$, where $V_1 = 1.5$ m³, $V_2 = 0.5$ m³ and $p_1 = 250$ kPa. Hence $(250)(1.5) = p_2(0.5)$, from which,

$$\textbf{pressure, } \boldsymbol{p_2} = \frac{(250)(1.5)}{(0.5)} = \textbf{750 kPa}$$

Pressure = (force)/(area), from which, force = pressure × area. Hence

force on the piston
$= (750 \times 10^3 \text{ Pa})(300 \times 10^{-4} \text{ m}^2) = \textbf{22.5 kN}$

24.3   Charles' law

Charles' law states:

for a given mass of gas at constant pressure, the volume V is directly proportional to its thermodynamic temperature T,

i.e. $V \propto T$ or $V = kT$ or

$$\boxed{\frac{V}{T} = k,}$$

at constant pressure, where T = thermodynamic temperature in kelvin (K).

A process which takes place at constant pressure is called an **isobaric** process.

The relationship between the Celsius scale of temperature and the thermodynamic or absolute scale is given by:

kelvin = degrees Celsius + 273

i.e.

$$\boxed{\textbf{K = °C + 273 or °C = K – 273}}$$

(as stated in Chapter 22).

If a given mass of gas at a constant pressure occupies a volume V_1 at a temperature T_1 and a volume V_2 at temperature T_2, then

$$\boxed{\frac{V_1}{T_1} = \frac{V_2}{T_2}}$$

Problem 4.   A gas occupies a volume of 1.2 litres at 20°C. Determine the volume it occupies at 130°C if the pressure is kept constant.

Since the change occurs at constant pressure (i.e. an isobaric process), Charles' law applies, i.e.

$$\frac{V_1}{T_1} = \frac{V_2}{T_2}$$

where $V_1 = 1.2$ l, $T_1 = 20°C = (20 + 273)$ K = 293 K and $T_2 = (130 + 273)$ K = 403 K. Hence

$$\frac{1.2}{293} = \frac{V_2}{403}$$

from which,

$$\textbf{volume at 130°C, } \boldsymbol{V_2} = \frac{(1.2)(403)}{293} = \textbf{1.65 litres}$$

Problem 5.   Gas at a temperature of 150°C has its volume reduced by one-third in an isobaric process. Calculate the final temperature of the gas.

Since the process is isobaric it takes place at constant pressure and hence Charles' law applies, i.e.

$$\frac{V_1}{T_1} = \frac{V_2}{T_2}$$

where $T_1 = (150 + 273)$ K = 423 K and $V_2 = \frac{2}{3} V_1$. Hence

$$\frac{V_1}{423} = \frac{\frac{2}{3} V_1}{T_2}$$

from which,

final temperature, $T_2 = \frac{2}{3}$ (423)

= **282 K** or (282 − 273)°C

i.e. **9°C**

24.4   The pressure law

The **pressure law** states:

the pressure p of a fixed mass of gas is directly proportional to its thermodynamic temperature T at constant volume.

i.e. $p \propto T$ or $p = kT$ or

$$\boxed{\frac{P}{T} = k}$$

When a fixed mass of gas at constant volume changes from pressure p_1 and temperature T_1, to pressure p_2 and temperature T_2 then:

$$\boxed{\frac{p_1}{T_1} = \frac{p_2}{T_2}}$$

Problem 6.   Gas initially at a temperature of 17°C and pressure 150 kPa is heated at constant volume until its temperature is 124°C. Determine the final pressure of the gas, assuming no loss of gas.

Since the gas is at constant volume, the pressure law applies, i.e.

$$\frac{p_1}{T_1} = \frac{p_2}{T_2}$$

where $T_1 = (17 + 273)$ K = 290 K,
$T_2 = (124 + 273)$ K = 397 K and
$p_1 = 150$ kPa. Hence

$$\frac{150}{290} = \frac{p_2}{397}$$

from which,

final pressure, p_2 $= \frac{(150)(397)}{290} = $ **205.3 kPa**

24.5   Dalton's law of partial pressure

Dalton's law of partial pressure states:

the total pressure of a mixture of gases occupying a given volume is equal to the sum of the pressures of each gas, considered separately, at constant temperature.

The pressure of each constituent gas when occupying a fixed volume alone is known as the **partial pressure** of that gas.

An **ideal gas** is one which completely obeys the gas laws given in sections 24.2 to 24.5. In practice no gas is an ideal gas, although air is very close to being one. For calculation purposes the difference between an ideal and an actual gas is very small.

Problem 7.   A gas R in a container exerts a pressure of 200 kPa at a temperature of 18°C. Gas Q is added to the container and the pressure increases to 320 kPa at the same temperature. Determine the pressure that gas Q alone exerts at the same temperature.

Initial pressure p_R = 200 kPa and pressure of gases R and Q together, $p = p_R + p_Q$ = 320 kPa. By Dalton's law of partial pressure, the pressure of gas Q alone is

$$p_Q = p - p_R = 320 - 200 = \mathbf{120\ kPa}$$

24.6   Characteristic gas equation

Frequently, when a gas is undergoing some change, the pressure, temperature and volume all vary simultaneously. Provided there is no change in the mass of a gas, the above gas laws can be combined, giving

$$\frac{p_1V_1}{T_1} = \frac{p_2V_2}{T_2} = k$$

where k is a constant.

For an ideal gas, constant $k = mR$, where m is the mass of the gas in kg, and R is the **characteristic gas constant**, i.e.

$$\frac{pV}{T} = mR$$

or

$$pV = mRT$$

This is called the **characteristic gas equation**. In this equation, p = absolute pressure in pascals, V = volume in m³, m = mass in kg, R = characteristic gas constant in J/(kg K), and T = thermodynamic temperature in kelvin.

Some typical values of the characteristic gas constant R include: air, 287 J/(kg K), hydrogen 4160 J/(kg K), oxygen 260 J/(kg K) and carbon dioxide 184 J/(kg K).

Standard temperature and pressure (i.e. **STP**) refers to a temperature of 0°C, i.e. 273 K, and normal atmospheric pressure of 101.325 kPa.

Problem 8. A gas occupies a volume of 2.0 m³ when at a pressure of 100 kPa and a temperature of 120°C. Determine the volume of the gas at 15°C if the pressure is increased to 250 kPa.

Using the combined gas law

$$\frac{p_1V_1}{T_1} = \frac{p_2V_2}{T_2}$$

where V_1 = 2.0 m³, p_1 = 100 kPa, p_2 = 250 kPa, T_1 = (120 + 273) K = 393 K and T_2 = (15 + 273) K = 288 K, gives:

$$\frac{(100)(2.0)}{393} = \frac{(250)V_2}{288}$$

from which,

$$\textbf{volume at 15°C, } V_2 = \frac{(100)(2.0)(288)}{(393)(250)}$$

$$= \textbf{0.586 m}^3$$

Problem 9. 20 000 mm³ of air initially at a pressure of 600 kPa and temperature 180°C is expanded to a volume of 70 000 mm³ at a pressure of 120 kPa. Determine the final temperature of the air, assuming no losses during the process.

Using the combined gas law,

$$\frac{p_1V_1}{T_1} = \frac{p_2V_2}{T_2}$$

where V_1 = 20 000 mm³, V_2 = 70 000 mm³, p_1 = 600 kPa, p_2 = 120 kPa, and T_1 = (180 + 273) K = 453 K, gives:

$$\frac{(600)(20\,000)}{453} = \frac{(120)(70\,000)}{T_2}$$

from which,

$$\textbf{final temperature, } T_2 = \frac{(120)(70\,000)(453)}{(600)(20\,000)}$$

$$= \textbf{317 K or 44°C}$$

Problem 10. Some air at a temperature of 40°C and pressure 4 bar occupies a volume of 0.05 m³. Determine the mass of the air assuming the characteristic gas constant for air to be 287 J/(kg K).

From above, $pV = mRT$, where p = 4 bar = 4×10^5 Pa (since 1 bar = 10^5 Pa – see Chapter 39), V = 0.05 m³, T = (40 + 273) K = 313 K, and R = 287 J/(kg K). Hence $(4 \times 10^5)(0.05 = m(287)(313)$ from which,

$$\textbf{mass of air, } m = \frac{(4 \times 10^5)(0.05)}{(287)(313)}$$

$$= \textbf{0.223 kg or 223 g}$$

Problem 11. A cylinder of helium has a volume of 600 cm³. The cylinder contains 200 g of helium at a temperature of 25°C. Determine the pressure of the helium if the characteristic gas constant for helium is 2080 J/(kg K).

From the characteristic gas equation, $pV = mRT$, where V = 600 cm³ = 600×10^{-6} m³, m = 200 g = 0.2 kg, T = (25 + 273) K = 298 K and

$R = 2080$ J/(kg K). Hence $(p)(600 \times 10^{-6}) = (0.2)(2080)(298)$ from which,

$$\text{pressure, } p = \frac{(0.2)(2080)(298)}{(600 \times 10^{-6})}$$

$$= 206\ 600\ 000\ \text{Pa}$$

$$= \textbf{206.6 MPa}$$

Problem 12. A spherical vessel has a diameter of 1.2 m and contains oxygen at a pressure of 2 bar and a temperature of $-20°C$. Determine the mass of oxygen in the vessel. Take the characteristic gas constant for oxygen to be 0.260 kJ/(kg K).

From the characteristic gas equation, $pV = mRT$ where V = volume of spherical vessel, and is given by

$$\frac{4}{3}\pi r^2 = \frac{4}{3}\pi \left(\frac{1.2}{2}\right)^3$$

$$= 0.905\ \text{m}^3$$

$p = 2$ bar $= 2 \times 10^5$ Pa, $T = (-20 + 273)$ K $= 253$ K and $R = 0.260$ kJ/(kg K) $= 260$ J/(kg K). Hence $(2 \times 10^5)(0.905) = m(260)(253)$ from which,

$$\textbf{mass of oxygen, } m = \frac{(2 \times 10^5)(0.905)}{(260)(253)}$$

$$= \textbf{2.75 kg}$$

Problem 13. Determine the characteristic gas constant of a gas which has a specific volume of 0.5 m³/kg at a temperature of 20°C and pressure 150 kPa.

From the characteristic gas equation, $pV = mRT$ from which,

$$R = \frac{pV}{mT}$$

where $p = 150 \times 10^3$ Pa, $T = (20 + 273)$ K $= 293$ K and specific volume, $V/m = 0.5$ m³/kg. Hence the **characteristic gas constant**,

$$R = \left(\frac{p}{T}\right)\left(\frac{V}{m}\right) = \left(\frac{150 \times 10^3}{293}\right)(0.5)$$

$$= \textbf{256 J/(kg K)}$$

Problem 14. A vessel has a volume of 0.80 m³ and contains a mixture of helium and hydrogen at a pressure of 450 kPa and a temperature of 17°C. If the mass of helium present is 0.40 kg determine (a) the partial pressure of each gas, and (b) the mass of hydrogen present. Assume the characteristic gas constant for helium to be 2080 J/(kg K) and for hydrogen 4160 J/(kg K).

(a) $V = 0.80$ m³, $p = 450$ kPa, $T = (17 + 273)$ K $= 290$ K, $m_{He} = 0.40$ kg, $R_{He} = 2080$ J/(kg K). If p_{He} is the partial pressure of the helium, then using the characteristic gas equation, $p_{He}V = m_{He}R_{He}T$ gives $(p_{He})(0.80) = (0.40)(2080)(290)$ from which,

the **partial pressure of the helium, p_{He}**

$$= \frac{(0.40)(2080)(290)}{(0.80)}$$

$$= \textbf{301.6 kPa}$$

By Dalton's law of partial pressure the total pressure p is given by the sum of the partial pressures, i.e. $p = p_H + p_{He}$, from which,

the **partial pressure of the hydrogen, p_H**

$$= p - p_{He} = 450 - 301.6$$

$$= \textbf{148.4 kPa}$$

(b) From the characteristic gas equation, $p_H V = m_H R_H T$.
Hence $(148.4 \times 10^3)(0.8) = m_H(4160)(290)$ from which,

mass of hydrogen, m_H

$$= \frac{(148.4 \times 10^3)(0.8)}{(4160)(290)}$$

$$= 0.098\ \text{kg or 98 g}$$

Problem 15. A compressed air cylinder has a volume of 1.2 m³ and contains air at a pressure of 1 MPa and a temperature of 25°C. Air is released from the cylinder until the pressure falls to 300 kPa and the temperature is 15°C. Determine (a) the mass of air released from the container, and (b) the volume it would occupy at STP. Assume the characteristic gas constant for air to be 287 J/(kg K).

$V_1 = 1.2 \text{ m}^3$ ($= V_2$, $p_1 = 1 \text{ MPa} = 10^6 \text{ Pa}$, $T_1 = (25 + 273) \text{ K} = 298 \text{ K}$, $T_2 = (15 + 273) \text{ K} = 288 \text{ K}$, $p_2 = 300 \text{ kPa} = 300 \times 10^3 \text{ Pa}$ and $R = 287 \text{ J/(kg K)}$.

(a) Using the characteristic gas equation, $p_1 V_1 = m_1 R T_1$, to find the initial mass of air in the cylinder gives:

$$(10^6)(1.2) = m_1 (287)(298)$$

from which,

mass $m_1 = \dfrac{(10^6)(1.2)}{(287)(288)} = 14.03 \text{ kg}$

Similarly, using $p_2 V_2 = m_2 R T_2$ to find the final mass of air in the cylinder gives

$$(300 \times 10^3)(1.2) = m_2 (287)(288)$$

from which,

mass $m_2 = \dfrac{(300 \times 10^3)(1.2)}{(287)(288)} = 4.36 \text{ kg}$

Mass of air released from cylinder

$= m_1 - m_2 = 14.03 - 4.36$

$= \mathbf{9.67 \text{ kg}}$

(b) At STP, $T = 273 \text{ K}$ and $p = 101.325 \text{ kPa}$. Using the characteristic gas equation $pV = mRT$

volume, V $= \dfrac{mRT}{p} = \dfrac{(9.67)(287)(273)}{101\,325}$

$= \mathbf{7.48 \text{ m}^3}$

Problem 16.   A vessel X contains gas at a pressure of 750 kPa at a temperature of 27°C. It is connected via a valve to vessel Y which is filled with a similar gas at a pressure of 1.2 MPa and a temperature of 27°C. The volume of vessel X is 2.0 m³ and that of vessel Y is 3.0 m³. Determine the final pressure at 27°C when the valve is opened and the gases are allowed to mix. Assume R for the gas to be 300 J/(kg K).

For vessel X:

$p_X = 750 \times 10^3 \text{ Pa}$

$T_X = (27 + 273) \text{ K} = 300 \text{ K}$

$V_X = 2.0 \text{ m}^3 \qquad R = 300 \text{ J/(kg K)}$

From the characteristic gas equation, $p_X V_X = m_X R T_X$. Hence

$$(750 \times 10^3)(2.0) = m_X (300)(300)$$

from which,

mass of gas in vessel X, m_X

$= \dfrac{(750 \times 10^3)(2.0)}{(300)(300)}$

$= 16.67 \text{ kg}$

For vessel Y:

$p_Y = 1.2 \times 10^6 \text{ Pa}$

$T_Y = (27 + 273) \text{ K} = 300 \text{ K}$

$V_Y = 3.0 \text{ m}^3 \qquad R = 300 \text{ J/(kg K)}$

From the characteristic gas equation, $p_Y V_Y = m_Y R T_Y$. Hence

$$(1.2 \times 10^6)(3.0) = m_Y (300)(300)$$

from which,

mass of gas in vessel Y,

$m_Y = \dfrac{(1.2 \times 10^6)(3.0)}{(300)(300)} = 40 \text{ kg}$

When the valve is opened, mass of mixture, $m = m_X + m_Y = 16.67 + 40 = 56.67 \text{ kg}$. Total volume, $V = V_X + V_Y = 2.0 + 3.0 = 5.0 \text{ m}^3$, $R = 300 \text{ J/(kg K)}$, $T = 300 \text{ K}$.

From the characteristic gas equation, $pV = mRT$

$$p(5.0) = (56.67)(300)(300)$$

from which,

final pressure, p $= \dfrac{(56.67)(300)(300)}{(5.0)}$

$= \mathbf{1.02 \text{ MPa}}$

24.7 Multi-choice questions on ideal gas laws

(Answers on page 356.)

1. Which of the following statements is false?
    (a) At constant temperature, Charles' law applies.
    (b) The pressure of a given mass of gas decreases as the volume is increased at constant temperature.
    (c) Isobaric changes are those which occur at constant pressure.
    (d) Boyle's law applies at constant temperature.

2. A gas occupies a volume of 4 m^3 at a pressure of 400 kPa. At constant temperature, the pressure is increased to 500 kPa. The new volume occupied by the gas is:
   (a) 5 m^3 (b) 0.3 m^3 (c) 0.2 m^3 (d) 3.2 m^3
3. A gas at a temperature of 27°C occupies a volume of 5 m^3. The volume of the same mass of gas at the same pressure but at a temperature of 57°C is:
   (a) 10.56 m^3 (b) 5.50 m^3 (c) 4.55 m^3
   (d) 2.37 m^3
4. Which of the following statements is false?
   (a) An ideal gas is one which completely obeys the gas laws.
   (b) Isothermal changes are those which occur at constant volume.
   (c) The volume of a gas increases when the temperature increases at constant pressure.
   (d) Changes which occur at constant pressure are called isobaric changes.

A gas has a volume of 0.4 m^3 when its pressure is 250 kPa and its temperature is 400 K. Use this data in questions 5 and 6.

5. The temperature when the pressure is increased to 400 kPa and the volume is increased to 0.8 m^3 is:
   (a) 400 K   (b) 80 K   (c) 1280 K   (d) 320 K
6. The pressure when the temperature is raised to 600 K and the volume is reduced to 0.2 m^3 is:
   (a) 187.5 kPa   (b) 250 kPa   (c) 333.3 kPa
   (d) 750 kPa
7. A gas has a volume of 3 m^3 at a temperature of 546 K and a pressure of 101.325 kPa. The volume it occupies at STP is:
   (a) 3 m^3 (b) 1.5 m^3 (c) 6 m^3
8. Which of the following statements is false?
   (a) A characteristic gas constant has units of J/(kg K).
   (b) STP conditions are 273 K and 101.325 kPa.
   (c) All gases are ideal gases.
   (d) An ideal gas is one which obeys the gas laws.

A mass of 5 kg of air is pumped into a container of volume 2.87 m^3. The characteristic gas constant for air is 287 J/(kg K). Use this data in questions 9 and 10.

9. The pressure when the temperature is 27°C is:
   (a) 1.6 kPa   (b) 6 kPa   (c) 150 kPa
   (d) 15 kPa

10. The temperature when the pressure is 200 kPa is:
    (a) 400°C   (b) 127°C   (c) 127 K   (d) 283 K

24.8 Short answer questions on ideal gas laws

1. State Boyle's law.
2. State Charles' law.
3. State the Pressure law.
4. State Dalton's law of partial pressures.
5. State the relationship between the Celsius and the thermodynamic scale of temperature.
6. What is (a) an isothermal change, and (b) an isobaric change?
7. Define an ideal gas.
8. State the characteristic gas equation.
9. What is meant by STP?

24.9 Further questions on ideal gas laws

1. The pressure of a mass of gas is increased from 150 kPa to 750 kPa at constant temperature. Determine the final volume of the gas, if its initial volume is 1.5 m^3.   [0.3 m^3]
2. Some gas initially at 16°C is heated to 96°C at constant pressure. If the initial volume of the gas is 0.8 m^3, determine the final volume of the gas.   [1.02 m^3]
3. In an isothermal process, a mass of gas has its volume reduced from 50 cm^3 to 32 cm^3. If the initial pressure of the gas is 80 kPa, determine its final pressure.   [125 kPa]
4. The piston of an air compressor compresses air to $\frac{1}{4}$ of its original volume during its stroke. Determine the final pressure of the air if the original pressure is 100 kPa, assuming an isothermal change.   [400 kPa]
5. A gas is contained in a vessel of volume 0.02 m^3 at a pressure of 300 kPa and a temperature of 15°C. The gas is passed into a vessel of volume 0.015 m^3. Determine to what temperature the gas must be cooled for the pressure to remain the same.   [−57°C]
6. In an isobaric process gas at a temperature of 120°C has its volume reduced by a sixth. Determine the final temperature of the gas.   [54.5°C]
7. Gas, initially at a temperature of 27°C and pressure 100 kPa, is heated at constant

volume until its temperature is 150°C. Assuming no loss of gas, determine the final pressure of the gas. [141 kPa]

8. A gas A in a container exerts a pressure of 120 kPa at a temperature of 20°C. Gas B is added to the container and the pressure increases to 300 kPa at the same temperature. Determine the pressure which gas B alone exerts at the same temperature.
[180 kPa]

9. A quantity of gas in a cylinder occupies a volume of 2 m^3 at a pressure of 300 kPa. A piston slides in the cylinder and compresses the gas, according to Boyle's law, until the volume is 0.5 m^3. If the area of the piston is 0.02 m^2, calculate the force on the piston when the gas is compressed. [24 kN]

10. A given mass of air occupies a volume of 0.5 m^3 at a pressure of 500 kPa and a temperature of 20°C. Find the volume of the air at STP. [2.30 m^3]

11. A spherical vessel has a diameter of 2.0 m and contains hydrogen at a pressure of 300 kPa and a temperature of –30°C. Determine the mass of hydrogen in the vessel. Assume the characteristic gas constant R for hydrogen is 4160 J/(kg K). [1.24 kg]

12. A gas occupies a volume of 1.20 m^3 when at a pressure of 120 kPa and a temperature of 90°C. Determine the volume of the gas at 20°C if the pressure is increased to 320 kPa.
[0.363 m^3]

13. A cylinder 200 mm in diameter and 1.5 m long contains oxygen at a pressure of 2 MPa and a temperature of 20°C. Determine the mass of oxygen in the cylinder. Assume the characteristic gas constant for oxygen is 260 J/(kg K). [1.24 kg]

14. A gas is pumped into an empty cylinder of volume 0.1 m^3 until the pressure is 5 MPa. The temperature of the gas is 40°C. If the cylinder mass increases by 5.32 kg when the gas has been added, determine the value of the characteristic gas constant. [300 J/(kg K)]

15. The mass of a gas is 1.2 kg and it occupies a volume of 13.45 m^3 at STP. Determine its characteristic gas constant. [4160 J/(kg K)]

16. A vessel P contains gas at a pressure of 800 kPa at a temperature of 25°C. It is connected via a valve to vessel Q which is filled with similar gas at a pressure of 1.5 MPa and a temperature of 25°C. The volume of vessel P is 1.5 m^3 and that of vessel R is 2.5 m^3. Determine the final pressure at 25°C

when the valve is opened and the gases are allowed to mix. Assume R for the gas to be 297 J/(kg K). [1.24 MPa]

17. 30 cm^3 of air initially at a pressure of 500 kPa and temperature 150°C is expanded to a volume of 100 cm^3 at a pressure of 200 kPa. Determine the final temperature of the air, assuming no losses during the process.
[291°C]

18. A vessel contains 4 kg of air at a pressure of 600 kPa and a temperature of 40°C. The vessel is connected to another by a short pipe and the air exhausts into it. The final pressure in both vessels is 250 kPa and the temperature in both is 15°C. If the pressure in the second vessel before the air entered was zero, determine the volume of each vessel. Assume R for air is 287 J/(kg K).
[0.60 m^3, 0.72 m^3]

19. A quantity of gas in a cylinder occupies a volume of 0.5 m^3 at a pressure of 400 kPa and a temperature of 27°C. It is compressed according to Boyle's law until its pressure is 1 MPa, and then expanded according to Charles' law until its volume is 0.03 m^3. Determine the final temperature of the gas.
[177°C]

20. Some air at a temperature of 35°C and pressure 2 bar occupies a volume of 0.08 m^3. Determine the mass of the air assuming the characteristic gas constant for air to be 287 J/(kg K). [0.181 kg]

21. Determine the characteristic gas constant R of a gas which has a specific volume of 0.267 m^3/kg at a temperature of 17°C and pressure 200 kPa. [184 J/(kg K)]

22. A vessel has a volume of 0.75 m^3 and contains a mixture of air and carbon dioxide at a pressure of 200 kPa and a temperature of 27°C. If the mass of air present is 0.5 kg determine (a) the partial pressure of each gas, and (b) the mass of carbon dioxide. Assume the characteristic gas constant for air to be 287 J/(kg K) and for carbon dioxide 184 J/(kg K).
[(a) 57.4 kPa, 142.6 kPa (b) 1.94 kg]

23. A cylinder contains 20 kg of air at a pressure of 2.5 MPa and a temperature of 27°C. Oxygen is now pumped into the cylinder until the pressure is increased to 4 MPa, the temperature remaining at 27°C. Calculate (a) the mass of oxygen pumped into the cylinder, and (b) the temperature to which the mixture must rise in order to increase the pressure to

5 MPa. Take R for air as 287 J/(kg K) and R for oxygen as 270 J/(kg K).

[(a) 13.25 kg (b) 102°C]

24. A mass of gas occupies a volume of 0.02 m³ when its pressure is 150 kPa and its temperature is 17°C. If the gas is compressed until its pressure is 500 kPa and its temperature is 57°C, determine (a) the volume it will occupy and (b) its mass, if the characteristic gas constant for the gas is 205 J/(kg K).

[(a) 0.0068 m³ (b) 0.052 kg]

25. A compressed air cylinder has a volume of 0.6 m³ and contains air at a pressure of 1.2 MPa absolute and a temperature of 37°C. After use the pressure is 800 kPa absolute and the temperature is 17°C. Calculate (a) the mass of air removed from the cylinder, and (b) the volume the mass of air removed would occupy at STP conditions. Take R for air as 287 J/(kg K) and atmospheric pressure as 100 kPa.

[(a) 2.33 kg (b) 1.826 m³]

25

Properties of water and steam

At the end of this chapter you should be able to:

- state the principle of conservation of energy
- understand the term 'internal energy, U'
- understand the terms 'enthalpy, H' and 'specific enthalpy, h'
- perform calculations using $H = U + pV$
- appreciate that sensible heat, $h_f = c\theta$
- understand the term 'saturated steam'
- perform calculations using $h_g = h_f + h_{fg}$ where h_g is the total specific enthalpy of steam at saturation temperature and h_{fg} is the specific latent heat of vaporization
- understand the terms 'dryness fraction, q' and 'wet steam'
- perform calculations using specific enthalpy of wet steam = $h_f + qh_{fg}$
- understand the term 'superheated steam'
- appreciate a typical temperature/specific enthalpy graph
- use steam tables

25.1 Principle of conservation of energy

When two systems are at different temperatures, the transfer of energy from one system to the other is called **heat transfer**. For a block of hot metal cooling in air, heat is transferred from the hot metal to the cool air. The **principle of conservation of energy** may be stated as

energy cannot be created nor can it be destroyed,

and since heat is a form of energy, this law applies to heat transfer problems.

A more convenient way of expressing this law when referring to heat transfer problems is:

$$\text{Initial energy of the system} + \begin{pmatrix} \text{energy entering} \\ \text{the system} \end{pmatrix}$$

$$= \begin{pmatrix} \text{final energy of} \\ \text{the system} \end{pmatrix} + \begin{pmatrix} \text{energy leaving} \\ \text{the system} \end{pmatrix}$$

or,

$$\begin{pmatrix} \text{energy entering} \\ \text{the system} \end{pmatrix} = \begin{pmatrix} \text{change of energy} \\ \text{within the system} \end{pmatrix}$$

$$+ \begin{pmatrix} \text{energy leaving} \\ \text{the system} \end{pmatrix}$$

Problem 1. Apply the principle of conservation of energy to a lamp connected to an electrical supply.

If the lamp is considered to be a system, then the energy entering the system is the electrical energy (watt seconds or joules). The change of energy of the system is an increase in the kinetic and potential energies of the molecules of the lamp. The energy leaving the system is in the form of light and heat energies. Thus

$$\begin{pmatrix} \text{electrical energy entering} \\ \text{the system} \end{pmatrix} =$$

$$\begin{pmatrix} \text{change of} \\ \text{internal energy} \end{pmatrix} + \begin{pmatrix} \text{light and heat energy} \\ \text{leaving the system} \end{pmatrix}$$

25.2 Internal energy

Fluids consist of a very large number of molecules moving in random directions within the fluid. When the fluid is heated, the speeds of the molecules are increased, increasing the kinetic energy of the molecules. There is also an increase in volume due to an increase in the average distance between molecules, causing the potential energy of the fluid to increase. The **internal energy**, U, of a fluid is the sum of the internal kinetic and potential energies of the molecules of a fluid, measured in joules. It is not usual to state the internal energy of a fluid as a particular value in heat transfer problems, since it is normally only the **change** in internal energy which is required.

Problem 2.   Discuss three factors affecting the internal energy of a fluid.

The amount of internal energy of a fluid depends on:

(a) The type of fluid; in gases the molecules are well separated and move with high velocities, thus a gaseous fluid has higher internal energy than the same mass of a liquid.
(b) The mass of a fluid; the greater the mass, the greater the number of molecules and hence the greater the internal energy.
(c) The temperature; the higher the temperature the greater the velocity of the molecules.

25.3 Enthalpy

The sum of the internal energy and the pressure energy of a fluid is called the **enthalpy** of the fluid, denoted by the symbol H and measured in joules. The pressure energy, or work done, is given by the product of pressure p and volume V, that is:

$$\text{pressure energy} = pV \text{ joules}$$

(i.e. pressure energy = work done = Fd. Since pressure $p = F/A$, $F = pA$. Hence work done = $(pA)d = pV$). Thus, enthalpy = internal energy + pressure energy (or work done), i.e.

$$\boxed{H = U + pV}$$

As for internal energy, the actual value of enthalpy is usually unimportant and it is the change in enthalpy which is usually required. In heat transfer problems involving steam and water, water is considered to have zero enthalpy at a standard pressure of 101 kPa and a temperature of 0°C. The word 'specific' associated with quantities indicates 'per unit mass'. Thus the **specific enthalpy** is obtained by dividing the enthalpy by the mass and is denoted by the symbol h. Thus:

$$\text{Specific enthalpy} = \frac{\text{enthalpy}}{\text{mass}} = \frac{H}{m} = h$$

The units of specific enthalpy are joules per kilogram (J/kg).

Problem 3.   In a closed system, that is, a system in which the mass of fluid remains a constant, the internal energy changes from 25 kJ to 50 kJ and the work done by the system is 55 kJ. Determine the heat transferred to the system to effect this change.

From above, $H = U + pV$, where H is the enthalpy (often taken as the heat energy added or taken from a system), U is the change of internal energy and pV is the pressure energy of work done. Thus

$$H = [(50 - 25) + 55] \text{ kJ} = (25 + 55) \text{ kJ} = 80 \text{ kJ}$$

That is, **heat transferred to the system is 80 kJ**.

25.4 Sensible heat

The specific enthalpy of water, h_f, at temperature θ °C is the quantity of heat needed to raise 1 kg of water from 0°C to θ°C, and is called the **sensible heat** of the water. Its value is given by: specific heat capacity of water (c) × temperature change (θ), that is,

$$\boxed{h_f = c\theta}$$

The specific heat capacity of water varies with temperature and pressure but is normally taken as 4.2 kJ/kg, thus $h_f = 4.2\ \theta$ **kJ/kg**.

25.5   Saturated steam

When water is heated at a uniform rate, a stage is reached (at 100°C at standard atmospheric pressure) where the addition of more heat does not result in a corresponding increase in temperature. The temperature at which this occurs is called the **saturation temperature, t_{SAT}**, and the water is called **saturated water.** As heat is added to saturated water, it is turned into **saturated steam**. The amount of heat required to turn 1 kg of saturated water into saturated steam is called the **specific latent heat of vaporization**, and is given the symbol, h_{fg}. The total specific enthalpy of steam at saturation temperature, h_g, is given by: the specific sensible heat + the specific latent heat of vaporization, i.e.

$$\boxed{h_g = h_f + h_{fg}}$$

25.6   Dryness factor

If the amount of heat added to saturated water is insufficient to turn all the water into steam, then the ratio

$$\frac{\text{mass of saturated steam}}{\text{total mass of steam and water}}$$

is called the **dryness fraction** of the steam, denoted by the symbol q. The steam is called **wet steam** and its total enthalpy h is given by: enthalpy of saturated water + (dryness fraction) × (enthalpy of latent heat of vaporization), that is

$$\boxed{h = h_f + qh_{fg}}$$

25.7   Superheated steam

When the amount of heat added to water at saturation temperature is sufficient to turn all the water into steam, it is called either saturated

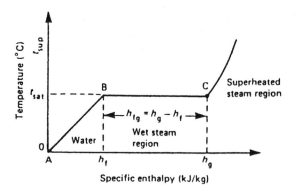

Figure 25.1

vapour or **dry saturated steam**. The addition of further heat results in the temperature of the steam rising and it is then called **superheated steam**. The specific enthalpy of superheated steam above that of dry saturated steam is given by $c(t_{SUP} - t_{SAT})$, where c is the specific heat capacity of the steam and t_{SUP} is the temperature of the superheated steam. The total specific enthalpy of the superheated steam is given by:

$$h_f + h_{fg} + c(t_{SUP} - t_{SAT}),\ \text{or}\ h_g + c(t_{SUP} - t_{SAT})$$

25.8   Temperature/specific enthalpy graph

The relationship between temperature and specific enthalpy can be shown graphically and a typical temperature/specific enthalpy diagram is shown in Fig. 25.1. In this figure, AB represents the sensible heat region where any increase in enthalpy results in a corresponding increase in temperature. BC is called the **evaporation line** and points between B and C represent the wet steam region (or latent region), point C representing dry saturated steam. Points to the right of C represent the superheated steam region.

25.9   Steam tables

The boiling point of water, t_{SAT} and the various specific enthalpies associated with water and steam [h_f, h_{fg} and $c(t_{SUP} - t_{SAT})$] all vary with pressure. These values at various pressures have been tabulated in **steam tables**, extracts from these being shown in Tables 25.1 and 25.2.

Table 25.1

Pressure		Saturation temperature t_{SAT} (°C)	Specific enthalpy (kJ/kg)		
(bar)	(kPa)		Saturated water h_f	Latent heat h_{fg}	Saturated vapour h_g
1	100	99.6	417	2258	2675
1.5	150	111.4	467	2226	2693
2	200	120.2	505	2202	2707
3	300	133.5	561	2164	2725
4	400	143.6	605	2134	2739
5	500	151.8	640	2109	2749
6	600	158.8	670	2087	2757
7	700	165.0	697	2067	2764
8	800	170.4	721	2048	2769
9	900	175.4	743	2031	2774
10	1000	179.9	763	2015	2778
15	1500	198.3	845	1947	2792
20	2000	212.4	909	1890	2799
30	3000	233.8	1008	1795	2803
40	4000	250.3	1087	1714	2801

Table 25.2

Pressure		Saturation temperature t_{SAT} (°C)	Saturated vapour h_g	Specific enthalpy (kJ/kg) Superheated steam at				
(bar)	(kPa)			200°C	250°C	300°C	350°C	400°C
1	100	99.6	2675	2876	2975	3075	3176	3278
1.5	150	111.4	2693	2873	2973	3073	3175	3277
2	200	120.2	2707	2871	2971	3072	3174	3277
3	300	133.5	2725	2866	2968	3070	3172	3275
4	400	143.6	2739	2862	2965	3067	3170	3274
5	500	151.8	2749	2857	2962	3065	3168	3272
6	600	158.8	2757	2851	2958	3062	3166	3270
7	700	165.0	2764	2846	2955	3060	3164	3269
8	800	170.4	2769	2840	2951	3057	3162	3267
9	900	175.4	2774	2835	2948	3055	3160	3266
10	1000	179.9	2778	2829	2944	3052	3158	3264
15	1500	198.3	2792	2796	2925	3039	3148	3256
20	2000	212.4	2799		2904	3025	3138	3248
30	3000	233.8	2803		2858	2995	3117	3231
40	4000	250.3	2801			2963	3094	3214

In Table 25.1, the pressure in both bar and kilopascals, and saturated water temperature, are shown in columns on the left. The columns on the right give the corresponding specific enthalpies of water (h_f) and dry saturated steam (h_g), together with the specific enthalpy of the latent heat of vaporization (h_{fg}).

The columns on the right of Table 25.2 give the specific enthalpies of dry saturated steam (h_g) and superheated steam at various temperatures. The values stated refer to zero enthalpy. However, if the degree of superheat is given, this refers to the saturation temperature. Thus at a pressure of 100 kPa, the column headed, say, 250°C has a degree of superheat of (250 – 99.6)°C, that is 150.4°C.

Problem 4. A steam-raising plant generates dry saturated steam at a pressure of 1.5 MPa. Use steam tables to find (a) the saturation temperature (b) the specific enthalpy of the dry saturated steam (c) the enthalpy of 1 t of the steam.

(a) From Table 25.1, at a pressure of 1.5 MPa (that is 1500 kPa), the saturation temperature t_{SAT} is **198.3°C**. That is, at a pressure of 1.5 MPa, the water boils at 198.3°C.
(b) The specific enthalpy of the dry saturated steam is also given in Table 25.1 (column 6), showing that h_g is **2792 kJ/kg**.
(c) The enthalpy is the total heat content of the steam and since, from part (b), 1 kg contains 2792 kJ, then 1 tonne (that is, 1000 kg) contains 1000 × 2792 kJ or **2792 MJ**.

Problem 5. Dry saturated steam at a pressure of 1.0 MPa is cooled at constant pressure until it has a dryness fraction of 0.6. Determine the change in the specific enthalpy of the steam.

From Table 25.1, the specific enthalpy of dry saturated steam h_g, at a pressure of 1.0 MPa (1000 kPa), is 2778 kJ/kg. From section 25.6, the specific enthalpy of wet steam is $h_f + q h_{fg}$. At a pressure of 1.0 MPa, h_f is 763 kJ/kg and h_{fg} is 2015 kJ/kg. Thus, the specific enthalpy of the wet steam is given by

$$763 + (0.6 \times 2015)$$

$$= 1972 \text{ kJ/kg}$$

The change in the specific enthalpy is (initial value minus the final value), that is, change in specific enthalpy = 2778 – 1972 = **806 kJ/kg**

Problem 6. The condenser of a turbine converts wet steam at the outlet of the turbine into water at the base of the condenser. The pressure in a condenser is 100 kPa and the dryness fraction of the steam entering the condenser is 0.7. The water is pumped from the condenser at a temperature of 27°C. Find the heat removed by the condenser cooling water per hour, if the mass of steam condensed is 120 t/h. Take the specific heat capacity of water as 4.2 kJ/kg.

From section 25.6, the specific enthalpy of wet steam is $h_f + q h_{fg}$. From Table 25.1, h_f at 100 kPa is 417 kJ/kg and h_{fg} is 2258 kJ/kg. Thus the specific enthalpy of the wet steam is 417 + (0.7 × 2258), i.e. 1997.6 kJ/kg. The specific enthalpy of the water leaving the condenser is given by 4.2 0 kJ/kg (see section 25.4), i.e. 4.2 × 27 = 113.4 kJ/kg. Thus specific heat removed by the cooling water is 1997.6 – 113.4, that is 1884.2 kJ/kg.

The total heat removed per hour is: mass × (specific heat removed by the cooling water) that is, 120 000 kg × 1884.2 kJ/kg, or **226.1 GJ/h**.

Problem 7. Determine the degree of superheat and the specific enthalpy of steam leaving a boiler at a pressure of 3.0 MPa and a temperature of 400°C.

Details of the specific enthalpies of superheated steam are given in Table 25.2. At a pressure of 3.0 MPa, i.e. 3000 kPa, the saturation temperature is 233.8°C, hence the degree of superheat is 400 – 233.8, or **166.2°C**.

The specific enthalpy of superheated steam at 3.0 MPa and 400°C is given in Table 25.2 as **3231 kJ/kg**.

Problem 8. Feed water is pumped into a boiler at a temperature of 30°C and steam leaves the boiler at a pressure of 2.0 MPa and having 87.6°C of superheat. Determine the heat supplied by the boiler for each kilogram of steam produced.

With reference to Table 25.2, at a pressure of 2.0 MPa, the saturation temperature is 212.4°C. Thus the temperature of the superheated steam is (212.4 + 87.6)°C, that is 300°C. At a pressure of 2.0 MPa and temperature of 300°C, the specific enthalpy of the steam leaving the boiler is 3025 kJ/kg. The specific enthalpy of the feed water is 4.2 × 30, that is, 126 kJ/kg (see section 25.4). Thus the heat supplied by the boiler is

$$3025 - 126, \text{ i.e. } \textbf{2899 kJ/kg}.$$

Problem 9. Draw the temperature/enthalpy diagram for 1 kg of water or steam at a pressure of 2 MPa when heated from 30°C to 400°C.

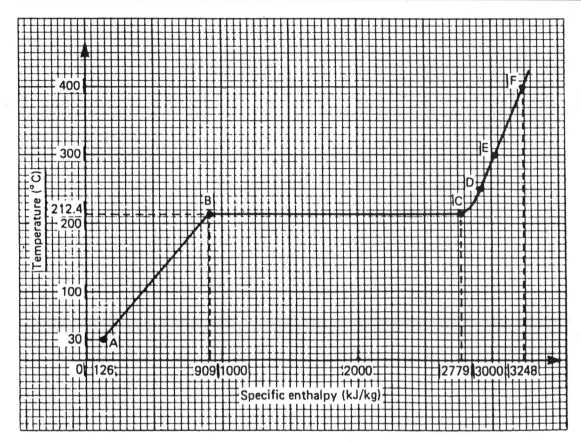

Figure 25.2

From section 25.4, the specific enthalpy of water at 30°C is 4.2 × 30, or 126 kJ/kg, shown as point A in Fig. 25.2. The saturation temperature at 2 MPa is given in Table 25.1 and is 212.4°C, the corresponding specific enthalpy, h_f, being 909 kJ/kg. This is shown as point B in Fig. 25.2. AB represents the sensible heat region. The specific enthalpy h_g of the saturated vapour (dry saturated steam) is 2799 kJ/kg at 212.4°C, shown as point C in Fig. 25.2. Thus BC represents the specific latent heat of vaporization region.

From Table 25.2, the specific enthalpies of superheated steam are: 2094 kJ/kg at 250°C; 3025 kJ/kg at 300°C; 3248 kJ/kg at 400°C, shown as points D, E and F, respectively, in Fig. 25.2. Thus CF represents the superheated steam region.

Superheated steam behaves very nearly as if it is an ideal gas and the gas laws introduced in Chapter 24 may be used to determine the relationship between pressure, volume and temperature.

> **Problem 10.**   From complete steam tables for superheated steam the following data for superheated steam at a pressure of 1 MPa and a temperature of 400°C is determined:
> Specific volume = 0.3065 m³/kg
> Specific gas constant, $R = 0.4562$ kJ/(kg K)
> Determine the percentage difference in volume, if it is assumed that superheated steam is an ideal gas.

The relationship between pressure, volume and temperature, assuming an ideal gas, is $pV = mRT$ (see Chapter 24). For the superheated steam $p = 1$ MPa $= 10^6$ Pa, $m = 1$ kg, $R = 456.2$ J/(kg K) and $T = (273 + 400)$ K $= 673$ K.
Since $pV = mRT$, then

$$V = \frac{mRT}{p} = \frac{1 \times 456.2 \times 673}{10^6} = 0.3070 \text{ m}^3$$

Since 0.3065 m^3 is the true value, the percentage difference is given by

$$\frac{0.3065 - 0.3070}{0.3065} \times 100 = \mathbf{-0.16\%}$$

If this principle is extended to other temperatures and pressures it may be shown that superheated steam does approximate closely to an ideal gas.

25.10   Multi-choice questions on properties of water and steam

(Answers on page 356.)

In questions 1 to 3, which of the statements are false?

1. The internal energy of a fluid:
   (a) is the sum of the kinetic and potential energies of the fluid
   (b) is a constant
   (c) depends on the mass of the fluid
   (d) depends on the temperature of the fluid
2. (a) The unit of specific enthalpy is the kilojoule.
   (b) Enthalpy = internal energy + pressure energy.
   (c) It is the change of enthalpy which is normally required in heat transfer problems.
   (d) Zero enthalpy is taken as 0°C and 101 kPa in heat transfer problems.
3. (a) A change of enthalpy causes a change of temperature for water in the sensible heat region.
   (b) A change of enthalpy causes no change of temperature in the latent heat of vaporization region.
   (c) A change of enthalpy causes a change of temperature for superheated steam.
   (d) The specific enthalpy of dry saturated steam is given by $h_f + h_g$.

In a steam raising plant operating at a pressure of 1.5 MPa, water is fed to a boiler at 30°C and steam leaves the boiler at 300°C. Use steam tables to find the quantities stated in questions 4 to 8, selecting the correct answer from those given in (a)–(j) below. Take the specific heat capacity of water as 4.2 kJ/kg.
(a) 2792 kJ/kg   (b) 420 kJ/kg   (c) 1102 kJ/kg
(d) 100°C   (e) 198.3°C   (f) 845 kJ/kg
(g) 126 kJ/kg   (h) 1947 kJ/kg   (i) 5584 kJ/kg
(j) 3039 kJ/kg.

4. The specific enthalpy of the feed water
5. The specific enthalpy of the saturated water
6. The specific latent heat of vaporization
7. The specific enthalpy of the steam when dry saturated.
8. The specific enthalpy of the steam leaving the boiler.

5 kg of wet steam at a pressure of 150 kPa and having a dryness fraction of 0.8 is condensed into water and cooled to a temperature of 50°C. Use steam tables to find the quantities stated in questions 9 and 10, selecting the correct answers from those given below:
(a) 111.4 kJ   (b) 467 kJ/kg   (c) 1050 kJ
(d) 2226 kJ/kg   (e) 11 130 kJ   (f) 11 239 kJ
(g) 2247.8 kJ/kg   (h) 9989 kJ   (i) 1997.8 kJ/kg

9. The enthalpy of the steam before being cooled
10. The enthalpy of the water after cooling has taken place

25.11   Short answer questions on properties of water and steam

1. State briefly the law of conservation of energy and how it is applied to heat transfer problems.
2. What is meant by the internal energy of a fluid?
3. What is meant by the enthalpy of a fluid? State its units.
4. Write down the relationship between enthalpy, internal energy and pressure energy.
5. State the values of pressure and temperature corresponding to zero enthalpy of water.
6. What is meant by specific enthalpy? State its units.
7. Write down the relationship for the specific enthalpy of water at a temperature of 0°C.
8. Define the dryness fraction of steam.
9. What are steam tables and why are they necessary?
10. State the relationship between superheated steam and an ideal gas.

25.12   Further questions on properties of water and steam

(Use steam tables as necessary. Take the specific heat capacity of water as 4.2 kJ/kg.)

1. In a closed system the internal energy is increased from 10 kJ to 70 kJ and the work

done by the system is 120 kJ. Find the energy entering the system. [180 kJ]

2. When the energy entering a closed system is 3 MJ, the work done by the system is 2.7 MJ. Determine the change in the internal energy of the system. [300 kJ]

3. Find the saturation temperature and specific enthalpy of saturated water at a pressure of 700 kPa. What will be the enthalpy of 100 kg of saturated water at this pressure?
[165°C, 697 kJ/kg, 69.7 MJ]

4. 50 kg of water is heated from a temperature of 15°C to its boiling point at a pressure of 150 kPa. Determine the increase in the enthalpy of the water. [20.2 MJ]

5. Saturated water at a pressure of 400 kPa is heated until its dryness fraction is 0.75. Determine the change of specific enthalpy.
[995.5 kJ/kg]

6. Water enters a steam raising plant at a pressure of 1 MPa and at a temperature of 30°C. If its specific enthalpy is increased by 2450.5 kJ/kg, determine the dryness fraction of the wet steam produced. [0.9]

7. 100 kg of dry saturated steam at a pressure of 2 MPa is superheated to a temperature of 400°C. Determine the increase in enthalpy of the steam. [44.9 MJ]

8. Wet steam, having a dryness fraction of 0.6 is heated to a temperature of 250°C at a pressure of 3.0 MPa. Determine the increase in the specific enthalpy. [773 kJ/kg]

9. 2 tonnes of steam at a pressure of 200 kPa and having a dryness fraction of 0.75, is cooled in a condenser to 27°C. Determine the amount of heat removed from the steam.
[4086 MJ]

10. A small boiler is used to convert 25 kg of water at 48°C to wet steam having a dryness fraction of 0.92, at a pressure of 600 kPa. Find the amount of heat added to the water by the boiler. [59.71 MJ]

11. A boiler delivers 200 t of steam per hour to a turbine at a pressure of 1.5 MPa and with 201.7°C of superheat. After passing through the turbine it has a dryness fraction of 0.3 at a pressure of 150 kPa. Find the amount of heat taken from the steam by the turbine.
[424.2 GJ]

12. Draw a temperature/enthalpy diagram for 1 kg of water/steam at a pressure of 1 MPa, when heated from 50°C to 300°C.
[The specific enthalpy/temperature coordinates are (201, 50), (763, 179.9), (2778, 179.9), (3052, 300)]

13. Water at a temperature of 40°C and at a pressure of 700 kPa is heated until it becomes superheated steam at a temperature of 400°C. Draw a temperature/specific enthalpy diagram for the water/steam.
[The specific enthalpy/temperature coordinates are (168, 40), (697, 165), (2764, 165), (3269, 400)]

14. From steam tables, superheated steam at a pressure of 2 MPa and a temperature of 350°C has a specific volume of 0.1386 m³/kg. Determine the volume of the superheated steam using the gas laws and comment on the result obtained. The specific gas constant of steam at this temperature and pressure is 0.4446 kJ/(kg K).
[0.1385 m³/kg, superheated steam approximates to an ideal gas]

15. Assuming superheated steam is an ideal gas, determine the pressure of 5 kg of superheated steam, if it occupies a volume of 2.613 m³ at a temperature of 300°C. Take R at this temperature and pressure as 0.455 kJ/(kg K). [499 kPa]

Part 3
Science for Electrical Engineering

26

An introduction to electric circuits

At the end of this chapter you should be able to:

- recognize common electrical circuit diagram symbols

- understand that electric current is the rate of movement of charge and is measured in amperes

- appreciate that the unit of charge is the coulomb

- calculate charge or quantity of electricity Q from $Q = It$

- understand that a potential difference between two points in a circuit is required for current to flow

- appreciate that the unit of p.d. is the volt

- understand that resistance opposes current flow and is measured in ohms

- appreciate what an ammeter, a voltmeter, an ohmmeter, a multimeter and a CRO measure

- distinguish between linear and non-linear devices

- state Ohm's law as $V = IR$ or $I = V/R$ or $R = V/I$

- use Ohm's law in calculations, including multiples and sub-multiples of units

- describe a conductor and an insulator, giving examples of each

- appreciate that electrical power P is given by $P = VI = I^2R = (V^2/R)$ watts

- calculate electrical power

- define electrical energy and state its unit

- calculate electrical energy

- state the three main effects of an electric current, giving practical examples of each

- explain the importance of fuses in electrical circuits

26.1 Standard symbols for electrical components

Symbols are used for components in electrical circuit diagrams and some of the more common ones are shown in Fig. 26.1.

26.2 Electrical current and quantity of electricity

All **atoms** consist of **protons**, **neutrons** and **electrons**. The protons, which have positive electrical charges, and the neutrons, which have

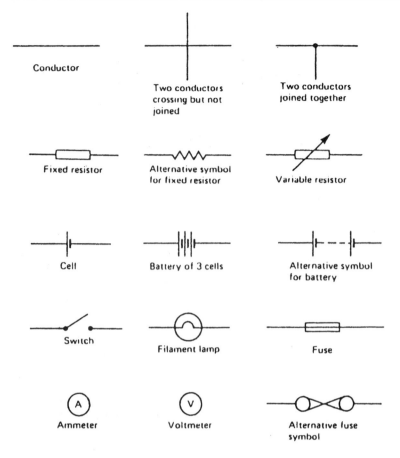

Figure 26.1

no electrical charge, are contained within the **nucleus**. Removed from the nucleus are minute negatively charged particles called electrons. Atoms of different materials differ from one another by having different numbers of protons, neutrons and electrons. An equal number of protons and electrons exist within an atom and it is said to be electrically balanced, as the positive and negative charges cancel each other out. When there are more than two electrons in an atom the electrons are arranged into **shells** at various distances from the nucleus.

All atoms are bound together by powerful forces of attraction existing between the nucleus and its electrons. Electrons in the outer shell of an atom, however, are attracted to their nucleus less powerfully than are electrons whose shells are nearer the nucleus.

It is possible for an atom to lose an electron; the atom, which is now called an **ion**, is not now electrically balanced, but is positively charged and is thus able to attract an electron to itself from another atom. Electrons that move from one atom to another are called free electrons and such random motion can continue indefinitely. However, if an electric pressure or **voltage** is applied across any material there is a tendency for electrons to move in a particular direction. This movement of free electrons, known as **drift**, constitutes an electric current flow.

Thus current is the rate of movement of charge

Conductors are materials that have electrons that are loosely connected to the nucleus and can easily move through the material from one atom to another.

Insulators are materials whose electrons are held firmly to their nucleus.

The unit used to measure the **quantity of electrical charge Q** is called the **coulomb C** (where 1 coulomb = 6.24×10^{18} electrons). If the drift of electrons in a conductor takes place at the rate of one coulomb per second the resulting current is said to be a current of one ampere. Thus,

1 ampere = 1 coulomb per second
or 1 A = 1 C/s

Hence,

1 coulomb = 1 ampere second or 1 C = 1 A s

Generally, if I is the current in amperes and t the time in seconds during which the current flows, then $I \times t$ represents the quantity of electrical charge in coulombs, i.e. quantity of electrical charge transferred,

$$Q = I \times t \text{ coulombs}$$

Problem 1.   What current must flow if 0.24 coulombs is to be transferred in 15 ms?

Since the quantity of electricity, $Q = It$, then

$$I = \frac{Q}{t} = \frac{0.24}{15 \times 10^{-3}} = \frac{0.24 \times 10^3}{15} = \frac{240}{15} = \textbf{16 A}$$

Problem 2.   If a current of 10 A flows for four minutes, find the quantity of electricity transferred.

Quantity of electricity, $Q = It$ coulombs. $I = 10$ A, $t = 4 \times 60 = 240$ s. Hence

$$Q = 10 \times 240 = \textbf{2400 C}$$

26.3   Potential difference and resistance

For a continuous current to flow between two points in a circuit a **potential difference (p.d.)** or **voltage, V**, is required between them; a complete conducting path is necessary to and from the source of electrical energy. The unit of p.d. is the **volt, V**.

Figure 26.2 shows a cell connected across a filament lamp. Current flow, by convention, is considered as flowing from the positive terminal of the cell around the circuit to the negative terminal.

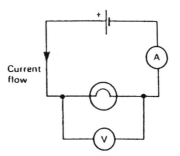

Figure 26.2

The flow of electric current is subject to friction. This friction, or opposition, is called **resistance R** and is the property of a conductor that limits current. The unit of resistance is the **ohm Ω**. 1 ohm is defined as the resistance which will have a current of 1 ampere flowing through it when 1 volt is connected across it, i.e.

$$\text{resistance } R = \frac{\textbf{potential difference}}{\textbf{current}}$$

26.4   Basic electrical measuring instruments

An **ammeter** is an instrument used to measure current and must be connected **in series** with the circuit. Figure 26.2 shows an ammeter connected in series with the lamp to measure the current flowing through it. Since all the current in the circuit passes through the ammeter it must have a very **low resistance**.

A **voltmeter** is an instrument used to measure p.d. and must be connected **in parallel** with the part of the circuit whose p.d. is required. In Fig. 26.2, a voltmeter is connected in parallel with the lamp to measure the p.d. across it. To avoid a significant current flowing through it a voltmeter must have a very **high resistance**.

An **ohmmeter** is an instrument for measuring resistance.

A **multimeter**, or universal instrument, may be used to measure voltage, current and resistance. An 'Avometer' is a typical example.

The **cathode ray oscilloscope** (CRO) may be used to observe waveforms and to measure voltages and currents. The display of a CRO involves a spot of light moving across a screen. The amount by which the spot is deflected from

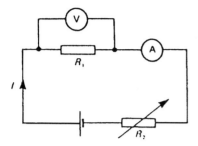

Figure 26.3

its initial position depends on the p.d. applied to the terminals of the CRO and the range selected. The displacement is calibrated in 'volts per cm'. For example, if the spot is deflected 3 cm and the volts/cm switch is on 10 V/cm then the magnitude of the p.d. is 3 cm × 10 V/cm, i.e. 30 V.

(See Chapter 36 for more detail about electrical measuring instruments and measurements.)

26.5   Linear and non-linear devices

Figure 26.3 shows a circuit in which current I can be varied by the variable resistor R_2. For various settings of R_2, the current flowing in resistor R_1, displayed on the ammeter, and the p.d. across R_1, displayed on the voltmeter, are noted and a graph is plotted of p.d. against current. The result is shown in Fig. 26.4(a) where the straight line graph passing through the origin indicates that current is directly proportional to the p.d. Since the gradient, i.e. (p.d.)/(current), is constant, resistance R_1 is constant. A resistor is thus an example of a **linear device**.

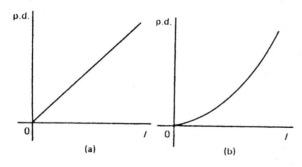

Figure 26.4

If the resistor R_1 in Fig. 26.3 is replaced by a component such as a lamp then the graph shown in Fig. 26.4(b) results when values of p.d. are noted for various current readings. Since the gradient is changing, the lamp is an example of a **non-linear device.**

26.6   Ohm's law

Ohm's law states that the current I flowing in a circuit is directly proportional to the applied voltage V and inversely proportional to the resistance R, provided the temperature remains constant. Thus,

$$I = \frac{V}{R} \text{ or } V = IR \text{ or } R = \frac{V}{I}$$

> Problem 3.   The current flowing through a resistor is 0.8 A when a p.d. of 20 V is applied. Determine the value of the resistance.

From Ohm's law,

$$\text{resistance } R = \frac{V}{I} = \frac{20}{0.8} = \frac{200}{8} = \textbf{25 } \Omega$$

26.7   Multiples and sub-multiples

Currents, voltages and resistances can often be very large or very small. Thus **multiples** and **sub-multiples** of units are often used, as stated in Chapter 7. The most common ones, with an example of each, are listed in Table 26.1.

> Problem 4.   Determine the p.d. which must be applied to a 2 kΩ resistor in order that a current of 10 mA may flow.

Resistance $R = 2 \text{ k}\Omega = 2 \times 10^3 = 2000 \text{ }\Omega$.
Current $I = 10 \text{ mA} = 10 \times 10^{-3} \text{ A}$ or $10/10^3$ or $(10/1000) \text{ A} = 0.01 \text{ A}$.
From Ohm's law, potential difference, $V = IR = (0.01)(2000) = \textbf{20 V}$.

> Problem 5.   A coil has a current of 50 mA flowing through it when the applied voltage is 12 V. What is the resistance of the coil?

Table 26.1

Prefix	Name	Meaning	Example
M	mega	multiply by 1 000 000 (i.e. $\times 10^6$)	$2\ M\Omega = 2\ 000\ 000$ ohms
k	kilo	multiply by 1000 (i.e. $\times 10^3$)	$10\ kV = 10\ 000$ volts
m	milli	divide by 1000 (i.e. $\times 10^{-3}$)	$25\ mA = \dfrac{25}{1000}\ A$ $= 0.025$ amperes
μ	micro	divide by 1 000 000 (i.e. $\times 10^{-3}$)	$50\ \mu V = \dfrac{50}{1\ 000\ 000}\ V$ $= 0.000\ 05$ volts

Resistance, $R = \dfrac{V}{I} = \dfrac{12}{50 \times 10^{-3}}$

$= \dfrac{12 \times 10^3}{50} = \dfrac{12\ 000}{50} = \textbf{240 }\boldsymbol{\Omega}$

> **Problem 6.** A 100 V battery is connected across a resistor and causes a current of 5 mA to flow. Determine the resistance of the resistor. If the voltage is now reduced to 25 V, what will be the new value of the current flowing?

Resistance $R = \dfrac{V}{I} = \dfrac{100}{5 \times 10^{-3}}$

$= \dfrac{100 \times 10^3}{5} = 20 \times 10^3 = \textbf{20 k}\boldsymbol{\Omega}$

Current when voltage is reduced to 25 V,

$I = \dfrac{V}{R} = \dfrac{25}{20 \times 10^3} = \dfrac{25}{20} \times 10^{-3} = \textbf{1.25 mA}$

> **Problem 7.** What is the resistance of a coil which draws a current of (a) 50 mA and (b) 200 μA from a 120 V supply?

(a) Resistance $R = \dfrac{V}{I} = \dfrac{120}{50 \times 10^{-3}}$

$= \dfrac{120}{0.05} = \dfrac{12\ 000}{5}$

$= \textbf{2400 }\boldsymbol{\Omega}$ or $\textbf{2.4 k}\boldsymbol{\Omega}$

(b) Resistance $R = \dfrac{120}{200 \times 10^{-6}} = \dfrac{120}{0.0002}$

$= \dfrac{1\ 200\ 000}{2}$

$= \textbf{600 000 }\boldsymbol{\Omega}$ or $\textbf{600 k}\boldsymbol{\Omega}$ or $\textbf{0.6 M}\boldsymbol{\Omega}$

26.8 Conductors and insulators

A **conductor** is a material having a low resistance which allows electric current to flow in it. All metals are conductors and some examples include copper, aluminium, brass, platinum, silver, gold and also carbon.

An **insulator** is a material having a high resistance which does not allow electric current to flow in it. Some examples of insulators include plastic, rubber, glass, porcelain, air, paper, cork, mica, ceramics and certain oils.

26.9 Electrical power and energy

Electrical Power

Power P in an electrical circuit is given by the product of potential difference V and current I. The unit of power is the **watt, W**. Hence

$$\boxed{P = V \times I \text{ watts}} \qquad (26.1)$$

From Ohm's law, $V = IR$. Substituting for V in equation (26.1) gives:

$$P = (IR) \times I$$

i.e. $\boxed{P = I^2R \text{ watts}}$

Also, from Ohm's law, $I = V/R$. Substituting for I in equation (26.1) gives:

$$P = V \times \left(\dfrac{V}{R}\right)$$

i.e. $\boxed{P = \dfrac{V^2}{R} \text{ watts}}$

There are thus three possible formulae which may be used for calculating power.

Problem 8. A 100 W electric light bulb is connected to a 250 V supply. Determine (a) the current flowing in the bulb, and (b) the resistance of the bulb.

Power $P = V \times I$, from which, current $I = P/V$.

(a) Current $I = \dfrac{100}{250} = \dfrac{10}{25} = \dfrac{2}{5} = \mathbf{0.4\ A}$

(b) Resistance $R = \dfrac{V}{I} = \dfrac{250}{0.4} = \dfrac{2500}{4} = \mathbf{625\ \Omega}$

Problem 9. Calculate the power dissipated when a current of 4 mA flows through a resistance of 5 kΩ.

Power $P = I^2R = (4 \times 10^{-3})^2(5 \times 10^3)$

$\quad\quad = 16 \times 10^{-6} \times 5 \times 10^3 = 80 \times 10^{-3}$

$\quad\quad = \mathbf{0.08\ W}$ or $\mathbf{80\ mW}$

Alternatively, since $I = 4 \times 10^{-3}$ and $R = 5 \times 10^3$ then from Ohm's law voltage $V = IR = 4 \times 10^{-3} \times 5 \times 10^3 = 20$ V. Hence

$\quad$ Power $P = V \times I = 20 \times 4 \times 10^{-3} = \mathbf{80\ mW}$

Problem 10. An electric kettle has a resistance of 30 Ω. What current will flow when it is connected to a 240 V supply? Find also the power rating of the kettle.

$\quad$ Current, $I = \dfrac{V}{R} = \dfrac{240}{30} = 8$ A

Power, $P = VI = 240 \times 8 = 1920$ W $= \mathbf{1.92\ kW} =$ power rating of kettle.

Problem 11. A current of 5 A flows in the winding of an electric motor, the resistance of the winding being 100 Ω. Determine (a) the p.d. across the winding, and (b) the power dissipated by the coil.

(a) Potential difference across winding, $V = IR$ $= 5 \times 100 = 500$ V.

(b) Power dissipated by coil is given by

$P = I^2R = 5^2 \times 100$

$\quad = \mathbf{2500\ W}$ or $\mathbf{2.5\ kW}$

(Alternatively, $P = V \times I = 500 \times 5 = \mathbf{2500\ W}$ or $\mathbf{2.5\ kW}$.)

Problem 12. The current/voltage relationship for two resistors A and B is as shown in Fig. 26.5. Determine the value of the resistance of each resistor.

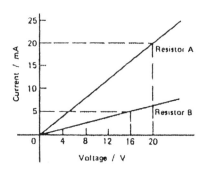

Figure 26.5

For resistor A,

$\quad R = \dfrac{V}{I} = \dfrac{20\ A}{20\ mA} = \dfrac{20}{0.02} = \dfrac{2000}{2}$

$\quad\quad = \mathbf{1000\ \Omega}$ or $\mathbf{1\ k\Omega}$

For resistor B,

$\quad R = \dfrac{V}{I} = \dfrac{16\ V}{5\ mA} = \dfrac{16}{0.005} = \dfrac{16\ 000}{5}$

$\quad\quad = \mathbf{3200\ \Omega}$ or $\mathbf{3.02\ k\Omega}$

Problem 13. The hot resistance of a 240 V filament lamp is 960 Ω. Find the current taken by the lamp and its power rating.

From Ohm's law,

$\quad$ current $I = \dfrac{V}{R} = \dfrac{240}{960} = \dfrac{24}{96} = \dfrac{1}{4}$ **A** or **0.25 A**

$\quad$ Power rating $P = VI = (240)\left(\dfrac{1}{4}\right) = \mathbf{60\ W}$

Electrical Energy

Electrical energy = power $\times$ time

If the power is measured in watts and the time in seconds then the unit of energy is watt-seconds or **joules**. If the power is measured in kilowatts and the time in hours then the unit of energy is **kilowatt-hours**, often called the '**unit of electricity**'. The 'electricity meter' in the home records the number of kilowatt-hours used and is thus an energy meter.

Problem 14. A 12 V battery is connected across a load having a resistance of 40 Ω. Determine the current flowing in the load, the power consumed and the energy dissipated in 2 minutes.

Current $I = \dfrac{V}{R} = \dfrac{12}{40} = \textbf{0.3 A}$

Power consumed, $P = VI = (12)(0.3) = \textbf{3.6 W}$.
Energy dissipated = power × time = (3.6 W)(2 × 60 s) = **432 J** (since 1 J = 1 W s).

Problem 15. A source of 15 V supplies a current of 2 A for six minutes. How much energy is provided in this time?

Energy = power × time, and power = voltage × current. Hence

energy = $VIt = 15 \times 2 \times (6 \times 60)$

= 10 800 W s or J

= **10.8 kJ**

Problem 16. Electrical equipment in an office takes a current of 13 A from a 240 V supply. Estimate the cost per week of electricity if the equipment is used for 30 hours each week and 1 kW h of energy costs 7p.

Power = VI watts = 240 × 13 = 3120 W = 3.12 kW.
Energy used per week = power × time = (3.12 kW) × (30 h) = 93.6 kW h. Cost at 7p per kW h = 93.6 × 7 = 655.2 p. Hence

weekly cost of electricity = £6.55

Problem 17. An electric heater consumes 3.6 MJ when connected to a 250 V supply for 40 minutes. Find the power rating of the heater and the current taken from the supply.

Power = $\dfrac{\text{energy}}{\text{time}} = \dfrac{3.6 \times 10^6}{40 \times 60} \dfrac{\text{J}}{\text{s}}$ (or W)

= 1500 W

i.e. power rating of heater = **1.5 kW**.
Power $P = VI$, thus

$I = \dfrac{P}{V} = \dfrac{1500}{250} = 6$ A

Hence the current taken from the supply is **6 A**.

Problem 18. Determine the power dissipated by the element of an electric fire of resistance 20 Ω when a current of 10 A flows through it. If the fire is on for 6 hours determine the energy used and the cost if 1 unit of electricity costs 7p.

Power $P = I^2R = 10^2 \times 20 = 100 \times 20 = \textbf{2000 W}$ or **2 kW**.
(Alternatively, from Ohm's law, $V = IR = 10 \times 20 = 200$ V, hence power $P = V \times I = 200 \times 10 = 2000$ W = 2 kW.)
Energy used in 6 hours = power × time = 2 kW × 6 h = **12 kW h**.
1 unit of electricity = 1 kW h. Hence the number of units used is 12. Cost of energy = 12 × 7 **= 84p**.

Problem 19. A business uses two 3 kW fires for an average of 20 hours each per week, and six 150 W lights for 30 hours each per week. If the cost of electricity is 7p per unit, determine the weekly cost of electricity to the business.

Energy = power × time.
Energy used by one 3 kW fire in 20 hours = 3 kW × 20 h = 60 kW h.
Hence weekly energy used by two 3 kW fires = 2 × 60 = 120 kW h.
Energy used by one 150 W light for 30 hours = 150 W × 30 h = 4500 W h = 4.5 kW h.
Hence weekly energy used by six 150 W lamps = 6 × 4.5 = 27 kW h.

Total energy used per week = 120 + 27 = 147 kW h.

1 unit of electricity = 1 kW h of energy, thus weekly cost of energy at 7p per kW h = 7 × 147 = 1029p = **£10.29**.

26.10   Main effects of electric current

The three main effects of an electric current are:

(a) magnetic effect;
(b) chemical effect;
(c) heating effect.

Some practical applications of the effects of an electric current include:

Magnetic effect: bells, relays, motors, generators, transformers, telephones, car ignition and lifting magnets

Chemical effect: primary and secondary cells and electroplating

Heating effect: cookers, water heaters, electric fires, irons, furnaces, kettles and soldering irons

26.11   Fuses

A **fuse** is used to prevent overloading of electrical circuits. The fuse, which is made of material having a low melting point, utilizes the heating effect of an electric current. A fuse is placed in an electrical circuit and if the current becomes too large the fuse wire melts and so breaks the circuit. A circuit diagram symbol for a fuse is shown in Fig. 26.1.

Problem 20.   If 5 A, 10 A and 13 A fuses are available, state which is most appropriate for the following appliances which are both connected to a 240 V supply.

(a)   Electric toaster having a power rating of 1 kW
(b)   Electric fire having a power rating of 3 kW

Power $P = VI$, from which,

$$\text{current } I = \frac{P}{V}$$

(a)   For the toaster,

$$\text{current } I = \frac{P}{V} = \frac{1000}{240} = \frac{100}{24} = 4\frac{1}{6} \text{ A}$$

Hence a **5 A** fuse is most appropriate.

(b)   For the fire,

$$\text{current } I = \frac{P}{V} = \frac{3000}{240} = \frac{300}{24} = 12\frac{1}{2} \text{ A}$$

Hence a **13 A** fuse is most appropriate.

26.12   Multi-choice questions on the introduction to electric circuits

(Answers on page 356.)

1. 60 µs is equivalent to:
   (a) 0.06 s   (b) 0.000 06 s   (c) 1000 minutes
   (d) 0.6 s
2. The current which flows when 0.1 coulomb is transferred in 10 ms is:
   (a) 1 A   (b) 10 A   (c) 10 mA   (d) 100 mA
3. The p.d. applied to a 1 kΩ resistance in order that a current of 100 µA may flow is:
   (a) 1 V   (b) 100 V   (c) 0.1 V   (d) 10 V
4. Which of the following formulae for electrical power is incorrect?
   (a) VI (b) V/I (c) I^2R (d) V^2/R
5. The power dissipated by a resistor of 4 Ω when a current of 5 A passes through it is
   (a) 6.25 W   (b) 20 W   (c) 80 W   (d) 100 W
6. Which of the following statements is true?
   (a) Electric current is measured in volts.
   (b) 200 kΩ resistance is equivalent to 0.2 MΩ.
   (c) An ammeter has a low resistance and must be connected in parallel with a circuit.
   (d) An electrical insulator has a high resistance.
7. A current of 3 A flows for 50 h through a 6 Ω resistor. The energy consumed by the resistor is:
   (a) 0.9 kW h   (b) 2.7 kW h   (c) 9 kW h
   (d) 27 kW h
8. What must be known in order to calculate the energy used by an electrical appliance?
   (a) voltage and current
   (b) current and time of operation
   (c) power and time of operation
   (d) current and resistance

26.13 Short answer questions on the introduction to electric circuits

1. Draw the preferred symbols for the following components used when drawing electrical circuit diagrams:
   (a) fixed resistor   (b) cell   (c) filament lamp
   (d) fuse   (e) voltmeter
2. State the unit of (a) current (b) potential difference (c) resistance.
3. State an instrument used to measure (a) current (b) potential difference (c) resistance.
4. What is a multimeter?
5. State Ohm's law.
6. Give one example of (a) a linear device (b) a non-linear device.
7. State the meaning of the following abbreviations of prefixes used with electrical units:
   (a) k (b) μ (c) m (d) M.
8. What is a conductor? Give four examples.
9. What is an insulator? Give four examples.
10. Complete the following statement:
    An ammeter has a .... resistance and must be connected .... with the load.
11. Complete the following statement:
    A voltmeter has a .... resistance and must be connected .... with the load.
12. State the unit of electrical power. State three formulae used to calculate power.
13. State two units used for electrical energy.
14. State the three main effects of an electric current and give two examples of each.
15. What is the function of a fuse in an electrical circuit?

26.14 Further questions on the introduction to electric circuits

1. In what time would a current of 10 A transfer a charge of 50 C?   [5 s]
2. A current of 6 A flows for 10 minutes. What charge is transferred?   [3600 C]
3. How long must a current of 100 mA flow so as to transfer a charge of 50 C? [8 min 20 s]
4. The current flowing through a heating element is 5 A when a p.d. of 35 V is applied across it. Find the resistance of the element.   [7 Ω]
5. A 60 W electric light bulb is connected to a 240 V supply. Determine (a) the current flowing in the bulb and (b) the resistance of the bulb.   [(a) 0.25 A (b) 960 Ω]

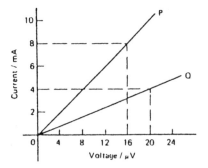

Figure 26.6

6. Graphs of current against voltage for two resistors P and Q are shown in Fig. 26.6. Determine the value of each resistor.
   [2 mΩ, 5 mΩ]
7. Determine the p.d. which must be applied to a 5 kΩ resistor such that a current of 6 mA may flow.   [30 V]
8. The hot resistance of a 250 V filament lamp is 625 Ω. Determine the current taken by the lamp and its power rating.   [0.4 A, 100 W]
9. Determine the resistance of a coil connected to a 150 V supply when a current of (a) 75 mA (b) 300 μA flows through it.
   [(a) 2 kΩ (b) 0.5 MΩ]
10. Determine the resistance of an electric fire which takes a current of 12 A from a 240 V supply. Find also the power rating of the fire and the energy used in 20 h.
    [20 Ω, 2.88 kW, 57.6 kW h]
11. Determine the power dissipated when a current of 10 mA flows through an appliance having a resistance of 8 kΩ.   [0.8 W]
12. 85.5 J of energy are converted into heat in nine seconds. What power is dissipated?
    [9.5 W]
13. A current of 4 A flows through a conductor and 10 W is dissipated. What p.d. exists across the ends of the conductor?   [2.5 V]
14. Find the power dissipated when:
    (a) a current of 5 mA flows through a resistance of 20 kΩ;
    (b) a voltage of 400 V is applied across a 120 kΩ resistor;
    (c) a voltage applied to a resistor is 10 kV and the current flow is 4 mA.
    [(a) 0.5 W (b) $1\frac{1}{3}$ W (c) 40 W]
15. A battery of e.m.f. 15 V supplies a current of 2 A for 5 min. How much energy is supplied in this time?   [9 kJ]

16. A d.c. electric motor consumes 72 MJ when connected to a 400 V supply for 2 h 30 min. Find the power rating of the motor and the current taken from the supply.

[8 kW, 20 A]

17. A p.d. of 500 V is applied across the winding of an electric motor and the resistance of the winding is 50 Ω. Determine the power dissipated by the coil. [5 kW]

18. In a household during a particular week three 2 kW fires are used on average 25 h each and eight 100 W light bulbs are used on average 35 h each. Determine the cost of electricity for the week if 1 unit of electricity costs 7p. [£12.46]

19. Calculate the power dissipated by the element of an electric fire of resistance 30 Ω when a current of 10 A flows in it. If the fire is on for 30 hours in a week determine the energy used. Determine also the weekly cost of energy if electricity costs 7.2p per unit.

[3 kW, 90 kW h, £6.48]

20. A television set having a power rating of 120 W and an electric lawnmower of power rating 1 kW are both connected to a 240 V supply. If 3 A, 5 A and 10 A fuses are available state which is the most appropriate for each appliance. [3 A, 5 A]

27

Resistance variation

At the end of this chapter you should be able to:

- appreciate that electrical resistance depends on four factors
- appreciate that resistance $R = \rho l/a$, where ρ is the resistivity
- recognize typical values of resistivity and its unit
- perform calculations using $R = \rho l/a$
- define the temperature coefficient of resistance, α
- recognize typical values for α
- perform calculations using $R_\theta = R_0(1 + \alpha\theta)$

27.1 Resistance and resistivity

The resistance of an electrical conductor depends on four factors, these being: (a) the length of the conductor, (b) the cross-sectional area of the conductor, (c) the type of material and (d) the temperature of the material.

Resistance, R, is directly proportional to length, l, of a conductor, i.e. $R \propto l$. Thus, for example, if the length of a piece of wire is doubled, then the resistance is doubled.

Resistance, R, is inversely proportional to cross-sectional area, a, of a conductor, i.e. $R \propto 1/a$. Thus, for example, if the cross-sectional area of a piece of wire is doubled then the resistance is halved.

Since $R \propto 1$ and $R \propto 1/a$ then $R \propto l/a$. By inserting a constant of proportionality into this relationship the type of material used may be taken into account. The constant of proportionality is known as the **resistivity** of the material and is given the symbol ρ (rho). Thus

resistance $\boxed{R = \dfrac{\rho l}{a} \text{ ohms}}$

ρ is measured in ohm metres (Ω m). The value of the resistivity is that resistance of a unit cube of the material measured between opposite faces of the cube.

Resistivity varies with temperature and some typical values of resistivities measured at about room temperature are given below:

Copper $1.7 \times 10^{-8}\,\Omega$ m (or $0.017\,\mu\Omega$ m)
Aluminium $2.6 \times 10^{-8}\,\Omega$ m (or $0.026\,\mu\Omega$ m)
Carbon (graphite) $10 \times 10^{-8}\,\Omega$ m ($0.10\,\mu\Omega$ m)
Glass $1 \times 10^{10}\,\Omega$ m (or $10^4\,\mu\Omega$ m)
Mica $1 \times 10^{13}\,\Omega$ m (or $10^7\,\mu\Omega$)

Note that good conductors of electricity have a low value of resistivity and good insulators have a high value of resistivity.

Problem 1. The resistance of a 5 m length of wire is 600 Ω. Determine (a) the resistance of an 8 m length of the same wire, and (b) the length of the same wire when the resistance is 420 Ω.

(a) Resistance, R, is directly proportional to length, l, i.e. $R \propto l$. Hence, 600 Ω $\propto$ 5 m or

$600 = (k)(5)$, where k is the coefficient of proportionality. Hence

$$k = \frac{600}{5} = 120$$

When the length l is 8 m, then

resistance $R = kl = (120)(8) = \mathbf{960\ \Omega}$

(b)   When the resistance is 420 Ω, $420 = kl$, from which,

$$\text{length } l \quad = \frac{420}{k} = \frac{420}{120} = \mathbf{3.5\ m}$$

Problem 2.   A piece of wire of cross-sectional area 2 mm^2 has a resistance of 300 Ω. Find (a) the resistance of a wire of the same length and material if the cross-sectional area is 5 mm^2, (b) the cross-sectional area of a wire of the same length and material of resistance 750 Ω.

Resistance R is inversely proportional to cross-sectional area, a, i.e. $R \propto 1/a$. Hence

$$300\ \Omega \propto \frac{1}{2\ \text{mm}^2} \text{ or } 300 = (k)\left(\frac{1}{2}\right)$$

from which, the coefficient of proportionality, $k = 300 \times 2 = 600$.

(a)   When the cross-sectional area $a = 5$ mm^2 then

$$R = (k)\left(\frac{1}{5}\right)$$

$$= (600)\left(\frac{1}{5}\right) = \mathbf{120\ \Omega}$$

(Note that resistance has decreased as the cross-sectional area is increased.)

(b)   When the resistance is 750 Ω then $750 = (k)(l/a)$, from which

$$\text{cross-sectional area, } a, = \frac{k}{750} = \frac{600}{750}$$

$$= \mathbf{0.8\ mm^2}$$

Problem 3.   A wire of length 8 m and cross-sectional area 3 mm^2 has a resistance of 0.16 Ω. If the wire is drawn out until its cross-sectional area is 1 mm^2, determine the resistance of the wire.

Resistance R is directly proportional to length, l, and inversely proportional to the cross-sectional area, a, i.e.

$$R \propto \frac{l}{a} \text{ or } R = k\,\frac{l}{a}$$

where k is the coefficient of proportionality. Since $R = 0.16$, $l = 8$ and $a = 3$, then $0.16 = (k)(8/3)$, from which

$$k = 0.16 \times \frac{3}{8} = 0.06$$

If the cross-sectional area is reduced to $\frac{1}{3}$ of its original area then the length must be tripled to 3×8, i.e. 24 m.

$$\text{New resistance } R = k\,\frac{l}{a} = 0.06\left(\frac{24}{1}\right)$$

$$= \mathbf{1.44\ \Omega}$$

Problem 4.   Calculate the resistance of a 2 km length of aluminium overhead power cable if the cross-sectional area of the cable is 100 mm^2. Take the resistivity of aluminium to be $0.03 \times 10^{-6}\ \Omega$ m.

Length $l = 2$ km $= 2000$ m; area, $a = 100$ mm$^2 = 100 \times 10^{-6}$ m^2; resistivity $= \rho = 0.03 \times 10^{-6}\ \Omega$ m.

$$\text{Resistance } R = \frac{\rho l}{a}$$

$$= \frac{(0.03 \times 10^{-6}\ \Omega\ \text{m})(2000\ \text{m})}{(100 \times 10^{-6}\ \text{m}^2)}$$

$$= \frac{0.03 \times 2000}{100}\ \Omega$$

$$= \mathbf{0.6\ \Omega}$$

Problem 5.   Calculate the cross-sectional area, in mm^2, of a piece of copper wire, 40 m in length and having a resistance of 0.25 Ω. Take the resistivity of copper as $0.02 \times 10^{-6}\ \Omega$ m.

Resistance $R = \rho l/a$, hence cross-sectional area is given by

$$a = \frac{\rho l}{R}$$

$$= \frac{(0.02 \times 10^{-6}\ \Omega\ \text{m})(40\ \text{m})}{0.25\ \Omega}$$

$$= 3.2 \times 10^{-6}\ \text{m}^2$$

$$= (3.2 \times 10^{-6}) \times 10^6\ \text{mm}^2$$

$$= \mathbf{3.2\ mm^2}$$

Problem 6. The resistance of 1.5 km of wire of cross-sectional area 0.17 mm^2 is 150 Ω. Determine the resistivity of the wire.

Resistance, $R = \rho l/a$. Hence

$$\text{resistivity} = \frac{Ra}{l} = \frac{(150\,\Omega)(0.17 \times 10^{-6}\,\text{m}^2)}{(1500\,\text{m})}$$

$$= \mathbf{0.017 \times 10^{-6}\,\Omega\,m}\ \text{or}\ \mathbf{0.017\,\mu\Omega\,m}$$

Problem 7. Determine the resistance of 1200 m of copper cable having a diameter of 12 mm if the resistivity of copper is $1.7 \times 10^{-8}\,\Omega$ m.

Cross-sectional area of cable, $a = \pi r^2$

$$= \pi \left(\frac{12}{2}\right)^2 = 36\pi\ \text{mm}^2$$

$$= 36\pi \times 10^{-6}\,\text{m}^2$$

Resistance $R = \dfrac{\rho l}{a}$

$$= \frac{(1.7 \times 10^{-8}\,\Omega\,\text{m})(1200\,\text{m})}{(36\pi \times 10^{-6}\,\text{m}^2)}$$

$$= \frac{1.7 \times 1200 \times 10^6}{10^8 \times 36\pi}\,\Omega$$

$$= \frac{1.7 \times 12}{36\pi}\,\Omega = \mathbf{0.180\,\Omega}$$

27.2 Temperature coefficient of resistance

In general, as the temperature of a material increases, most conductors increase in resistance, insulators decrease in resistance, while the resistance of some special alloys remain almost constant.

The **temperature coefficient of resistance** of a material is the increase in the resistance of a 1 Ω resistor of that material when it is subjected to a rise of temperature of 1°C. The symbol used for the temperature coefficient of resistance is α (alpha). Thus, if some copper wire of resistance 1 Ω is heated through 1°C and its resistance is

then measured as 1.0043 Ω then $\alpha = 0.0043\ \Omega/\Omega$ °C for copper. The units are usually expressed only as 'per °C', . i.e. $\alpha = 0.0043$/°C for copper. If the 1 Ω resistor of copper is heated through 100°C then the resistance at 100°C would be $1 + 100 \times 0.0043 = 1.43\ \Omega$.

Some typical values of temperature coefficient of resistance measured at 0°C are given below:

Copper	0.0043/°C	Aluminium	0.0038/°C
Nickel	0.0062/°C	Carbon	−0.000 48/°C
Constantan	0	Eureka	0.000 01/°C

(Note that the negative sign for carbon indicates that its resistance falls with increase of temperature.)

If the resistance of a material at 0°C is known the resistance at any other temperature can be determined from:

$$\boxed{R_\theta = R_0(1 + \alpha_0\theta)}$$

where R_0 = resistance at 0°C
R_θ = resistance at temperature θ°C
α_0 = temperature coefficient of resistance at 0°C.

Problem 8. A coil of copper wire has a resistance of 100 Ω when its temperature is 0°C. Determine its resistance at 100°C if the temperature coefficient of resistance of copper at 0°C is 0.0043/°C.

Resistance $R_\theta = R_0(1 + \alpha_0\theta)$.
Hence resistance at 100°C is given by

$$R_{100} = 100[1 + (0.0043)(100)]$$

$$= 100[1 + 0.43] = 100(1.43) = \mathbf{143\ \Omega}$$

Problem 9. An aluminium cable has a resistance of 27 Ω at a temperature of 35°C. Determine its resistance at 0°C. Take the temperature coefficient of resistance at 0°C to be 0.0038/°C.

Resistance at θ°C, $R_\theta = R_0 (1 + \alpha_0\theta)$.
Hence resistance at 0°C is given by

$$R_0 = \frac{R_\theta}{(1 + \alpha_0\theta)} = \frac{27}{[1 + (0.0038)(35)]}$$

$$= \frac{27}{1 + 0.133} = \frac{27}{1.133} = \mathbf{23.83\ \Omega}$$

Problem 10. A carbon resistor has a resistance of 1 kΩ at 0°C. Determine its resistance at 80°C. Assume that the temperature coefficient of resistance for carbon at 0°C is –0.0005.

Resistance at temperature θ°C is given by

$$R_\theta = R_0 (1 + \alpha_0\theta)$$

i.e.

$$R_\theta = 1000[1 + (-0.0005)(80)]$$

$$= 1000(1 - 0.040) = 1000(0.96)$$

$$= \mathbf{960 \ \Omega}$$

If the resistance of a material at room temperature (approximately 20°C), R_{20}, and the temperature coefficient of resistance at 20°C, α_{20}, are known then the resistance R_θ at temperature θ°C is given by:

$$\boxed{R_\theta = R_{20}[1 + \alpha_{20}(\theta - 20)]}$$

Problem 11. A coil of copper wire has a resistance of 10 Ω at 20°C. If the temperature coefficient of resistance of copper at 20°C is 0.004/°C determine the resistance of the coil when the temperature rises to 100°C.

Resistance at θ°C, $R_\theta = R_{20} [1 + \alpha_{20}(\theta - 20)]$. Hence resistance at 100°C is given by

$$R_{100} = 10[1 + (0.004)(100 - 20)]$$

$$= 10[1 + (0.004)(80)]$$

$$= 10[1 + 0.32]$$

$$= 10(1.32) = \mathbf{13.2 \ \Omega}$$

Problem 12. The resistance of a coil of aluminium wire at 18°C is 200 Ω. The temperature of the wire is increased and the resistance rises to 240 Ω. If the temperature coefficient of resistance of aluminium is 0.0039/°C at 18°C determine the temperature to which the coil has risen.

Let the temperature rise to θ°C. Resistance at θ°C, $R_\theta = R_{18}[1 + \alpha_{18}(\theta - 18)]$, i.e.

$$240 = 200[1 + (0.0039)(\theta - 18)]$$

$$240 = 200 + (200)(0.0039)(\theta - 18)$$

$$240 - 200 = 0.78(\theta - 18)$$

$$40 = 0.78(\theta - 18)$$

$$\frac{40}{0.78} = \theta - 18$$

$$51.28 = \theta - 18, \text{ from which, } \theta = 51.28 + 18$$
$$= 69.28°C$$

Hence the temperature of the coil increases to 69.28°C.

If the resistance at 0°C is not known, but is known at some other temperature θ_1, then the resistance at any temperature can be found as follows:

$$R_1 = R_0(1 + \alpha_0\theta_1) \text{ and } R_2 = R_0(1 + \alpha_0\theta_2)$$

Dividing one equation by the other gives:

$$\boxed{\frac{R_1}{R_2} = \frac{1 + \alpha_0\theta_1}{1 + \alpha_0\theta_2}}$$

where R_2 = resistance at temperature θ_2.

Problem 13. Some copper wire has a resistance of 200 Ω at 20°C. A current is passed through the wire and the temperature rises to 90°C. Determine the resistance of the wire at 90°C, correct to the nearest ohm, assuming that the temperature coefficient of resistance is 0.004/°C at 0°C.

$R_{20} = 200 \ \Omega$, $\alpha_0 = 0.004/°C$

$$\frac{R_{20}}{R_{90}} = \frac{[1 + \alpha_0(20)]}{[1 + \alpha_0(90)]}$$

Hence

$$R_{90} = \frac{R_{20}[1 + 90\alpha_0]}{[1 + 20\alpha_0]}$$

$$= \frac{200[1 + 90(0.004)]}{[1 + 20(0.004)]}$$

$$= \frac{200[1 + 0.36]}{[1 + 0.08]}$$

$$= \frac{200(1.36)}{(1.08)} = 251.85 \ \Omega$$

i.e. **the resistance of the wire at 90°C is 252 Ω.**

27.3 Multi-choice questions on resistance variation

(Answers on page 356.)

1. The unit of resistivity is:
   (a) ohms   (b) ohm millimetre
   (c) ohm metre   (d) ohm/metre
2. The length of a certain conductor of resistance 100 Ω is doubled and its cross-sectional area is halved. Its new resistance is:
   (a) 100 Ω (b) 200 Ω (c) 50 Ω (d) 400 Ω
3. The resistance of a 2 km length of cable of cross-sectional area 2 m^2 and resistivity of 2×10^{-8} Ω m is:
   (a) 0.02 Ω (b) 20 Ω (c) 0.02 m Ω
   (d) 200 Ω
4. A piece of graphite has a cross-sectional area of 10 mm^2. If its resistance is 0.1 Ω and its resistivity 10×10^{-8} Ω m, its length is:
   (a) 10 km   (b) 10 cm   (c) 10 mm   (d) 10 m
5. The symbol for the unit of temperature coefficient of resistance is:
   (a) Ω/°C   (b) Ω (c) °C   (d) Ω/Ω°C
6. A coil of wire has a resistance of 10 Ω at 0°C. If the temperature coefficient of resistance for the wire is 0.004/°C its resistance at 100°C is:
   (a) 0.4 Ω (b) 1.4 Ω (c) 14 Ω (d) 10 Ω
7. A nickel coil has a resistance of 13 Ω at 50°C. If the temperature coefficient of resistance at 0°C is 0.006/°C, the resistance at 0°C is:
   (a) 16.9 Ω (b) 10 Ω (c) 43.3 Ω (d) 0.1 Ω

27.4 Short answer questions on resistance variation

1. Name four factors which can affect the resistance of a conductor.
2. If the length of a piece of wire of constant cross-sectional area is halved, the resistance of the wire is ....
3. If the cross-sectional area of a certain length of cable is trebled, the resistance of the cable is ....
4. What is resistivity? State its unit and the symbol used.
5. Complete the following:
   Good conductors of electricity have a .... value of resistivity and good insulators have a .... value of resistivity.

6. What is meant by the 'temperature coefficient of resistance'? State its units and the symbols used.
7. If the resistance of a metal at 0°C is R_0, R_θ is the resistance at θ°C and α_0 is the temperature coefficient of resistance at 0°C then:
   $R_\theta = $ ....

27.5 Further questions on resistance variation

1. The resistance of a 2 m length of cable is 2.5 Ω. Determine
   (a) the resistance of a 7 m length of the same cable, and
   (b) the length of the same wire when the resistance is 6.25 Ω.
   [(a) 8.75 Ω (b) 5 m]
2. Some wire of cross-sectional area 1 mm^2 has a resistance of 20 Ω. Determine (a) the resistance of a wire of the same length and material if the cross-sectional area is 4 mm^2, and (b) the cross-sectional area of a wire of the same length and material if the resistance is 32 Ω.
   [(a) 5 Ω (b) 0.625 mm^2]
3. Some wire of length 5 m and cross-sectional area 2 mm^2 has a resistance of 0.08 Ω. If the wire is drawn out until its cross-sectional area is 1 mm^2, determine the resistance of the wire.
   [0.32 Ω]
4. Find the resistance of 800 m of copper cable of cross-sectional area 20 mm^2. Take the resistivity of copper as 0.02 $\mu\Omega$ m.
   [0.8 Ω]
5. Calculate the cross-sectional area, in mm^2, of a piece of aluminium wire 100 m long and having a resistance of 2 Ω. Take the resistivity of aluminium as 0.03×10^{-6} Ω m.
   [1.5 mm^2]
6. (a) What does the resistivity of a material mean?
   (b) The resistance of 500 m of wire of cross-sectional area 2.6 mm^2 is 5 Ω. Determine the resistivity of the wire in $\mu\Omega$ m.
   [0.026 $\mu\Omega$ m]
7. Find the resistance of 1 km of copper cable having a diameter of 10 mm if the resistivity of copper is 0.017×10^{-6} Ω m.   [0.216 Ω]
8. A coil of aluminium wire has a resistance of 50 Ω when its temperature is 0°C. Determine its resistance at 100°C if the temperature coefficient of resistance of aluminium at 0°C is 0.0038/°C.   [69 Ω]

9. A copper cable has a resistance of 30 Ω at a temperature of 50°C. Determine its resistance at 0°C. Take the temperature coefficient of resistance of copper at 0°C as 0.0043/°C. [24.69 Ω]

10. The temperature coefficient of resistance for carbon at 0°C is –0.00048/°C. What is the significance of the minus sign? A carbon resistor has a resistance of 500 Ω at 0°C. Determine its resistance at 50°C. [488 Ω]

11. A coil of copper wire has a resistance of 20 Ω at 18°C. If the temperature coefficient of resistance of copper at 18°C is 0.004/°C, determine the resistance of the coil when the temperature rises to 98°C. [26.4 Ω]

12. The resistance of a coil of nickel wire at 20°C is 100 Ω. The temperature of the wire is increased and the resistance rises to 130 Ω. If the temperature coefficient of resistance of nickel is 0.006/°C at 20°C, determine the temperature to which the coil has risen. [70°C]

13. Some aluminium wire has a resistance of 50 Ω at 20°C. The wire is heated to a temperature of 100°C. Determine the resistance of the wire at 100°C, assuming that the temperature coefficient of resistance at 0°C is 0.004/°C. [64.8 Ω]

14. A copper cable is 1.2 km long and has a cross-sectional area of 5 mm². Find its resistance at 80°C if at 20°C the resistivity of copper is 0.02 × 10⁻⁶ Ω m and its temperature coefficient of resistance is 0.004/°C. [5.952 Ω]

28

Chemical effects of electricity

At the end of this chapter you should be able to:

- understand electrolysis and its applications, including electroplating

- appreciate the purpose and construction of a simple cell

- explain polarization and local action

- explain corrosion and its effects

- define the terms e.m.f., E, and internal resistance, r, of a cell

- perform calculations using $V = E - Ir$

- determine the total e.m.f. and total internal resistance for cells connected in series and in parallel

- distinguish between primary and secondary cells

- explain the construction and practical applications of the Leclanché cell, the mercury cell, the lead–acid cell and the alkaline cell

- list the advantages and disadvantages of alkaline cells over lead–acid cells

- understand the term 'cell capacity' and state its unit

28.1 Introduction

A material must contain **charged particles** to be able to conduct electric current. In **solids**, the current is carried by **electrons**. Copper, lead, aluminium, iron and carbon are some examples of solid conductors. In **liquids** and **gases**, the current is carried by the part of a molecule which has acquired an electric charge, called **ions**. These can possess a positive or negative charge, and examples include hydrogen ion H^+, copper ion Cu^{++} and hydroxyl ion OH^-. Distilled water contains no ions and is a poor conductor of electricity whereas salt water contains ions and is a fairly good conductor of electricity.

28.2 Electrolysis

Electrolysis is the decomposition of a liquid compound by the passage of electric current through it. Practical applications of electrolysis include the electroplating of metals (see section 28.3), the refining of copper and the extraction of aluminium from its ore.

An **electrolyte** is a compound which will undergo electrolysis. Examples include salt water, copper sulphate and sulphuric acid.

The **electrodes** are the two conductors carrying current to the electrolyte. The positive-connected electrode is called the **anode** and the negative-connected electrode the **cathode**.

When two copper wires connected to a battery are placed in a beaker containing a salt water solution, then current will flow through the solution. Air bubbles appear around the wires as the water is changed into hydrogen and oxygen by electrolysis.

28.3 Electroplating

Electroplating uses the principle of electrolysis to apply a thin coat of one metal to another metal. Some practical applications include the tin-plating of steel, silver-plating of nickel alloys and chromium-plating of steel. If two copper electrodes connected to a battery are placed in a beaker containing copper sulphate as the electrolyte it is found that the cathode (i.e. the electrode connected to the negative terminal of the battery) gains copper while the anode loses copper.

28.4 The simple cell

The purpose of an **electric cell** is to convert chemical energy into electrical energy.

A **simple cell** comprises two dissimilar conductors (electrodes) in an electrolyte. Such a cell is shown in Fig. 28.1, comprising copper and zinc electrodes. An electric current is found to flow between the electrodes. Other possible electrode pairs exist, including zinc–lead and zinc–iron. The electrode potential (i.e. the p.d. measured between the electrodes) varies for each pair of metals. By knowing the e.m.f. of each metal with respect to some standard electrode the e.m.f. of any pair of metals may be determined. The standard used is the hydrogen electrode. The **electrochemical series** is a way of listing elements in order of electrical potential, and Table 28.1 shows a number of elements in such a series. In a simple cell two faults exist – those due to polarization and those due to local action.

Table 28.1 Part of the electrochemical series

potassium
sodium
aluminium
zinc
iron
lead
hydrogen
copper
silver
carbon

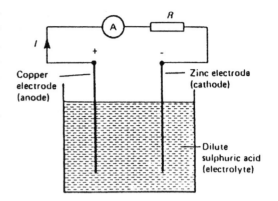

Figure 28.1

Polarization

If the simple cell shown in Fig. 28.1 is left connected for some time, the current I decreases fairly rapidly. This is because of the formation of a film of hydrogen bubbles on the copper anode. This effect is known as the polarization of the cell. The hydrogen prevents full contact between the copper electrode and the electrolyte and this increases the internal resistance of the cell. The effect can be overcome by using a chemical depolarizing agent or depolarizer, such as potassium dichromate which removes the hydrogen bubbles as they form. This allows the cell to deliver a steady current.

Local action

When commercial zinc is placed in dilute sulphuric acid, hydrogen gas is liberated from it and the zinc dissolves. The reason for this is that impurities, such as traces of iron, are present in the zinc which set up small primary cells with the zinc. These small cells are short-circuited by the electrolyte, with the result that localized currents flow causing corrosion. This action is known as local action of the cell. This may be prevented by rubbing a small amount of mercury on the zinc surface, which forms a protective layer on the surface of the electrode.

When two metals are used in a simple cell the electrochemical series may be used to predict the behaviour of the cell:

(i) The metal that is higher in the series acts as the negative electrode, and vice versa. For example, the zinc electrode in the cell shown in Fig. 28.1 is negative and the copper electrode is positive.

(ii) The greater the separation in the series between the two metals the greater is the e.m.f. produced by the cell.

The electrochemical series is representative of the order of reactivity of the metals and their compounds:

(i) The higher metals in the series react more readily with oxygen and vice-versa.

(ii) When two metal electrodes are used in a simple cell the one that is higher in the series tends to dissolve in the electrolyte.

28.5   Corrosion

Corrosion is the gradual destruction of a metal in a damp atmosphere by means of simple cell action. In addition to the presence of moisture and air required for rusting, an electrolyte, an anode and a cathode are required for corrosion. Thus, if metals widely spaced in the electrochemical series are used in contact with each other in the presence of an electrolyte, corrosion will occur. For example, if a brass valve is fitted to a heating system made of steel, corrosion will occur.

The **effects of corrosion** include the weakening of structures, the reduction of the life of components and materials, the wastage of materials and the expense of replacement.

Corrosion may be **prevented** by coating with paint, grease, plastic coatings and enamels, or by plating with tin or chromium. Also, iron may be galvanized, i.e. plated with zinc, the layer of zinc helping to prevent the iron from corroding.

28.6   E.m.f. and internal resistance of a cell

The **electromotive force (e.m.f.)**, E, of a cell is the p.d. between its terminals when it is not connected to a load (i.e. the cell is on 'no-load').

The e.m.f. of a cell is measured by using a **high resistance voltmeter** connected in parallel with the cell. The voltmeter must have a high resistance otherwise it will pass current and the cell will not be on no-load. For example, if the resistance of a cell is 1 Ω and that of a voltmeter 1 MΩ then the equivalent resistance of the circuit

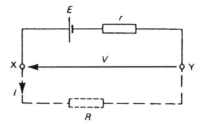

Figure 28.2

is 1 MΩ + 1 Ω, i.e. approximately 1 MΩ, hence no current flows and the cell is not loaded.

The voltage available at the terminals of a cell falls when a load is connected. This is caused by the **internal resistance** of the cell which is the opposition of the material of the cell to the flow of current. The internal resistance acts in series with other resistances in the circuit. Figure 28.2 shows a cell of e.m.f. E volts and internal resistance, r, and XY represents the terminals of the cell.

When a load (shown as resistance R) is not connected, no current flows and the terminal p.d., $V = E$. When R is connected a current I flows which causes a voltage drop in the cell, given by Ir. The p.d. available at the cell terminals is less than the e.m.f. of the cell and is given by:

$$V = E - Ir$$

Thus if a battery of e.m.f. 12 volts and internal resistance 0.1 Ω delivers a current of 100 A, the terminal p.d. is given by

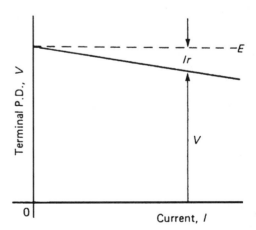

Figure 28.3

$$V = 12 - (100)(0.01)$$

$$= 12 - 1 = 11 \text{ V}$$

When different values of potential difference V across a cell or power supply are measured for different values of current I, a graph may be plotted as shown in Fig. 28.3. Since the e.m.f. E of the cell or power supply is the p.d. across its terminals on no-load (i.e. when $I = 0$), then E is as shown by the broken line. Since $V = E - Ir$ then the internal resistance may be calculated from

$$r = \frac{E - V}{I}$$

When a current is flowing in the direction shown in Fig. 28.2 the cell is said to be **discharging** ($E > V$).

When a current flows in the opposite direction to that shown in Fig. 28.2 the cell is said to be **charging** ($V > E$).

A battery is a combination of more than one cell. The cells in a battery may be connected in series or in parallel.

(i) For cells connected in series:

Total e.m.f. = sum of cell's e.m.f.s
Total internal resistance = sum of cell's internal resistances

(ii) For cells connected in parallel:

If each cell has the same e.m.f. and internal resistance:
Total e.m.f. = e.m.f. of one cell
Total internal resistance of n cells
$= \dfrac{1}{n} \times$ internal resistance of one cell

Problem 1.   Eight cells, each with an internal resistance of 0.2 Ω and an e.m.f. of 2.2 V are connected (a) in series, (b) in parallel. Determine the e.m.f. and the internal resistance of the batteries so formed.

(a) When connected in series, total e.m.f.
= sum of cell's e.m.f.
= 2.2 × 8 = **17.6 V**
Total internal resistance
= sum of cell's internal resistance
= 0.2 × 8 = **1.6 Ω**

(b) When connected in parallel, total e.m.f.
= e.m.f. of one cell
= **2.2 V**
Total internal resistance of 8 cells

$= \dfrac{1}{8} \times$ internal resistance of one cell

$= \dfrac{1}{8} \times 0.2 = \textbf{0.025 Ω}$

Problem 2.   A cell has an internal resistance of 0.02 Ω and an e.m.f. of 2.0 V. Calculate its terminal p.d. if it delivers (a) 5 A, (b) 50 A.

(a) Terminal p.d., $V = E - Ir$ where E = e.m.f. of cell, I = current flowing and r = internal resistance of cell. $E = 2.0$ V, $I = 5$ A and $r = 0.02$ Ω, hence

$$V = 2.0 - (5)(0.2) = 2.0 - 0.1 = \textbf{1.9 V}$$

(b) When the current is 50 A, terminal p.d. is given by

$$V = E - Ir = 2.0 - 50(0.02)$$

i.e.

$$V = 2.0 - 1.0 = \textbf{1.0 V}$$

Thus the terminal p.d. decreases as the current drawn increases.

Problem 3.   The p.d. at the terminals of a battery is 25 V when no-load is connected and 24 V when a load taking 10 A is connected. Determine the internal resistance of the battery.

When no-load is connected the e.m.f. of the battery, E, is equal to the terminal p.d., V, i.e. $E = 25$ V. When current $I = 10$ A and terminal p.d. $V = 24$ V, then $V = E - Ir$, i.e.

$$24 = 25 - (10)r$$

Hence, rearranging, gives

$$10r = 25 - 24 = 1$$

and the internal resistance

$$r = \frac{1}{10} = \textbf{0.1 Ω}$$

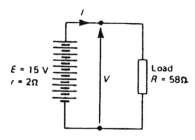

Figure 28.4

Problem 4.   Ten 1.5 V cells, each having an internal resistance of 0.2 Ω, are connected in series to a load of 58 Ω. Determine (a) the current flowing in the circuit and (b) the p.d. at the battery terminals.

(a)   For ten cells, battery e.m.f., $E = 10 \times 1.5 = 15$ V, and the total internal resistance, $r = 10 \times 0.2 = 2$ Ω. When connected to a 58 Ω load the circuit is as shown in Fig. 28.4.

$$\text{Current } I = \frac{\text{e.m.f.}}{\text{total resistance}} = \frac{15}{58 + 2}$$

$$= \frac{15}{60} = \textbf{0.25 A}$$

(b)   p.d. to battery terminals is $V = E - Ir$, i.e.

$$V = 15 - (0.25)(2) = \textbf{14.5 V}$$

28.7   Primary cells

Primary cells cannot be recharged, that is, the conversion of chemical energy to electrical energy is irreversible and the cell cannot be used once the chemicals are exhausted. Examples of primary cells include the Leclanché cell and the mercury cell.

Leclanché cell

A typical dry Leclanché cell is shown in Fig. 28.5. Such a cell has an e.m.f. of about 1.5 V when new, but this falls rapidly due to polarization if in continuous use. The hydrogen film on the carbon electrode forms faster than can be dissipated by the depolarizer. The Leclanché cell is suitable only for intermittent use, applications including torches, transistor radios, bells, indicator circuits, gas lighters, controlling switch-gear, and so on. The cell is the most commonly used of primary cells, is cheap, requires little maintenance and has a shelf life of about 2 years.

Mercury cell

A typical mercury cell is shown in Fig. 28.6. Such a cell has an e.m.f. of about 1.3 V which remains constant for a relatively long time. Its main advantages over the Leclanché cell is its smaller size and its long shelf life. Typical practical

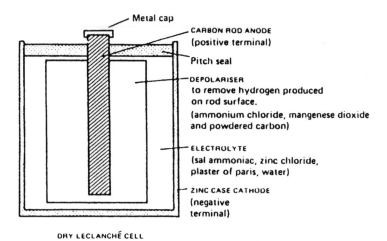

DRY LECLANCHÉ CELL

Figure 28.5

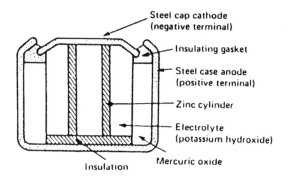

Figure 28.6

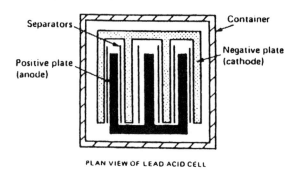

Figure 28.7

applications include hearing aids, medical electronics, cameras, and for guided missiles.

28.8 Secondary cells

Secondary cells can be recharged after use, that is, the conversion of chemical energy to electrical energy is reversible and the cell may be used many times. Examples of secondary cells include the lead–acid cell and alkaline cells. Practical applications of such cells include car batteries, telephone circuits and for traction purposes – such as milk delivery vans and fork lift trucks.

Lead–acid cell

A typical lead-acid cell is constructed of:

(i) A container made of glass, ebonite or plastic.
(ii) **Lead plates**
     (a) the negative plate (cathode) consists of spongy lead,
     (b) the positive plate (anode) is formed by pressing lead peroxide into the lead grid.
     The plates are interleaved, as shown in the plan view of Fig. 28.7, to increase their effective cross-sectional area and to minimize internal resistance.
(iii) **Separators** made of glass, celluloid or wood.
(iv) An **electrolyte** which is a mixture of sulphuric acid and distilled water.

The relative density (or specific gravity) of a lead–acid cell, which may be measured using a hydrometer, varies between about 1.26 when the cell is fully charged to about 1.19 when the cell is

discharged. The terminal p.d. of a lead–acid cell is about 2 V.

When a cell supplies current to a load it is said to be **discharging**. During discharge:

(i) the lead peroxide (positive plate) and the spongy lead (negative plate) are converted into lead sulphate, and
(ii) the oxygen in the lead peroxide combines with hydrogen in the electrolyte to form water. The electrolyte is therefore weakened and the relative density falls.

The terminal p.d. of a lead–acid cell when fully discharged is about 1.8 V.

A cell is **charged** by connecting a d.c. supply to its terminals, the positive terminal of the cell being connected to the positive terminal of the supply. The charging current flows in the reverse direction to the discharge current and the chemical action is reversed. During charging:

(i) the lead sulphate on the positive and negative plates is converted back to lead peroxide and lead respectively, and
(ii) the water content of the electrolyte decreases as the oxygen released from the electrolyte combines with the lead of the positive plate. The relative density of the electrolyte thus increases.

The colour of the positive plate when fully charged is dark brown and when discharged is light brown. The colour of the negative plate when fully charged is grey and when discharged is light grey.

Alkaline cell

There are two main types of alkaline cell – the nickel–iron cell and the nickel–cadmium cell. In

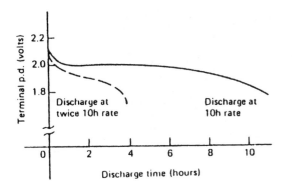

Figure 28.8

both types the positive plate is made of nickel hydroxide enclosed in finely perforated steel tubes, the resistance being reduced by the addition of pure nickel or graphite. The tubes are assembled into nickel–steel plates.

In the nickel–iron cell (sometimes called the Edison cell or nife cell), the negative plate is made of iron oxide, with the resistance being reduced by a little mercuric oxide, the whole being enclosed in perforated steel tubes and assembled in steel plates. In the nickel–cadmium cell the negative plate is made of cadmium. The electrolyte in each type of cell is a solution of potassium hydroxide which does not undergo any chemical change and thus the quantity can be reduced to a minimum. The plates are separated by insulating rods and assembled in steel containers which are then enclosed in a non-metallic crate to insulate the cells from one another. The average discharge p.d. of an alkaline cell is about 1.2 V.

Advantages of an alkaline cell (for example, a nickel–cadmium cell or a nickel–iron cell) over a lead–acid cell include:

(i)   More robust construction
(ii) Capable of withstanding heavy charging and discharging currents without damage
(iii) Has a longer life
(iv) For a given capacity is lighter in weight
(v)   Can be left indefinitely in any state of charge or discharge without damage
(vi) Is not self-discharging.

Disadvantages of an alkaline cell over a lead–acid cell include:

(i) Is relatively more expensive
(ii) Requires more cells for a given e.m.f.

(iii) Has a higher internal resistance
(iv) Must be kept sealed
(v)   Has a lower efficiency.

Alkaline cells may be used in extremes of temperature, in conditions where vibration is experienced or where duties require long idle periods or heavy discharge currents. Practical examples include traction and marine work, lighting in railway carriages, military portable radios and for starting diesel and petrol engines. However, the lead–acid cell is the most common one in practical use.

28.9   Cell capacity

The **capacity** of a cell is measured in ampere-hours (Ah). A fully charged 50 Ah battery rated for 10 h discharge can be discharged at a steady current of 5 A for 10 h, but if the load current is increased to 10 A then the battery is discharged in 3–4 h, since the higher the discharge current, the lower is the effective capacity of the battery. Typical discharge characteristics for a lead–acid cell are shown in Fig. 28.8.

28.10   Multi-choice questions on the chemical effects of electricity

(Answers on page 356.)

1. The terminal p.d. of a cell of e.m.f. 2 V and internal resistance 0.1 Ω when supplying a current of 5 A will be:
   (a) 1.5 V   (b) 2 V   (c) 1.9 V   (d) 2.5 V
2. Five cells, each with an e.m.f. of 2 V and internal resistance 0.5 Ω, are connected in series. The resulting battery will have:
   (a) an e.m.f. of 2 V and an internal resistance of 0.5 Ω
   (b) an e.m.f. of 10 V and an internal resistance of 2.5 Ω
   (c) an e.m.f. of 2 V and an internal resistance of 0.1 Ω
   (d) an e.m.f. of 10 V and an internal resistance of 0.1 Ω
3. If the five cells of question 2 are connected in parallel the resulting battery will have:
   (a) an e.m.f. of 2 V and an internal resistance of 0.5 Ω
   (b) an e.m.f. of 10 V and an internal resistance of 2.5 Ω

(c) an e.m.f. of 2 V and an internal resistance of 0.1 Ω

(d) an e.m.f. of 10 V and an internal resistance of 0.1 Ω

4. Which of the following statements is false?
(a) A Leclanché cell is suitable for use in torches.
(b) A nickel–cadmium cell is an example of a primary cell.
(c) When a cell is being charged its terminal p.d. exceeds the cell e.m.f.
(d) A secondary cell may be recharged after use.

5. Which of the following statements is false? When two metal electrodes are used in a simple cell, the one that is higher in the electrochemical series:
(a) tends to dissolve in the electrolyte
(b) is always the negative electrode
(c) reacts most readily with oxygen
(d) acts as the anode

6. Five 2 V cells, each having an internal resistance of 0.2 Ω, are connected in series to a load of resistance 14 Ω. The current flowing in the circuit is:
(a) 10 A   (b) 1.4 A   (c) 1.5 A   (d) $\frac{2}{3}$ A

7. For the circuit of question 6, the p.d. at the battery terminals is:
(a) 10 V   (b) $9\frac{1}{3}$ V   (c) 0 V   (d) $10\frac{2}{3}$ V

8. Which of the following statements is true?
(a) The capacity of a cell is measured in volts.
(b) A primary cell converts electrical energy into chemical energy.
(c) Galvanizing iron helps to prevent corrosion.
(d) A positive electrode is termed the cathode.

28.11   Short answer questions on the chemical effects of electricity

1. What is electrolysis?
2. What is an electrolyte?
3. Conduction in electrolytes is due to ....
4. A positive-connected electrode is called the .... and the negative-connected electrode the ....
5. Name two practical applications of electrolysis.
6. The purpose of an electric cell is to convert .... to ....

7. Make a labelled sketch of a simple cell.
8. What is the electrochemical series?
9. What is corrosion?
10. Name two effects of corrosion and state how they may be prevented.
11. What is meant by the e.m.f. of a cell? How may the e.m.f. of a cell be measured?
12. Define internal resistance.
13. If a cell has an e.m.f. of E volts, an internal resistance of r ohms and supplies a current I amperes to a load, the terminal p.d. V volts is given by: $V = $ ....
14. Name the two main types of cells.
15. Explain briefly the difference between primary and secondary cells.
16. Name two types of primary cells.
17. Name two types of secondary cells.
18. State three typical applications of primary cells.
19. State three typical applications of secondary cells.
20. In what units are the capacity of a cell measured?

28.12   Further questions on the chemical effects of electricity

1. Twelve cells, each with an internal resistance of 0.24 Ω and an e.m.f. of 1.5 V are connected (a) in series, (b) in parallel. Determine the e.m.f. and internal resistance of the batteries so formed.
[(a) 18 V, 2.88 Ω (b) 1.5 V, 0.02 Ω]

2. A piece of chromium and a piece of iron are placed in an electrolyte in a container. A d.c. supply is connected between the pieces of metal, the positive terminal being connected to the chromium. Explain what is likely to happen and why this happens.

3. With reference to conduction in electrolytes, explain briefly how silver plating of nickel alloys is achieved.

4. A cell has an internal resistance of 0.03 Ω and an e.m.f. of 2.2 V. Calculate its terminal p.d. if it delivers (a) 1 A, (b) 20 A, (c) 50 A.
[(a) 2.17 V (b) 1.6 V (c) 0.7 V]

5. The p.d. at the terminals of a battery is 16 V when no load is connected and 14 V when a load taking 8 A is connected. Determine the internal resistance of the battery.
[0.25 Ω]

6. A battery of e.m.f. 20 V and internal resistance 0.2 Ω supplies a load taking 10 A.

Determine the p.d. at the battery terminals and the resistance of the load.

[18 V, 1.8 Ω]

7. Ten 2.2 V cells, each having an internal resistance of 0.1 Ω are connected in series to a load of 21 Ω. Determine (a) the current flowing in the circuit, and (b) the p.d. at the battery terminals. [(a) 1 A (b) 21 V]

8. For the circuits shown in Fig. 28.9 the resistors represent the internal resistance of the batteries. Find, in each case:
(a) the total e.m.f. across PQ,
(b) the total equivalent internal resistances of the batteries.
[(a)(i) 6 V (ii) 2 V (b)(i) 4 V (ii) 0.25 Ω]

9. The voltage at the terminals of a battery is 52 V when no load is connected and 48.8 V when a load taking 80 A is connected. Find the internal resistance of the battery. What would be the terminal voltage when a load taking 20 A is connected?

[0.04 Ω, 51.2 V]

10. Define electrolysis and use an example of electrolysis to show the meaning of the terms (a) electrode, (b) electrolyte.

11. The simple cell has two main faults – polarization and local action. Explain these two phenomena.

12. What is corrosion? State its effects and how it may be prevented.

13. Explain the difference between primary and secondary cells. Make a fully labelled sketch of one of each type of cell.

14. Describe the charging and discharging sequence of a simple lead–acid secondary cell.

15. Compare the performances and uses of the Leclanché dry cell and the mercury cell.

16. State three practical applications for each of the following cells:
(a) dry Leclanché (b) mercury (c) lead–acid (d) alkaline.

17. How may the electrochemical series be used to predict the behaviour of a cell? Also, how is the series representative of the order of reactivity of the metals and their compounds?

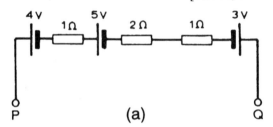

(a)

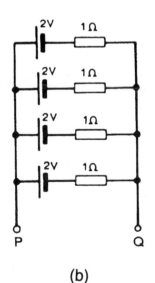

(b)

Figure 28.9

29

Series and parallel networks

At the end of this chapter you should be able to:

- calculate unknown voltages, current and resistances in a series circuit
- understand voltage division in a series circuit
- calculate unknown voltages, currents and resistances in a parallel network
- calculate unknown voltages, currents and resistances in series–parallel networks
- understand current division in a two-branch parallel network
- describe the advantages and disadvantages of series and parallel connection of lamps

29.1 Series circuits

Figure 29.1 shows three resistors R_1, R_2 and R_3 connected end to end, i.e. in series, with a battery source of V volts. Since the circuit is closed a current I will flow and the p.d. across each resistor may be determined from the voltmeter readings V_1, V_2 and V_3.

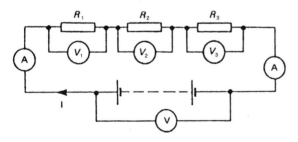

Figure 29.1

In a series circuit

(a) the current I is the same in all parts of the circuit and hence the same reading is found on each of the ammeters shown, and

(b) the sum of the voltages V_1, V_2 and V_3 is equal to the total applied voltage, V, i.e.

$$V = V_1 + V_2 + V_3$$

From Ohm's law:

$V_1 = IR_1$, $V_2 = IR_2$, $V_3 = IR_3$ and $V = IR$

where R is the total circuit resistance.
Since $V = V_1 + V_2 + V_3$, then $IR = IR_1 + IR_2 + IR_3$. Dividing throughout by I gives

$$R = R_1 + R_2 + R_3$$

Thus for a series circuit, the total resistance is obtained by adding together the values of the separate resistances.

Problem 1. For the circuit shown in Fig. 29.2, determine (a) the battery voltage V, (b) the total resistance of the circuit, and (c) the values of resistance of resistors R_1, R_2 and R_3, given that the p.d.s across R_1, R_2 and R_3 are 5 V, 2 V and 6 V, respectively.

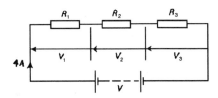

Figure 29.2

(a)

Battery voltage $V = V_1 + V_2 + V_3$

$$= 5 + 2 + 6 = \textbf{13 V}$$

(b)

Total circuit resistance $R = \dfrac{V}{I} = \dfrac{13}{4} = \textbf{3.25 }\Omega$

(c)

Resistance $R_1 = \dfrac{V_1}{I} = \dfrac{5}{4} = \textbf{1.25 }\Omega$

Resistance $R_2 = \dfrac{V_2}{I} = \dfrac{2}{4} = \textbf{0.5 }\Omega$

Resistance $R_3 = \dfrac{V_3}{I} = \dfrac{6}{4} = \textbf{1.5 }\Omega$

(Check: $R_1 + R_2 + R_3 = 1.25 + 0.5 + 1.5 = 3.25\ \Omega = R$)

Problem 2. For the circuit shown in Fig. 29.3, determine the p.d. across resistor R_3. If the total resistance of the circuit is $100\ \Omega$, determine the current flowing through resistor R_1. Find also the value of resistor R_2.

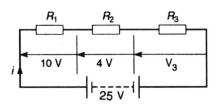

Figure 29.3

---

p.d. across R_3, $V_3 = 25 - 10 - 4 = \textbf{11 V}$.

Current $I = \dfrac{V}{R} = \dfrac{25}{100} = \textbf{0.25 A}$

which is the current flowing in each resistor.

Resistance $R_2 = \dfrac{V_2}{I} = \dfrac{4}{0.25} = \textbf{16 }\Omega$

Problem 3. A 12 V battery is connected in a circuit having three series-connected resistors having resistance of $4\ \Omega$, $9\ \Omega$ and $11\ \Omega$. Determine the current flowing through, and the p.d. across, the $9\ \Omega$ resistor. Find also the power dissipated in the $11\ \Omega$ resistor.

The circuit diagram is shown in Fig. 29.4.

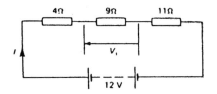

Figure 29.4

Total resistance $R = 4 + 9 + 11 = 24\ \Omega$.

Current $I = \dfrac{V}{R} = \dfrac{12}{24} = \textbf{0.5 A}$

which is the current in the $9\ \Omega$ resistor.
p.d. across the $9\ \Omega$ resistor, $V_1 = I \times 9 = 0.5 \times 9 = \textbf{4.5 V}$.
Power dissipated in the $11\ \Omega$ resistor is given by

$$P = I^2 R = 0.5^2(11)$$

$$= (0.25)(11) = \textbf{2.75 W}$$

29.2 Potential divider

The voltage distribution for the circuit shown in Fig. 29.5(a) is given by:

$$V_1 = \left(\frac{R_1}{R_1 + R_2}\right)V$$

$$V_2 = \left(\frac{R_2}{R_1 + R_2}\right)V$$

The circuit shown in Fig. 29.5(b) is often referred to as a **potential divider** circuit. Such a circuit can consist of a number of similar elements in series

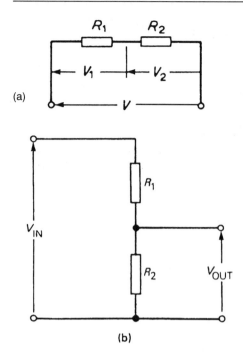

Figure 29.5

connected across a voltage source, voltages being taken from connections between the elements. Frequently the divider consists of two resistors as shown in Fig. 29.5(b), where

$$V_{OUT} = \left(\frac{R_2}{R_1 + R_2}\right)V_{IN}$$

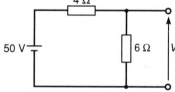

Figure 29.6

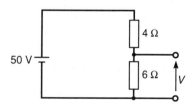

Figure 29.7

Problem 4.   Determine the value of voltage V shown in Fig. 29.6.

Figure 29.6 may be redrawn as shown in Fig. 29.7, and

$$\text{voltage } V = \left(\frac{6}{6 + 4}\right)(50) = \textbf{30 V}$$

Problem 5.   Two resistors are connected in series across a 24 V supply and a current of 3 A flows in the circuit. If one of the resistors has a resistance of 2 Ω determine (a) the value of the other resistor, and (b) the p.d. across the 2 Ω resistor. If the circuit is connected for 50 hours, how much energy is used?

The circuit diagram is shown in Fig. 29.8.

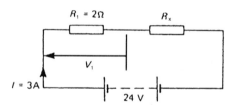

Figure 29.8

(a)   Total circuit resistance

$$R = \frac{V}{I} = \frac{24}{3} = 8\ \Omega$$

Value of unknown resistance

$$R_x = 8 - 2 = \textbf{6}\ \boldsymbol{\Omega}$$

(b)   p.d. across 2 Ω resistor

$$V_1 = IR_1 = 3 \times 2 = \textbf{6 V}$$

Alternatively, from above,

$$V_1 = \left(\frac{R_1}{R_1 + R_x}\right)V = \left(\frac{2}{2 + 6}\right)24 = 6\text{ V}$$

Energy used = power × time

$$= (V \times I) \times t$$
$$= (24 \times 3 \text{ W})(50 \text{ h})$$
$$= 3600 \text{ W h} = \textbf{3.6 kW h}$$

29.3 Parallel networks

Figure 29.9 shows three resistors, R_1, R_2 and R_3 connected across each other, i.e. in parallel, across a battery source of V volts.

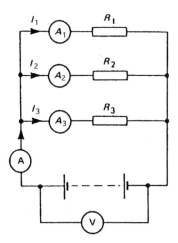

Figure 29.9

In a parallel circuit:

(a) the sum of the currents I_1, I_2 and I_3 is equal to the total circuit current, I, i.e.

$$\boxed{I = I_1 + I_2 + I_3}$$

and

(b) the source p.d., V volts, is the same across each of the resistors.

From Ohm's law:

$$I_1 = \frac{V}{R_1}, \; I_2 = \frac{V}{R_2}, \; I_3 = \frac{V}{R_3} \text{ and } I = \frac{V}{R}$$

where R is the total circuit resistance. Since

$$I = I_1 + I_2 + I_3$$

then

$$\frac{V}{R} = \frac{V}{R_1} + \frac{V}{R_2} + \frac{V}{R_3}$$

Dividing throughout by V gives

$$\boxed{\frac{1}{R} = \frac{1}{R_1} + \frac{1}{R_2} + \frac{1}{R_3}}$$

This equation must be used when finding the total resistance R of a parallel circuit. For the special case of **two resistors in parallel**

$$\frac{1}{R} = \frac{1}{R_1} + \frac{1}{R_2} = \frac{R_2 + R_1}{R_1 R_2}$$

Hence

$$\boxed{R = \frac{R_1 R_2}{R_1 + R_2}} \quad \left(\text{i.e. } \frac{\text{product}}{\text{sum}}\right)$$

Problem 6. For the circuit shown in Fig. 29.10, determine (a) the reading on the ammeter, and (b) the value of resistor R_2.

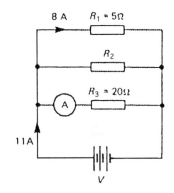

Figure 29.10

---

p.d. across R_1 is the same as the supply voltage V. Hence supply voltage, $V = 8 \times 5 = 40$ V.

(a) Reading on ammeter

$$I = \frac{V}{R_3} = \frac{40}{20} = \textbf{2 A}$$

(b) Current flowing through $R_2 = 11 - 8 - 2 = 1$ A, hence

$$R_2 = \frac{V}{I_2} = \frac{40}{1} = \textbf{40 } \Omega$$

Problem 7. Two resistors, of resistance 3 Ω and 6 Ω, are connected in parallel across a battery having a voltage of 12 V. Determine (a) the total circuit resistance and (b) the current flowing in the 3 Ω resistor.

The circuit diagram is shown in Fig. 29.11.

(a) The total circuit resistance R is given by

$$\frac{1}{R} = \frac{1}{R_1} + \frac{1}{R_2} = \frac{1}{3} + \frac{1}{6}$$

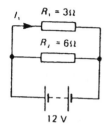

Figure 29.11

$$\frac{1}{R} = \frac{2+1}{6} = \frac{3}{6}$$

Hence, $R = (6/3) = \textbf{2 } \Omega$.

(Alternatively,

$$R = \frac{R_1 R_2}{R_1 + R_2} = \frac{3 \times 6}{3 + 6} = \frac{18}{9} = 2\Omega)$$

(b)   Current in the 3 Ω resistance,

$$I_1 = \frac{V}{R_1} = \frac{12}{3} = \textbf{4 A}$$

Problem 8.   For the circuit shown in Fig. 29.12, find (a) the value of the supply voltage V and (b) the value of current I.

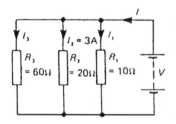

Figure 29.12

(a)   p.d. across 20 Ω resistor $= I_2 R_2 = 3 \times 20$
$= 60$ V, hence supply voltage $V = \textbf{60 V}$ since the circuit is connected in parallel.

(b)

Current $I_1 = \dfrac{V}{R_1} = \dfrac{60}{10} = 6$ A; $I_2 = \textbf{3 A}$

$$I_3 = \frac{V}{R_3} = \frac{60}{60} = 1 \text{ A}$$

Current $I = I_1 + I_2 + I_3$ and hence $I = 6 + 3 + 1 = \textbf{10 A.}$

Alternatively

$$\frac{1}{R} = \frac{1}{60} + \frac{1}{20} + \frac{1}{10} = \frac{1 + 3 + 6}{60} = \frac{10}{60}$$

Hence total resistance

$$R = \frac{60}{10} = 6 \; \Omega$$

Current

$$I = \frac{V}{R} = \frac{60}{6} = \textbf{10 A}$$

Problem 9.   Given four $1 \; \Omega$ resistors, state how they must be connected to give an overall resistance of (a) $\frac{1}{4} \Omega$ (b) $1 \; \Omega$ (c) $1\frac{1}{3} \; \Omega$ (d) $2\frac{1}{2} \; \Omega$, all four resistors being connected in each case.

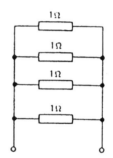

Figure 29.13

(a)   **All four in parallel** (see Fig. 29.13), since

$$\frac{1}{R} = \frac{1}{1} + \frac{1}{1} + \frac{1}{1} + \frac{1}{1} = \frac{4}{1}$$

$$R = \frac{1}{4} \; \Omega$$

(b)   **Two in series, in parallel with another two in series** (see Fig. 29.14), since $1 \; \Omega$ and $1 \; \Omega$ in series gives $2 \; \Omega$, and $2 \; \Omega$ in parallel with $2 \; \Omega$ gives:

$$\frac{2 \times 2}{2 + 2} = \frac{4}{4} = 1 \; \Omega$$

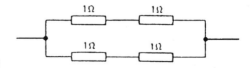

Figure 29.14

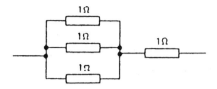

Figure 29.15

(c) **Three in parallel in series with one** (see Fig. 29.15), since for the three in parallel

$$\frac{1}{R} = \frac{1}{1} + \frac{1}{1} + \frac{1}{1} = \frac{3}{1}$$

i.e. $R = \frac{1}{3}\,\Omega$ and $\frac{1}{3}\,\Omega$ in series with $1\,\Omega$ gives $1\frac{1}{3}\,\Omega$.

(d) **Two in parallel, in series with two in series** (see Fig. 29.16), since for the two in parallel

$$R = \frac{1 \times 1}{1 + 1} = \frac{1}{2}\,\Omega$$

and $\frac{1}{2}\,\Omega$, $1\,\Omega$ and $1\,\Omega$ in series gives $2\frac{1}{2}\,\Omega$.

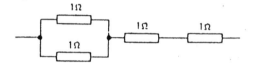

Figure 29.16

Problem 10.   Find the equivalent resistance for the circuit shown in Fig. 29.17.

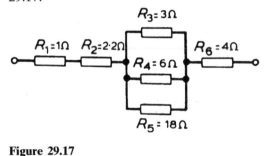

Figure 29.17

R_3, R_4 and R_5 are connected in parallel and their equivalent resistance R is given by:

$$\frac{1}{R} = \frac{1}{3} + \frac{1}{6} + \frac{1}{18} = \frac{6 + 3 + 1}{18} = \frac{10}{18}$$

Hence $R = (18/10) = 1.8\,\Omega$. The circuit is now equivalent to four resistors in series and the

equivalent circuit resistance $= 1 + 2.2 + 1.8 + 4 = 9\,\Omega$.

Problem 11.   Resistances of $10\,\Omega$, $20\,\Omega$ and $30\,\Omega$ are connected (a) in series and (b) in parallel to a 240 V supply. Calculate the supply current in each case.

(a) The series circuit is shown in Fig. 29.18. The equivalent resistance $R_T = 10\,\Omega + 20\,\Omega + 30\,\Omega = 60\,\Omega$. Supply current

$$I = \frac{V}{R_T} = \frac{240}{60} = \textbf{4 A}$$

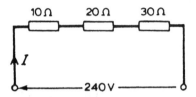

Figure 29.18

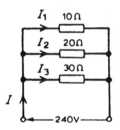

Figure 29.19

(b) The parallel circuit is shown in Fig. 29.19. The equivalent resistance R_T of $10\,\Omega$, $20\,\Omega$ and $30\,\Omega$ resistances connected in parallel is given by:

$$\frac{1}{R_T} = \frac{1}{10} + \frac{1}{20} + \frac{1}{30} = \frac{6 + 3 + 2}{60} = \frac{11}{60}$$

Hence $R_T = (60/11)\,\Omega$. Supply current

$$I = \frac{V}{R_T} = \frac{240}{60/11} = \frac{240 \times 11}{60} = \textbf{44 A}$$

(Check:

$$I_1 = \frac{V}{R_1} = \frac{240}{10} = 24\,\text{A}; \quad I_2 = \frac{V}{R_2} = \frac{240}{20} = 12\,\text{A}$$

$$I_3 = \frac{V}{R_3} = \frac{240}{30} = 8 \text{ A}$$

For a parallel circuit $I = I_1 + I_2 + I_3 = 24 + 12 + 8 = \mathbf{44 \text{ A}}$, as above.)

29.4   Current division

For the circuit shown in Fig. 29.20, the total circuit resistance, R_T, is given by:

$$R_T = \frac{R_1 R_2}{R_1 + R_2}$$

and

$$V = IR_T = I\left(\frac{R_1 R_2}{R_1 + R_2}\right)$$

$$\text{Current } I_1 = \frac{V}{R_1} = \frac{I}{R_1}\left(\frac{R_1 R_2}{R_1 + R_2}\right)$$

$$= \left(\frac{R_2}{R_1 + R_2}\right)(I)$$

Similarly

$$\text{current } I_2 = \frac{V}{R_2} = \frac{I}{R_2}\left(\frac{R_1 R_2}{R_1 + R_2}\right)$$

$$= \left(\frac{R_1}{R_1 + R_2}\right)(I)$$

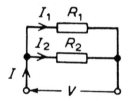

Figure 29.20

Summarizing, with reference to Fig. 29.20

$$\boxed{I_1 = \left(\frac{R_2}{R_1 + R_2}\right)(I)} \text{ and } \boxed{I_2 = \left(\frac{R_1}{R_1 + R_2}\right)(I)}$$

> **Problem 12.** For the series–parallel arrangement shown in Fig. 29.21, find (a) the supply current, (b) the current flowing through each resistor and (c) the p.d. across each resistor.

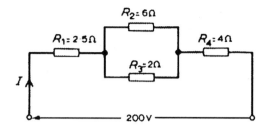

Figure 29.21

(a)   The equivalent resistance R_x of R_2 and R_3 in parallel is:

$$R_x = \frac{6 \times 2}{6 + 2} = \frac{12}{8} = 1.5 \ \Omega$$

The equivalent resistance R_T of R_1, R_x and R_4 in series is:

$$R_T = 2.5 + 1.5 + 4 = 8 \ \Omega$$

Supply current

$$I = \frac{V}{R_T} = \frac{200}{8} = \mathbf{25 \text{ A}}$$

(b)   The current flowing through R_1 and R_4 is 25 A.
The current flowing through R_2 is

$$\left(\frac{R_3}{R_2 + R_3}\right) I = \left(\frac{2}{6 + 2}\right) 25$$

$$= \mathbf{6.25 \text{ A}}$$

The current flowing through R_3 is

$$\left(\frac{R_3}{R_2 + R_3}\right) I = \left(\frac{6}{6 + 2}\right) 25$$

$$= \mathbf{18.75 \text{ A}}$$

(Note that the currents flowing through R_2 and R_3 must add up to the total current flowing into the parallel arrangement, i.e. 25 A.)

(c)   The equivalent circuit of Fig. 29.21 is shown in Fig. 29.22.

p.d. across R_1, i.e. $V_1 = IR_1 = (25)(2.5) = \mathbf{62.5 \text{ V}}$

p.d. across R_x, i.e. $V_x = IR_x$ (25)(1.5) = $\mathbf{37.5 \text{ V}}$

p.d. across R_4, i.e. $V_4 = IR_4 = (25)(4) = \mathbf{100 \text{ V}}$

Hence the p.d. across R_2 = p.d. across R_3 = $\mathbf{37.5 \text{ V}}$.

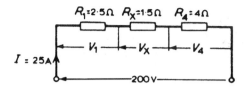

Figure 29.22

Problem 13.   For the circuit shown in Fig. 29.23 calculate (a) the value of resistor R_x such that the total power dissipated in the circuit is 2.5 kW, (b) the current flowing in each of the four resistors.

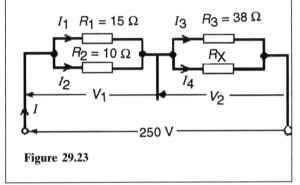

Figure 29.23

(a)   Power dissipated $P = VI$ watts, hence 2500 = (250)(I), i.e.

$$I = \frac{2500}{250} = 10 \text{ A}$$

From Ohm's law

$$R_T = \frac{V}{I} = \frac{250}{10} = 25 \text{ }\Omega$$

where R_T is the equivalent circuit resistance. The equivalent resistance of R_1 and R_2 in parallel is

$$\frac{15 \times 10}{15 + 10} = \frac{150}{25} = 6 \text{ }\Omega$$

The equivalent resistance of resistors R_3 and R_x in parallel is equal to 25 Ω – 6 Ω, i.e. 19 Ω.
There are three methods whereby R_x can be determined.

Method 1
The voltage $V_1 = IR$, where R is 6 Ω, from above, i.e. $V_1 = (10)(6) = 60$ V. Hence

$V_2 = 250 \text{ V} – 60 \text{ V} = 190 \text{ V}$

$\qquad$ = p.d. across R_3

$\qquad$ = p.d. across R_x

$$I_3 = \frac{V_2}{R_3} = \frac{190}{38} = 5 \text{ A}$$

Thus $I_4 = 5$ A also, since $I = 10$ A. Thus

$$R_x = \frac{V_2}{I_4} = \frac{190}{5} = 38 \text{ }\Omega$$

Method 2
Since the equivalent resistance of R_3 and R_x in parallel is 19 Ω,

$$19 = \frac{38R_x}{38 + R_x} \left(\text{i.e. } \frac{\text{product}}{\text{sum}}\right)$$

Hence

$19(38 + R_x) = 38R_x$

$722 + 19R_x = 38R_x$

$722 \qquad\quad = 38R_x – 19R_x = 19R_x$

Thus

$$R_x = \frac{722}{19} = 38 \text{ }\Omega$$

Method 3
When two resistors having the same value are connected in parallel the equivalent resistance is always half the value of one of the resistors. Thus, in this case, since $R_T = 19$ Ω and $R_3 = 38$ Ω, then $R_x = 38$ Ω could have been deduced on sight.

(b)   Current

$$I_1 = \left(\frac{R_2}{R_1 + R_2}\right) I = \left(\frac{10}{15 + 10}\right) 10$$

$$= \left(\frac{2}{5}\right) 10 = \textbf{4 A}$$

Current

$$I_2 = \left(\frac{R_1}{R_1 + R_2}\right) I = \left(\frac{15}{15 + 10}\right) 10$$

$$= \left(\frac{3}{5}\right) 10 = \textbf{6 A}$$

From part (a), method 1, $I_3 = I_4 = 5$ A.

Problem 14.   For the arrangement shown in Fig. 29.24, find the current I_x.

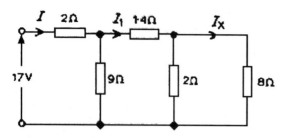

Figure 29.24

Commencing at the right-hand side of the arrangement shown in Fig. 29.24, the circuit is gradually reduced in stages as shown in Fig. 29.25(a)–(d).

From Fig. 29.25(d)

$$I = \frac{17}{4.25} = 4 \text{ A}$$

From Fig. 29.25(b)

$$I_1 = \left(\frac{9}{9+3}\right) I = \left(\frac{9}{12}\right) 4 = 3 \text{ A}$$

From Fig. 29.24

$$I_x = \left(\frac{2}{2+8}\right) I_1 = \left(\frac{2}{10}\right) 3 = \textbf{0.6 A}$$

29.5 Wiring lamps in series and in parallel

Series connection

Figure 29.26 shows three lamps, each rated at 240 V, connected in series across a 240 V supply.

(i) Each lamp has only (240/3) V, i.e. 80 V across it and thus each lamp glows dimly.

(ii) If another lamp of similar rating is added in series with the other three lamps, then each lamp now has (240/4) V, i.e. 60 V across it and each now glows even more dimly.

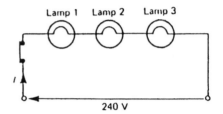

Figure 29.26

(iii) If a lamp is removed from the circuit or if a lamp develops a fault (i.e. an open circuit)

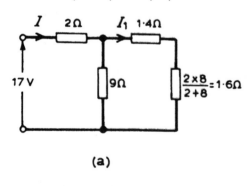

(a)

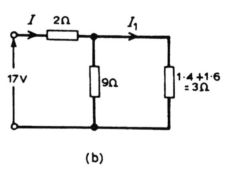

(b)

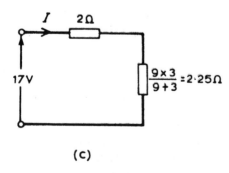

(c)

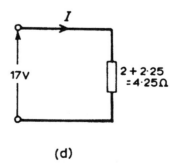

(d)

Figure 29.25

or if the switch is opened then the circuit is broken, no current flows, and the remaining lamps will not light up.

(iv) Less cable is required for a series connection than for a parallel one.

The series connection of lamps is usually limited to decorative lighting such as for Christmas tree lights.

Parallel connection

Fig. 29.27 shows three similar lamps, each rated at 240 V, connected in parallel across a 240 V supply.

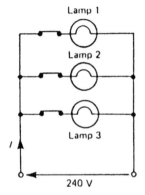

Figure 29.27

(i) Each lamp has 240 V across it and thus each will glow brilliantly at their rated voltage.

(ii) If any lamp is removed from the circuit or develops a fault (open circuit) or a switch is opened, the remaining lamps are unaffected.

(iii) The addition of further similar lamps in parallel does not affect the brightness of the other lamps.

(iv) More cable is required for parallel connection than for a series one.

The parallel connection of lamps is the most widely used in electrical installations.

Problem 15. If three identical lamps are connected in parallel and the combined resistance is 150 Ω, find the resistance of one lamp.

Let the resistance of one lamp be R, then

$$\frac{1}{150} = \frac{1}{R} + \frac{1}{R} + \frac{1}{R} = \frac{3}{R}$$

from which, $R = 3 \times 150 = \textbf{450 } \Omega$.

Problem 16. Three identical lamps A, B and C are connected in series across a 150 V supply. State (a) the voltage across each lamp, and (b) the effect of lamp C failing.

(a) Since each lamp is identical and they are connected in series there is (150/3) V, i.e. 50 V across each.

(b) If lamp C fails, i.e. open circuits, no current will flow and **lamps A and B will not operate**.

29.6 Multi-choice questions on series and parallel networks

(Answers on page 356.)

1. If two 4 Ω resistors are connected in series the effective resistance of the circuit is:
   (a) 8 Ω (b) 4 Ω (c) 2 Ω (d) 1 Ω
2. If two 4 Ω resistors are connected in parallel the effective resistance of the circuit is:
   (a) 8 Ω (b) 4 Ω (c) 2 Ω (d) 1 Ω
3. With the switch in Fig. 29.28 closed, the ammeter reading will indicate:
   (a) $1\frac{2}{3}$ A (b) 75 A (c) $\frac{1}{3}$ A (d) 3 A

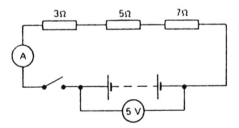

Figure 29.28

4. The effect of connecting an additional parallel load to an electrical supply source is to increase the
   (a) resistance of load
   (b) voltage of the source
   (c) current taken from the source
   (d) p.d. across the load
5. The equivalent resistance when a resistor of

$\frac{1}{3}\,\Omega$ is connected in parallel with a $\frac{1}{4}\,\Omega$ resistor is
(a) $\frac{1}{7}\,\Omega$ (b) $7\,\Omega$

6. With the switch in Fig. 29.29 closed the ammeter reading will ¡indicate:
(a) 108 A   (b) $\frac{1}{3}$ A   (c) 3 A   (d) $4\frac{3}{5}$ A

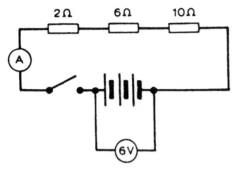

Figure 29.29

7. A $6\,\Omega$ resistor is connected in parallel with the three resistors of Fig. 29.29. With the switch closed the ammeter reading will indicate:
(a) $\frac{3}{4}$ A   (b) 4 A   (c) $\frac{1}{4}$ A   (d) $1\frac{1}{3}$ A

8. A $10\,\Omega$ resistor is connected in parallel with a $15\,\Omega$ resistor and the combination is connected in series with a $12\,\Omega$ resistor. The equivalent resistance of the circuit is:
(a) $37\,\Omega$ (b) $18\,\Omega$ (c) $27\,\Omega$ (d) $4\,\Omega$

29.7   Short answer questions on series and parallel networks

1. Name three characteristics of a series circuit.
2. Show that for three resistors R_1, R_2 and R_3 connected in series the equivalent resistance R is given by $R = R_1 + R_2 + R_3$.
3. Name three characteristics of a parallel circuit.
4. Show that for three resistors R_1, R_2 and R_3 connected in parallel the equivalent resistance R is given by

$$\frac{1}{R} = \frac{1}{R_1} + \frac{1}{R_2} + \frac{1}{R_3}$$

5. Explain the potential divider circuit.
6. Compare the merits of wiring lamps in (a) series, (b) parallel.

29.8   Further questions on series and parallel networks

1. The p.d.s measured across three resistors connected in series are 5 V, 7 V and 10 V, and the supply current is 2 A. Determine (a) the supply voltage, (b) the total circuit resistance and (c) the values of the three resistors.
   [(a) 22 V (b) 11 Ω (c) 2.5 Ω, 3.5 Ω, 5 Ω]
2. For the circuit shown in Fig. 29.30, determine the value of V_1. If the total circuit resistance is 36 Ω determine the supply current and the value of the resistors R_1, R_2 and R_3.
   [10 V, 0.5 A, 20 Ω, 10 Ω, 6 Ω]

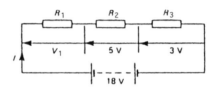

Figure 29.30

3. When the switch in the circuit in Fig. 29.31 is closed the reading on voltmeter 1 is 30 V and that on voltmeter 2 is 10 V. Determine the reading on the ammeter and the value of resistor R_x.   [4 A, 2.5 Ω]

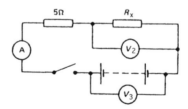

Figure 29.31

4. Two resistors are connected in series across an 18 V supply and a current of 5 A flows. If one of the resistors has a value of 2.4 Ω determine (a) the value of the other resistor and (b) the p.d. across the 2.4 Ω resistor.
   [(a) 1.2 Ω (b) 12 V]
5. Resistances of 4 Ω and 12 Ω are connected in parallel across a 9 V battery. Determine (a) the equivalent circuit resistance, (b) the supply current, and (c) the current in each resistor.
   [(a) 3 Ω (b) 3 A (c) $2\frac{1}{4}$ A, $\frac{3}{4}$ A]

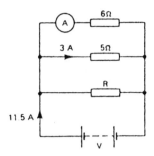

Figure 29.32

6. For the circuit shown in Fig. 29.32 determine (a) the reading on the ammeter, and (b) the value of resistor R. [2.5 A, 2.5 Ω]
7. Find the equivalent resistance when the following resistances are connected (a) in series, (b) in parallel:
   (i) 3 Ω and 2 Ω   (ii) 20 kΩ and 40 kΩ
   (iii) 4 Ω, 8 Ω and 16 Ω   (iv) 800 Ω, 4 kΩ and 1500 Ω.
   $$\left[\begin{array}{c} \text{(a)(i) } 5\,\Omega \text{ (ii) } 60\,\text{k}\Omega \text{ (iii) } 28\,\Omega \text{ (iv) } 6.3\,\text{k}\Omega \\ \text{(b)(i) } 1.2\,\Omega \text{ (ii) } 13\tfrac{1}{3}\,\text{k}\Omega \text{ (iii) } 2\tfrac{2}{7}\,\Omega \\ \text{(iv) } 461.5\,\text{k}\Omega \end{array}\right]$$
8. Find the total resistance between terminals A and B of the circuit shown in Fig. 29.33(a). [8 Ω]

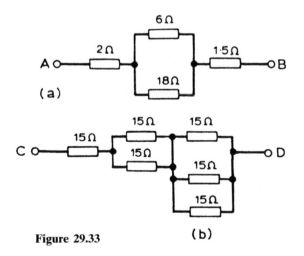

Figure 29.33

(a)

(b)

9. Find the equivalent resistance between terminals C and D of the circuit shown in Fig. 29.33(b). [27.5 Ω]
10. Resistors of 20 Ω, 20 Ω and 30 Ω are connected in parallel. What resistance must

be added in series with the combination to obtain a total resistance of 10 Ω. If the complete circuit expends a power of 0.36 kW, find the total current flowing.
[2.5 Ω, 6 A]
11. (a) Calculate the current flowing in the 30 Ω resistor shown in Fig. 29.34.
    (b) What additional value of resistance would have to be placed in parallel with the 20 Ω and 30 Ω resistors to change the supply current to 8 A, the supply voltage remaining constant.
    [(a) 1.6 A (b) 6 Ω]

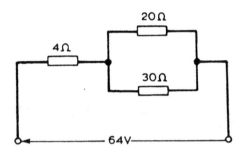

Figure 29.34

12. For the circuit shown in Fig. 29.35, find (a) V_1, (b) V_2, without calculating the current flowing. [(a) 30 V (b) 42 V]

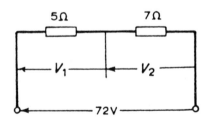

Figure 29.35

13. Determine the currents and voltages indicated in the circuit shown in Fig. 29.36.
    $$\left[\begin{array}{l} I_1 = 5\,\text{A}, I_2 = 2.5\,\text{A}, I_3 = 1\tfrac{2}{3}\,\text{A}, I_4 = \tfrac{5}{6}\,\text{A}, \\ I_5 = 3\,\text{A}, I_6 = 2\,\text{A}, V_1 = 20\,\text{V}, V_2 = 5\,\text{V}, \\ V_3 = 6\,\text{V} \end{array}\right]$$
14. Find the current I in Fig. 29.37. [1.8 A]
15. If four identical lamps are connected in parallel and the combined resistance is 100 Ω, find the resistance of one lamp.
[400 Ω]

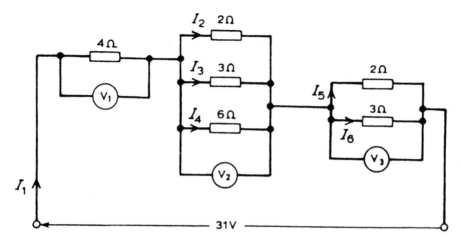

Figure 29.36

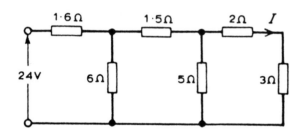

Figure 29.37

16. Three identical filament lamps are connected (a) in series, (b) in parallel across a 210 V supply. State for each connection the p.d. across each lamp.   [(a) 70 V (b) 210 V]

30

Kirchhoff's laws

At the end of this chapter you should be able to:

- state Kirchhoff's current and voltage laws
- evaluate unknown currents and e.m.f.s in series/parallel circuits, using Kirchhoff's laws

30.1 Introduction

More complex d.c. circuits cannot be solved by Ohm's law and the formulae for series and parallel resistors alone. Kirchhoff (a German physicist) developed two laws which further help the determination of unknown currents and voltages in d.c. series/parallel networks.

30.2 Kirchhoff's current and voltage laws

Current law

At any junction in an electric circuit the total current flowing towards that junction is equal to the total current flowing away from the junction, i.e. $\Sigma I = 0$.

Thus referring to Fig. 30.1:

$$I_1 + I_2 + I_3 = I_4 + I_5$$

or

$$I_1 + I_2 + I_3 - I_4 - I_5 = 0$$

Voltage law

In any closed loop in a network, the algebraic sum of the voltage drops (i.e. products of current and resistance) taken around the loop is equal to the resultant e.m.f. acting in that loop.

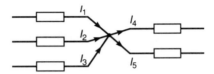

Figure 30.1

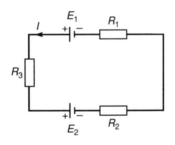

Figure 30.2

Thus referring to Fig. 30.2:

$$E_1 - E_2 = IR_2 + IR_2 + IR_3$$

(Note that if current flows away from the positive terminal of a source, that source is considered by convention to be positive. Thus moving anticlockwise around the loop of Fig. 30.2, E_1 is positive and E_2 is negative.)

Problem 1. Determine the value of the unknown currents marked in Fig. 30.3.

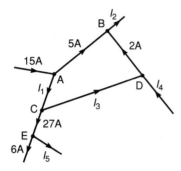

Figure 30.3

Applying Kirchhoff's current law to each junction in turn gives:
For junction A:

$$15 = 5 + I_1$$

Hence

$I_1 = 10$ A

For junction B:

$$5 + 2 = I_2$$

Hence

$I_2 = 7$ A

For junction C:

$$I_1 = 27 + I_3$$

i.e.

$$10 = 27 + I_3$$

Hence

$I_3 = 10 - 27 = -17$ A

(i.e. in the opposite direction to that shown in Fig. 30.3).
For junction D:

$$I_3 + I_4 = 2$$

i.e.

$$-17 + I_4 = 2$$

Hence

$I_4 = 17 + 2 = 19$ A

For junction E:

$$27 = 6 + I_5$$

Hence

$I_5 = 27 - 6 = 21$ A

Problem 2. Determine the value of e.m.f. E in Fig. 30.4.

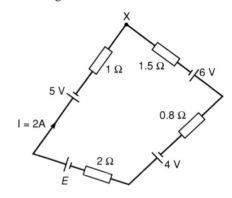

Figure 30.4

Applying Kirchhoff's voltage law and moving clockwise around the loop of Fig. 30.4 starting at point X gives:

$$6 + 4 + E - 5 = I(1.5) + I(0.8) + I(2) + I(1)$$
$$5 + E = I(5.3) = 2(5.3)$$

since current I is 2 A.
Hence

$$5 + E = 10.6$$

and

e.m.f. $E = 10.6 - 5 = 5.6$ V

Problem 3. Use Kirchhoff's laws to determine the current flowing in the 4 Ω resistance of the network shown in Fig. 30.5.

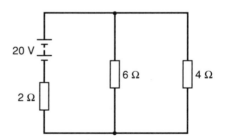

Figure 30.5

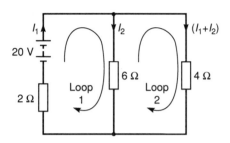

Figure 30.6

Step 1
Label current I_1 flowing from the positive terminal of the 20 V source and current I_2 flowing through the 6 Ω resistance as shown in Fig. 30.6. By **Kirchhoff's current law**, the current flowing in the 4 Ω resistance must be $(I_1 - I_2)$.

Step 2
Label loops 1 and 2 as shown in Fig. 30.6 (both loops have been shown clockwise, although they do not need to be in the same direction). **Kirchhoff's voltage law** is now applied to each loop in turn:
For loop 1:
$$20 = 2I_1 + 6I_2 \qquad (1)$$
For loop 2:
$$0 = 4(I_1 - I_2) - 6I_2 \qquad (2)$$
Note the zero on the left-hand side of equation (2) since there is no voltage source in loop 2. Note also the minus sign in front of $6I_2$. This is because loop 2 is moving through the 6 Ω resistance in the opposite direction to current I_2.
Equation (2) simplifies to:
$$0 = 4I_1 - 10I_2 \qquad (3)$$

Step 3
Solve the **simultaneous equations** (1) and (3) for currents I_1 and I_2 (see Chapter 2):
$$20 = 2I_1 + 6I_2 \qquad (1)$$
$$0 = 4I_1 - 10I_2 \qquad (3)$$
$2 \times$ equation (1) gives:
$$40 = 4I_1 + 12I_2 \qquad (4)$$
Equation (4) − equation (3) gives:
$$40 = 0 + (12I_2 - -10I_2)$$

i.e.
$$40 = 22I_2$$
Hence
$$\text{current } I_2 = \frac{40}{22} = \textbf{1.818 A}$$

Substituting $I_2 = 1.818$ into equation (1) gives:
$$20 = 2I_1 + 6(1.818)$$
$$20 = 2I_1 + 10.908$$
and
$$20 - 10.908 = 2I_1$$
from which
$$I_1 = \frac{20 - 10.908}{2} = \frac{9.092}{2} = \textbf{4.546 A}$$

Hence the current flowing in the 4 Ω resistance is $I_1 - I_2$, i.e. $(4.546 - 1.818) = \textbf{2.718 A}$.
The currents and their directions are as shown in Fig. 30.7.

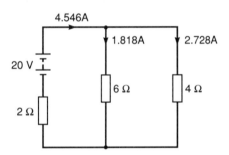

Figure 30.7

Problem 4. Use Kirchhoff's laws to determine the current flowing in each branch of the network shown in Fig. 30.8.

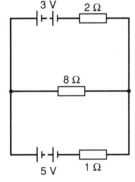

Figure 30.8

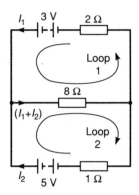

Figure 30.9

Step 1
The currents I_1 and I_2 are labelled as shown in Fig. 30.9 and by Kirchhoff's current law the current in the 8 Ω resistance is $(I_1 + I_2)$.

Step 2
Loops 1 and 2 are labelled as shown in Fig. 30.9. Kirchhoff's voltage law is now applied to each loop in turn.
For loop 1:

$$3 = 2I_1 + 8(I_1 + I_2) \qquad (1)$$

For loop 2:

$$5 = 8(I_1 + I_2) + (1)(I_2) \qquad (2)$$

Equation (30.1) simplifies to:

$$3 = 10I_1 + 8I_2 \qquad (3)$$

Equation (30.2) simplifies to:

$$5 = 8I_1 + 9I_2 \qquad (4)$$

4 × equation (30.3) gives:

$$12 = 40I_1 + 32I_2 \qquad (5)$$

5 × equation (4) gives:

$$25 = 40I_1 + 45I_2 \qquad (6)$$

Equation (6) – equation (5) gives:

$$13 = 13I_2$$

and

$$I_2 = \mathbf{1\,A}$$

Substituting $I_2 = 1$ in equation (3) gives:

$$3 = 10I_1 + 8(1)$$

$$3 - 8 = 10I_1$$

and

$$I_1 = \frac{-5}{10} = \mathbf{-0.5\,A}$$

(i.e. I_1 is flowing in the opposite direction to that shown in Fig. 30.9).
The current in the 8 Ω resistance is $(I_1 + I_2) = (-0.5 + 1) = \mathbf{0.5\,A}$.

30.3 Multi-choice questions on Kirchhoff's laws

(Answers on page 356.)

1. Which of the following statements is true? For the junction in the network shown in Fig. 30.10:
   (a) $I_5 - I_4 = I_3 - I_2 + I_1$
   (b) $I_1 + I_2 + I_3 = I_4 + I_5$
   (c) $I_2 + I_3 + I_5 = I_1 + I_4$
   (d) $I_1 - I_2 - I_3 - I_4 + I_5 = 0$

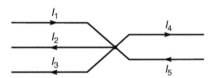

Figure 30.10

2. Which of the following statements is true? For the circuit shown in Fig. 30.11:
   (a) $E_1 + E_2 + E_3 = Ir_1 + Ir_2 + Ir_3$
   (b) $E_2 + E_3 - E_1 - I(r_1 + r_2 + r_3) = 0$
   (c) $I(r_1 + r_2 + r_3) = E_1 - E_2 - E_3$
   (d) $E_2 + E_3 - E_1 = Ir_1 + Ir_2 + Ir_3$

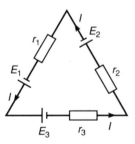

Figure 30.11

3. The current I flowing in resistor R in the circuit shown in Fig. 30.12 is:

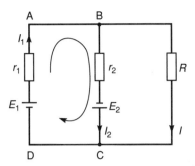

Figure 30.12

(a) $I_2 - I_1$ (b) $I_1 - I_2$ (c) $I_1 + I_2$ (d) I_1
4. Applying Kirchhoff's voltage law clockwise around loop ABCD in the circuit shown in Fig. 30.12 gives:
(a) $E_1 - E_2 = I_1 r_1 + I_2 r_2$
(b) $E_2 = E_1 + I_1 r_1 + I_2 r_2$
(c) $E_1 + E_2 = I_1 r_1 + I_2 r_2$
(d) $E_1 + I_1 r_1 = E_2 + I_2 r_2$

30.4 Short answer questions on Kirchhoff's laws

1. State Kirchhoff's current law.
2. State Kirchhoff's voltage law.

30.5 Further questions on Kirchhoff's laws

1. Find currents I_3, I_4 and I_6 in Fig. 30.13.
   [$I_3 = 2$ A, $I_4 = -1$ A, $I_6 = 3$ A]

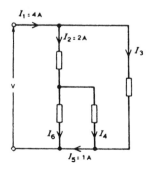

Figure 30.13

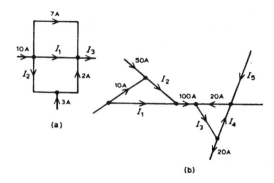

Figure 30.14

2. For the networks shown in Fig. 30.14, find the values of the currents marked.
   (a) $I_1 = 4$ A, $I_2 = -1$, $I_3 = 13$ A (b) $I_1 = 40$ A, $I_2 = 60$ A, $I_3 = 120$ A, $I_4 = 100$ A, $I_5 = -80$ A
3. Use Kirchhoff's laws to find the current flowing in the 6 Ω resistor of Fig. 30.15 and the power dissipated in the 4 Ω resistor.
   [2.162 A, 42.07 W]

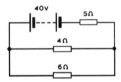

Figure 30.15

4. Find the current flowing in the 3 Ω resistor for the network shown in Fig. 30.16(a). Find also the p.d. across the 10 Ω and 2 Ω resistors. [2.715 A, 7.410 V, 3.948 V]

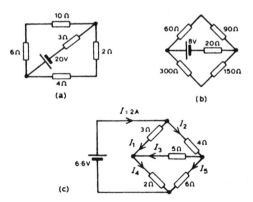

Figure 30.16

5. For the network shown in Fig. 30.16(b), find: (a) the current in the battery, (b) the current in the 300 Ω resistor, (c) the current in the 90 Ω resistor, and (d) the power dissipated in the 150 Ω resistor.

$$\left[\begin{array}{l} \text{(a) } 60.38 \text{ mA (b) } 15.10 \text{ mA} \\ \text{(c) } 45.28 \text{ mA (d) } 34.20 \text{ mW} \end{array} \right]$$

6. For the bridge network shown in Fig. 30.16(c), find the currents I_1 to I_5.

$$\left[\begin{array}{l} I_1 = 1.25 \text{ A, } I_2 = 0.75 \text{ A, } I_3 = 0.15 \text{ A} \\ I_4 = 1.40 \text{ A, } I_5 = 0.60 \text{ A} \end{array} \right]$$

31

Electromagnetism

At the end of this chapter you should be able to:

- understand the form of a magnetic field around a magnet

- understand that magnetic fields are produced by electric currents

- apply the screw rule to determine direction of magnetic field

- recognize that the magnetic field around a solenoid is similar to a magnet

- apply the screw rule or grip rule to a solenoid to determine magnetic field direction

- recognize and describe practical applications of an electromagnet

- define magnetic flux and state its unit

- define magnetic flux density and state its unit

- perform calculations using $B = \Phi/A$

- appreciate factors upon which the force F on a current-carrying conductor depends

- perform calculations using $F = BIl$ and $F = BIl \sin \theta$

- use Fleming's left-hand rule to predetermine direction of force in a current-carrying conductor

- describe the principle of operation of a simple d.c. motor

- describe the principle of operation and construction of a moving coil instrument

- appreciate the force F on a charge in a magnetic field is given by $F = QvB$

- perform calculations using $F = QvB$

31.1 Magnetic fields

A **permanent magnet** is a piece of ferromagnetic material (such as iron, nickel or cobalt) which has properties of attracting other pieces of these materials.

The area around a magnet is called the **magnetic field** and it is in this area that the effects of the **magnetic force** produced by the magnet can be detected. The magnetic field of a bar magnet can be represented pictorially by the 'lines of force' (or lines of 'magnetic flux' as they are called) as shown in Fig. 31.1. Such a field pattern can be produced by placing iron filings in the vicinity of the magnet.

The field direction at any point is taken as that in which the north-seeking pole of a compass needle points when suspended in the field. External to the magnet the direction of the field is north to south.

The laws of magnetic attraction and repulsion can be demonstrated by using two bar magnets. In Fig. 36.2(a), with **unlike pole** adjacent,

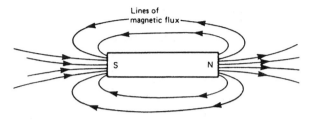

Figure 31.1

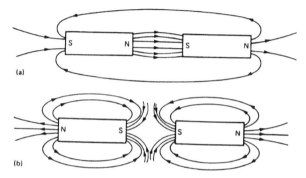

Figure 31.2

attraction occurs. In Fig. 36.2(b), with **like poles** adjacent, **repulsion** occurs.

Magnetic fields are produced by electric currents as well as by permanent magnets. The field forms a circular pattern with the current-carrying conductor at the centre. The effect is portrayed in Fig. 31.3 where the convention adopted is:

(a) current flowing **away** from the viewer is shown by $\times$ and can be thought of as the feathered end of the shaft of an arrow;

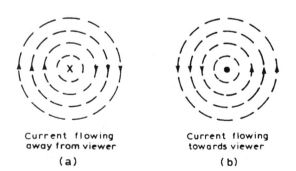

Figure 31.3

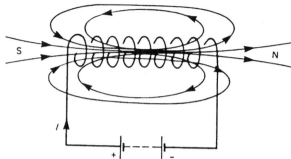

Figure 31.4

(b) current flowing **towards** the viewer is shown by ● and can be thought of as the tip of an arrow.

The direction of the fields in Fig. 31.3 is remembered by the **screw rule** which states: 'If a normal right-hand thread screw is screwed along the conductor in the direction of the current, the direction of rotation of the screw is in the direction of the magnetic field.'

A magnetic field produced by a long coil, or **solenoid**, is shown in Fig. 31.4 and is seen to be similar to that of a bar magnet shown in Fig. 31.1. If the solenoid is wound on an iron bar an even stronger field is produced. The **direction** of the field produced by current I is determined by a compass and is remembered by either:

(a) the **screw rule**, which states that if a normal right-hand thread screw is placed along the axis of the solenoid and is screwed in the direction of the current it moves in the direction of the solenoid (i.e. points in the direction of the north pole); or

(b) the **grip rule**, which states that if the coil is gripped with the right hand with the fingers pointing in the direction of the current, then the thumb, outstretched parallel to the axis of the solenoid, points in the direction of the magnetic field inside the solenoid (i.e. points in the direction of the north pole).

Problem 1.   Figure 31.5 shows a coil of wire wound on an iron core connected to a battery. Sketch the magnetic field pattern associated with the current-carrying coil and determine the polarity of the field.

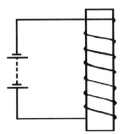

Figure 31.5

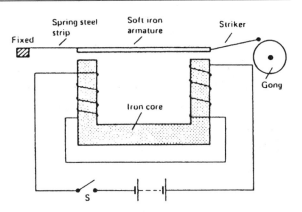

Figure 31.7

The magnetic field associated with the solenoid in Fig. 31.5 is similar to the field associated with a bar magnet and is as shown in Fig. 31.6. The polarity of the field is determined either by the screw rule or by the grip rule. Thus the north pole is at the bottom and the south pole at the top.

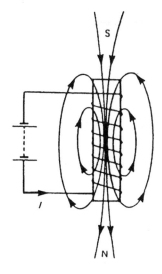

Figure 31.6

31.2 Electromagnets

An **electromagnet**, which is a solenoid wound on an iron core, provides the basis of many items of electrical equipment, examples including an electric bell, relays, lifting magnets and telephone receivers.

Problem 2. State what happens when the switch S is closed and then opened in the circuit shown in Fig. 31.7.

When the switch S is closed a current passes through the coil. The iron-cored solenoid is energized, the soft iron armature is attracted to the electromagnet and the striker hits the gong. When the switch S is opened the coil becomes demagnetized and the spring steel strip pulls the armature back to its original position. This is the principle of operation of an **electric bell**.

Problem 3. For the relay circuit shown in Fig. 31.8 state what happens when the switch S is closed.

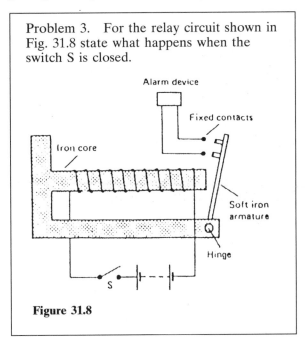

Figure 31.8

When the switch S is closed a current passes through the coil and the iron-cored solenoid is energized. The hinged soft iron armature is attracted to the electromagnet and pushes against

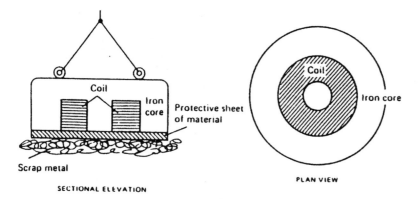

Figure 31.9

the two fixed contacts so that they are connected together, thus closing the electric circuit to be controlled – in this case, an alarm circuit. The alarm sounds for as long as the current flows in the coil.

> Problem 4.   Make a sketch showing the main parts of a lifting magnet that would be suitable for use in a scrap-metal yard and state its principle of operation.

A typical lifting magnet showing the plan and elevation is shown in Fig. 31.9. When current is passed through the coil, the iron core becomes magnetized (i.e. an electromagnet) and thus will attract to it other pieces of magnetic material. When the circuit is broken the iron core becomes demagnetized which releases the materials being lifted.

31.3   Magnetic flux and flux density

Magnetic flux is the amount of magnetic field (or the number of lines of force) produced by a magnetic source. The symbol for magnetic flux is Φ (Greek letter phi). The unit of magnetic flux is the **weber, Wb**.

Magnetic flux density is the amount of flux passing through a defined area that is perpendicular to the direction of the flux.

$$\textbf{Magnetic flux density} = \frac{\textbf{magnetic flux}}{\textbf{area}}$$

The symbol for magnetic flux density is B. The unit of magnetic flux density is the tesla, T, where $1\,\text{T} = 1\,\text{Wb/m}^2$. Hence

$$\boxed{B = \frac{\Phi}{A}\ \text{tesla}}$$

where $A\ \text{m}^2$ is the area.

> Problem 5.   A magnetic pole face has a rectangular section having dimensions 200 mm by 100 mm. If the total flux emerging from the pole is 150 μWb, calculate the flux density.

Flux $\Phi = 150\,\mu\text{Wb} = 150 \times 10^{-6}\,\text{Wb}$. Cross-sectional area $A = 200 \times 100 = 20\,000\,\text{mm}^2 = 20\,000 \times 10^{-6}\,\text{m}^2$. Flux density

$$B = \frac{\Phi}{A} = 150 \times 1\frac{0^{-6}}{20\,000 \times 10^{-6}}$$

$$= \textbf{0.0075 T}\ \text{or}\ \textbf{7.5 mT}$$

> Problem 6.   The maximum working flux density of a lifting electromagnet is 1.8 T and the effective area of a pole face is circular in cross-section. If the total magnetic flux produced is 353 mWb, determine the radius of the pole face.

Flux density $B = 1.8\,\text{T}$; flux $\Phi = 353\,\text{mWb} = 353 \times 10^{-3}\,\text{Wb}$. Since $B = \Phi/A$, cross-sectional area

$$A = \frac{\Phi}{B} = \frac{353 \times 10^{-3}}{1.8}\ \text{m}^2$$

$$= 0.1961\,\text{m}^2$$

The pole face is circular, hence area $= \pi r^2$, where r is the radius. Hence $\pi r^2 = 0.1961$, from which $\pi r^2 = 0.1961/\pi$ and

$$\text{radius } r = \sqrt{\left(\frac{0.1961}{\pi}\right)} = 0.250 \text{ m}^2$$

i.e. the radius of the pole face is **250 mm**.

31.4 Force on a current-carrying conductor

If a current-carrying conductor is placed in a magnetic field produced by permanent magnets, then the fields due to the current-carrying conductor and the permanent magnets interact and cause a force to be exerted on the conductor. The force on the current-carrying conductor in a magnetic field depends upon:

(a) the flux density of the field, B teslas,
(b) the strength of the current, I amperes,
(c) the length of the conductor perpendicular to the magnetic field, l metres, and
(d) the directions of the field and the current.

When the magnetic field, the current and the conductor are mutually at right angles then:

Force $F = BIl$ newtons

When the conductor and the field are at an angle $\theta°$ to each other then:

Force $F = BIl \sin \theta$ newtons

Since when the magnetic field, current and conductor are mutually at right angles, $F = BIl$, the magnetic flux density B may be defined by $B = F/Il$, i.e. the flux density is 1 T if the force exerted on 1 m of a conductor when the conductor carries a current of 1 A is 1 N.

Problem 7. A conductor carries a current of 20 A and is at right angles to a magnetic field having a flux density of 0.9 T. If the length of the conductor in the field is 30 cm, calculate the force acting on the conductor. Determine also the value of the force if the conductor is inclined at an angle of 30° to the direction of the field.

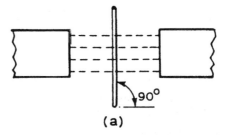

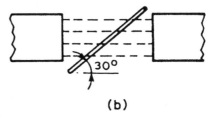

(a)

(b)

Figure 31.10

$B = 0.9$ T; $I = 20$ A; $l = 30$ cm $= 0.30$ m. Force $F = BIl = (0.9)(20)(0.30)$ newtons when the conductor is at right angles to the field, as shown in Fig. 31.10(a), i.e. **$F = 5.4$ N**. When the conductor is inclined at 30° to the field, as shown in Fig. 31.10(b), then

$$\text{Force } F = BIl \sin \theta$$

$$F = (0.9)(20)(0.30) \sin 30°$$

$$F = 2.7 \text{ N}$$

If the current-carrying conductor shown in Fig. 31.3(a) is placed in the magnetic field shown in

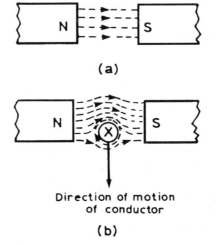

Figure 31.11

Fig. 31.11(a), then the two fields interact and cause a force to be exerted on the conductor as shown in Fig. 31.11(b). The field is strengthened above the conductor and weakened below, thus tending to move the conductor downwards. This is the basic principle of operation of the electric motor (see section 31.5) and the moving-coil instrument (see section 31.6).

The direction of the force exerted on a conductor can be predetermined by using Fleming's left-hand rule (often called the motor rule) which states:

Let the thumb, first finger and second finger of the left hand be extended such that they are all at right-angles to each other [as shown in Fig. 31.12]. If the first finger points in the direction of the magnetic field, the second finger points in the direction of the current, then the thumb will point in the direction of the motion of the conductor.

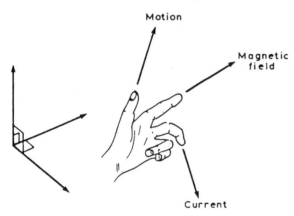

Figure 31.12

Summarizing:

First finger – Field

SeCond finger – Current

ThuMb – Motion

Problem 8.   Determine the current required in a 400 mm length of conductor of an electric motor, when the conductor is situated at right angles to a magnetic field of flux density 1.2 T, if a force of 1.92 N is to be exerted on the conductor. If the conductor is vertical, the current flowing downwards and the direction of the magnetic field is from left to right, what is the direction of the force?

$F = 1.92$ N; $l = 400$ mm $= 0.40$ m; $B = 1.2$ T. Since $F = BIl$, $I = F/Bl$. Hence

$$\text{current } I = \frac{1.92}{(1.2)(0.4)} = 4\,\textbf{A}$$

If the current flows downwards, the direction of its magnetic field due to the current alone will be clockwise when viewed from above. The lines of flux will reinforce (i.e. strengthen) the main magnetic field at the back of the conductor and will be in opposition in the front (i.e. weaken the field). **Hence the force on the conductor will be from back to front (i.e. toward the viewer).** This direction may also have been deduced using Fleming's left-hand rule.

Problem 9.   A conductor 350 mm long carries a current of 10 A and is at right angles to a magnetic field lying between two circular pole faces each of radius 60 mm. If the total flux between the pole faces is 0.5 mWb, calculate the magnitude of the force exerted on the conductor.

$l = 350$ mm $= 0.35$ m; $I = 10$ A; area of pole face $A = \pi r^2 = \pi(0.06)^2$ m^2; $\Phi = 0.5$ mWb $= 0.5 \times 10^{-3}$ Wb. Force $F = BIl$, and $B = \Phi/A$, hence

$$\text{Force} = \left(\frac{\Phi}{A}\right) Il$$

$$= \frac{(0.5 \times 10^{-3})}{\pi(0.06)^2}\,(10)\,(0.35)\text{ newtons}$$

i.e. **force = 0.155 N**.

Problem 10.   With reference to Fig. 31.13 determine (a) the direction of the force on the conductor in Fig. 31.13(a), (b) the direction of the force on the conductor in Fig. 31.13(b), (c) the direction of the current in Fig. 31.13(c), (d) the polarity of the magnetic system in Fig. 31.13(d).

(a) The direction of the main magnetic field is from north to south, i.e. left to right. The current is flowing towards the viewer, and using the screw rule, the direction of the field is anticlockwise. Hence either by Fleming's left-hand rule, or by sketching the interacting magnetic field as shown in Fig. 31.14(a), the direction of the force on the conductor is seen to be upward.

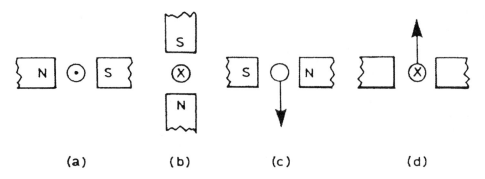

(a)   **(b)**   **(c)**   **(d)**

Figure 31.13

(b) Using a similar method to part (a) it is seen that the force on the conductor is to the right, see Fig. 31.14(b)
(c) Using Fleming's left-hand rule, or by sketching as in Fig. 31.14(c), it is seen that the current is toward the viewer, i.e. out of the paper.
(d) Similar to part (c), the polarity of the magnetic system is as shown in Fig. 31.14(d).

Problem 11. A coil is wound on a rectangular former of width 24 mm and length 30 mm. The former is pivoted about an axis passing through the middle of the two shorter sides and is placed in a uniform magnetic field of flux density 0.8 T, the axis being perpendicular to the field. If the coil carries a current of 50 mA, determine the force on each coil side (a) for a single-turn coil, (b) for a coil wound with 300 turns.

(a) Flux density $B = 0.8$ T; length of conductor lying at right angles to field $l = 30$ mm $= 30 \times 10^{-3}$ m; current $I = 50$ mA $= 50 \times 10^{-3}$ A. For a single-turn coil, force on each coil side $F = BIl = 0.8 \times 50 \times 10^{-3} \times 30 \times 10^{-3}$

$$= 1.2 \times 10^{-3} \text{ N, or } 0.0012 \text{ N}$$

(b) When there are 300 turns on the coil there are effectively 300 parallel conductors each carrying a current of 50 mA. Thus the total force produced by the current is 300 times that for a single-turn coil. Hence force on coil side $F = 300\, BIl$

$$= 300 \times 0.0012 = \textbf{0.36 N}$$

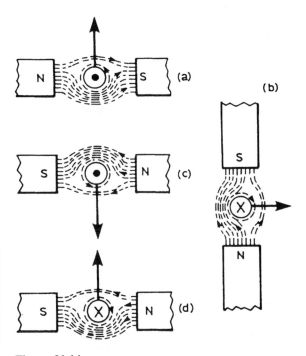

Figure 31.14

31.5 Principle of operation of a simple d.c. motor

A rectangular coil which is free to rotate about a fixed axis is shown placed inside a magnetic field produced by permanent magnets in Fig. 31.15. A direct current is fed into the coil via carbon brushes bearing on a commutator, which consists of a metal ring split into two halves separated by insulation. When current flows in the coil a magnetic field is set up around the coil which interacts with the magnetic field produced by the

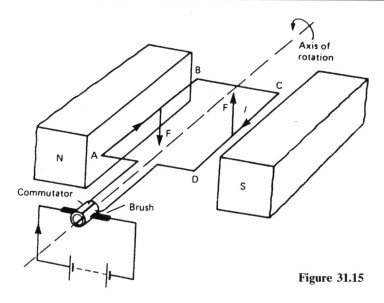

Figure 31.15

magnets. This causes a force F to be exerted on the current-carrying conductor which, by Fleming's left-hand rule, is downwards between points A and B and upward between C and D for the current direction shown. This causes a torque and the coil rotates anticlockwise. When the coil has turned through 90° from the position shown in Fig. 31.15 the brushes connected to the positive and negative terminals of the supply make contact with different halves of the commutator ring, thus reversing the direction of the current flow in the conductor. If the current is not reversed and the coil rotates past this position the forces acting on it change direction and it rotates in the opposite direction thus never making more than half a revolution. The current direction is reversed every time the coil swings through the vertical position and thus the coil rotates anticlockwise for as long as the current flows. This is the principle of operation of a d.c. motor which is thus a device that takes in electrical energy and converts it into mechanical energy.

31.6 Principle of operation of a moving-coil instrument

A moving-coil instrument operates on the motor principle. When a conductor-carrying current is placed in a magnetic field, a force F is exerted on the conductor, given by $F = BIl$. If the flux density B is made constant (by using permanent magnets) and the conductor is a fixed length (say, a coil) then the force will depend only on the current flowing in the conductor.

In a moving-coil instrument a coil is placed centrally in the gap between shaped pole pieces as shown by the front elevation in Fig. 31.16(a). (The airgap is kept as small as possible, although for clarity it is shown exaggerated in Fig. 31.16(b).) The coil is supported by steel pivots, resting in jewel bearings, on a cylindrical iron core. Current is led into and out of the coil by two phosphor bronze spiral hairsprings which are wound in opposite directions to minimize the effect of temperature change and to limit the coil swing (i.e. to **control** the movement) and return the movement to zero position when no current flows. Current flowing in the coil produces forces as shown in Fig. 31.16(b), the directions being obtained by Fleming's left-hand rule. The two forces, F_A and F_B, produce a torque which will move the coil in a clockwise direction, i.e. move the pointer from left to right. Since force is proportional to current the scale is linear.

When the aluminium frame, on which the coil is wound, is rotated between the poles of the magnet, small currents (called eddy currents) are induced into the frame, and this provides automatically the necessary **damping** of the system due to the reluctance of the former to move within the magnetic field.

The moving-coil instrument will measure only direct current or voltage and the terminals are marked positive and negative to ensure that the

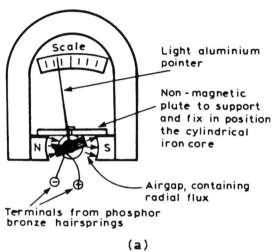

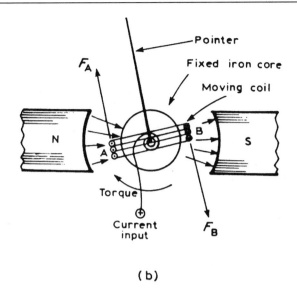

Figure 31.16

current passes through the coil in the correct direction to deflect the pointer 'up the scale'.

The range of this sensitive instrument is extended by using shunts and multipliers (see Chapter 36).

31.7 Force on a charge

When a charge of Q coulombs is moving at a velocity of v m/s in a magnetic field of flux density B teslas, the charge moving perpendicular to the field, then the magnitude of the force F exerted on the charge is given by:

$$F = QvB \text{ newtons}$$

> Problem 12. An electron in a television tube has a charge of 1.6×10^{-19} coulombs and travels at 3×10^7 m/s perpendicular to a field of flux density 18.5 μT. Determine the force exerted on the electron in the field.

From above, force $F = QvB$ newtons, where $Q =$ charge in coulombs $= 1.6 \times 10^{-19}$C; $v =$ velocity of charge $= 3 \times 10^7$ m/s; and $B =$ flux density $= 18.5 \times 10^{-6}$ T. Hence force on electron $F = 1.6 \times 10^{-19} \times 3 \times 10^7 \times 18.5 \times 10^{-6}$

$$= 1.6 \times 3 \times 18.5 \times 10^{-18}$$

$$= 88.8 \times 10^{-18} = \mathbf{8.88 \times 10^{-17} \, N}$$

31.8 Multi-choice questions on electromagnetism

(Answers on page 356.)

1. The unit of magnetic flux density is the:
   (a) weber   (b) weber per metre
   (c) ampere per metre   (d) tesla
2. The total flux in the core of an electrical machine is 20 mWb and its flux density is 1 T. The cross-sectional area of the core is:
   (a) 0.05 m² (b) 0.02 m² (c) 20 m² (d) 50 m²
3. A conductor carries a current of 10 A at right angles to a magnetic field having a flux density of 500 mT. If the length of the conductor in the field is 20 cm, the force on the conductor is:
   (a) 100 kN   (b) 1 kN   (c) 100 N   (d) 1 N
4. If a conductor is horizontal, the current flowing from left to right and the direction of the surrounding magnetic field is from above to below, the force exerted on the conductor is:
   (a) from left to right   (b) from below to above   (c) away from the viewer   (d) towards the viewer
5. For the current-carrying conductor lying in the magnetic field shown in Fig. 31.17(a), the direction of the force on the conductor is:
   (a) to the left   (b) upwards
   (c) to the right   (d) downwards

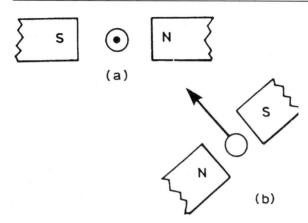

(a)

(b)

Figure 31.17

6. For the current-carrying conductor lying in the magnetic field shown in Fig. 31.17(b) the direction of the current in the conductor is (a) towards the viewer (b) away from the viewer.
7. Figure 31.18 shows a rectangular coil of wire placed in a magnetic field and free to rotate about axis AB. If current flows into the coil at C, the coil will:
   (a) commence to rotate anticlockwise
   (b) commence to rotate clockwise
   (c) remain in the vertical position
   (d) experience a force towards the north pole
8. The force on an electron travelling at 10^7 m/s in a magnetic field of density $10\,\mu T$ is 1.6×10^{-17} N. The electron has a charge of:
   (a) 1.6×10^{-28} C   (b) 1.6×10^{-15}
   (c) 1.6×10^{-19} C   (d) 1.6×10^{-25} C

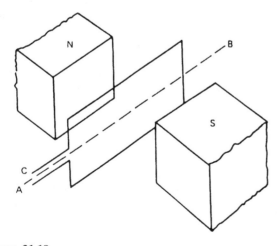

Figure 31.18

31.9 Short answer questions on electromagnetism

1. What is a permanent magnet?
2. Sketch the pattern of the magnetic field associated with a bar magnet. Mark the direction of the field.
3. The direction of the magnetic field around a current-carrying conductor may be remembered using the .... rule.
4. Sketch the magnetic field pattern associated with a solenoid connected to a battery and wound on an iron bar. Show the direction of the field.
5. Name three applications of electromagnetism.
6. State what happens when a current-carrying conductor is placed in a magnetic field between two magnets.
7. Define magnetic flux.
8. The symbol for magnetic flux is .... and the unit of flux is the ....
9. Define magnetic flux density.
10. The symbol for magnetic flux density is .... and the unit of flux density is ....
11. The force on a current-carrying conductor in a magnetic field depends on four factors: name them.
12. The direction of the force on a conductor in a magnetic field may be predetermined using Fleming's .... rule.

31.10 Further questions on electromagnetism

1. Determine the flux density in a magnetic field of cross-sectional area 20 cm^2 having a flux of 3 mWb.   [1.5 T]
2. Find the total flux emerging from a magnetic pole face having dimensions 50 mm by 60 mm, if the flux density is 0.9 T. [2.7 mWb]
3. An electromagnet of square cross-section produces a flux density of 0.45 T. If the magnetic flux is 720 μWb find the dimensions of the electromagnet cross-section.
                                                [40 mm by 40 mm]
4. The pole core of an electrical machine has a circular cross-section of diameter 120 mm. Determine the flux density if the flux in the core is 7.5 mWb.   [0.663 T]
5. A conductor carries a current of 70 A at right angles to a magnetic field having a flux

density of 1.5 T. If the length of the conductor in the field is 200 mm calculate the force acting on the conductor. What is the force when the conductor and field are at an angle of 45°?   [210 N, 14.8 N]

6. Calculate the current required in a 240 mm length of conductor of a d.c. motor when the conductor is situated at right angles to the magnetic field of flux density 1.25 T, if a force of 1.20 N is to be exerted on the conductor.   [4.0 A]

7. A conductor 30 cm long is situated at right angles to a magnetic field. Calculate the strength of the magnetic field if a current of 15 A in the conductor produces a force on it of 3.6 N.   [0.80 T]

8. A conductor 300 mm long carries a current of 13 A and is at right angles to a magnetic field between two circular pole faces, each of diameter 80 mm. If the total flux between the pole faces is 0.75 mWb calculate the force exerted on the conductor.   [0.582 N]

9. (a) A 400 mm length of conductor carrying a current of 25 A is situated at right angles to a magnetic field between two poles of an electric motor. The poles have a circular cross-section. If the force exerted on the conductor is 80 N and the total flux between the pole faces is 1.27 mWb, determine the diameter of a pole face.
    (b) If the conductor in part (a) is vertical, the current flowing downwards and the direction of the magnetic field is from left to right, what is the direction of the 80 N force?
                    [(a) 14.2 mm (b) towards the viewer]

10. A coil is wound uniformly on a former having a width of 18 mm and a length of 25 mm. The former is pivoted about an axis passing through the middle of the two shorter sides and is placed in a uniform magnetic field of flux density 0.75 T, the axis being perpendicular to the field. If the coil carries a current of 120 mA, determine the force exerted on each coil side, (a) for a single-turn coil, (b) for a coil wound with 400 turns.
                    [(a) 2.25×10^{-3} N (b) 0.9 N]

11. Figure 31.19 shows a simplified diagram of a section through the coil of a moving-coil instrument. For the direction of current flow shown in the coil determine the direction that the pointer will move.   [To the right]

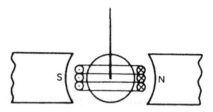

Figure 31.19

12. Explain, with the aid of a sketch, the action of a simplified d.c. motor.

13. Sketch and label the movement of a moving-coil instrument. Briefly explain the principle of operation of such an instrument.

14. Calculate the force exerted on a charge of 2×10^{-18} C travelling at 2×10^{6} m/s perpendicular to a field of density 2×10^{-7} T.
                    [8×10^{-19} N]

15. Determine the speed of a 10^{-19} C charge travelling perpendicular to a field of flux density 10^{-7} T, if the force on the charge is 10^{-20} N.   [10^{6} m/s]

32

Electromagnetic induction

At the end of this chapter you should be able to:

- understand how an e.m.f. may be induced in a conductor
- state Faraday's laws of electromagnetic induction
- state Lenz's law
- use Fleming's right-hand rule for relative directions
- appreciate that the induced e.m.f., $E = Blv$ or $E = Blv \sin \theta$
- calculate induced e.m.f. given B, l, v and θ and determine relative directions
- define inductance L and state its unit
- define mutual inductance
- appreciate that e.m.f. $E = N(\Delta \Phi / t = L(\Delta I / t)$
- calculate induced e.m.f., given N, t, L, change of flux or change of current
- calculate inductance L of a coil, given $L = N\Phi / I$
- calculate mutual inductance using $E_2 = M(\Delta I_1 / t)$
- describe the principle of operation of a transformer
- perform calculations using $(V_1/V_2) = (N_1/N_2) = (I_2/I_1)$ for an ideal transformer
- define the 'rating' of a transformer

32.1 Introduction to electromagnetic induction

When a conductor is moved across a magnetic field so as to cut through the lines of force (flux), an electromotive force (e.m.f.) is produced in the conductor. If the conductor forms part of a closed circuit then the e.m.f. produced causes an electric current to flow round the circuit. Hence an e.m.f. (and thus current) is 'induced' in the conductor as a result of its movement across the magnetic field. This effect is known as **'electromagnetic induction'**.

Figure 32.1(a) shows a coil of wire connected to a centre-zero galvanometer, which is a sensi-

tive ammeter with the zero-current position in the centre of the scale.

(a) When the magnet is moved at constant speed towards the coil (Fig. 32.1(a)), a deflection is noted on the galvanometer showing that a current has been produced in the coil.
(b) When the magnet is moved at the same speed as in (a) but away from the coil the same deflection is noted but is in the opposite direction (see Fig. 32.1(b)).
(c) When the magnet is held stationary even within the coil no deflection is recorded.
(d) When the coil is moved at the same speed as in (a) and the magnet held stationary the same galvanometer deflection is noted.

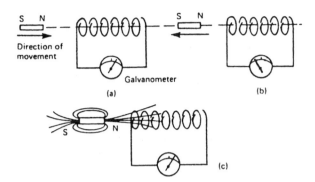

Figure 32.1

(e) When the relative speed is, say, doubled, the galvanometer deflection is doubled.
(f) When a stronger magnet is used, a greater galvanometer deflection is noted.
(g) When the number of turns of wire of the coil is increased, a greater galvanometer deflection is noted.

Figure 32.1(c) shows the magnetic field associated with the magnet. As the magnet is moved towards the coil, the magnetic flux of the magnet moves across, or cuts, the coil. **It is the relative movement of the magnetic flux and the coil that causes an e.m.f. and thus current to be induced in the coil.** This effect is known as electromagnetic induction. The laws of electromagnetic induction stated in section 32.2 evolved from experiments such as those described above.

32.2 Laws of electromagnetric induction

Faraday's laws of electromagnetic induction state:

(i) *An induced e.m.f. is set up whenever the magnetic field linking that circuit changes.*
(ii) *The magnitude of the induced e.m.f. in any circuit is proportional to the rate of change of the magnetic flux linking the circuit.*

Lenz's law states:

The direction of an induced e.m.f. is always such that it tends to set up a current opposing the motion or the change of flux responsible for inducing that e.m.f.

An alternative method to Lenz's law of determining relative directions is given by **Fleming's**

Right-hand rule (often called the geneRator rule) which states:

Let the thumb, first finger and second finger of the right hand be extended such that they are all at right angles to each other [as shown in Fig. 32.2]. If the first finger points in the direction of the magnetic field, the thumb points in the direction of motion of the conductor relative to the magnetic field, then the second finger will point in the direction of the induced e.m.f.

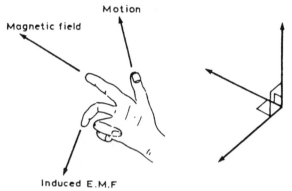

Figure 32.2

Summarizing:

First finger – Field

ThuMb – Motion

SEcond finger – E.m.f.

In a generator, conductors forming an electric circuit are made to move through a magnetic field. By Faraday's law an e.m.f. is induced in the conductors and thus a source of e.m.f. is created. A generator converts mechanical energy into electrical energy. (The action of a simple a.c. generator is described in Chapter 33.) The induced e.m.f. E set up between the ends of the conductor shown in Fig. 32.3 is given by:

$$E = Blv \text{ volts}$$

where B the flux density is measured in teslas, l the length of conductor in the magnetic field is measured in metres, and v the conductor velocity is measured in metres per second.

If the conductor moves at an angle $\theta°C$ to the magnetic field (instead of at 90° as assumed above) then

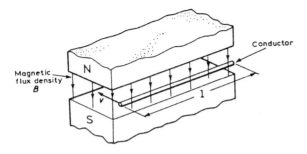

Figure 32.3

$$E = Blv \sin \theta$$

Problem 1. A conductor 300 mm long moves at a uniform speed of 4 m/s at right angles to a uniform magnetic field of flux density 1.25 T. Determine the current flowing in the conductor when (a) its ends are open-circuited, (b) its ends are connected to a load of 20 Ω resistance.

When a conductor moves in a magnetic field it will have an e.m.f. induced in it but this e.m.f. can only produce a current if there is a closed circuit. Induced e.m.f. $E = Blv$

$$= (1.25) \left(\frac{300}{1000} \right) (4) = 1.5 \text{ V}$$

(a) If the ends of the conductor are open-circuited **no current** will flow even though 1.5 V has been induced.
(b) From Ohm's law

$$I = \frac{E}{R} = \frac{1.5}{20} = \textbf{0.075 A or 75 mA}$$

Problem 2. At what velocity must a conductor 75 mm long cut a magnetic field of flux density 0.6 T if an e.m.f. of 9 V is to be induced in it? Assume the conductor, the field and the direction of motion are mutually perpendicular.

Induced e.m.f. $E = Blv$, hence

$$\text{velocity } v = \frac{E}{Bl}$$

i.e.

$$v = \frac{9}{(0.6)(75 \times 10^{-3})} = \frac{9 \times 10^3}{0.6 \times 75} = \textbf{200 m/s}$$

Problem 3. A conductor moves with a velocity of 15 m/s at an angle of (a) 90°, (b) 60°, and (c) 30° to a magnetic field produced between two square-faced poles of side length 2 cm. If the flux leaving a pole face is 5 μWb, find the magnitude of the induced e.m.f. in each case.

$v = 15$ m/s; length of conductor in magnetic field, $l = 2$ cm $= 0.02$ m; $A = (2 \times 2)$ cm$^2 = 4 \times 10^{-4}$ m^2, $\Phi = 5 \times 10^{-6}$ Wb.

(a) $E_{90} = Blv \sin 90°$

$$= \frac{\Phi}{A} lv = \frac{(5 \times 10^{-6})}{(4 \times 40^{-4})} (0.02)(15) = \textbf{3.75 mV}$$

(b) $E_{60} = Blv \sin 60°$

$$= E_{90} \sin 60° = 3.75 \sin 60° = \textbf{3.25 mV}$$

(c) $E_{30} = Blv \sin 30°$

$$= E_{90} \sin 30° = 3.75 \sin 30 = \textbf{1.875 mV}$$

Problem 4. The wing span of a metal aeroplane is 36 m. If the aeroplane is flying at 400 km/h, determine the e.m.f. induced between its wing tips. Assume the vertical component of the earth's magnetic field is 40 μT.

Induced e.m.f. across wing tips, $E = Blv$. $B = 40$ μT $= 40 \times 10^{-6}$ T; $l = 36$ m.

$$v = 400 \frac{\text{km}}{\text{h}} \times 1000 \frac{\text{m}}{\text{km}} \times \frac{1 \text{ h}}{60 \times 60 \text{ s}}$$

$$= \frac{(400)(1000)}{3600} = \frac{4000}{36} \text{ m/s}$$

Hence

$$E = (40 \times 10^{-6})(36) \left(\frac{4000}{36} \right) = \textbf{0.16 V}$$

Problem 5. The diagram shown in Fig. 32.4 represents the generation of e.m.f.s. Determine (i) the direction in which the conductor has to be moved in Fig. 32.4(a), (ii) the direction of the induced e.m.f. in Fig. 32.4(b), (iii) the polarity of the magnetic system in Fig. 32.4(c).

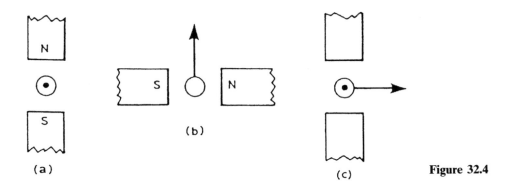

Figure 32.4

The direction of the e.m.f., and thus the current due to the e.m.f., may be obtained by either Lenz's law or Fleming's Right-hand rule (i.e. Gene**R**ator rule).

(i) Using Lenz's law: The field due to the magnet and the field due to the current-carrying conductor are shown in Fig. 32.5(a) and are seen to reinforce to the left of the conductor. Hence the force on the conductor is to the right. However Lenz's law says that the direction of the induced e.m.f. is always such as to oppose the effect producing it. **Thus the conductor will have to be moved to the left.**

(ii) Using Fleming's right-hand rule:
First finger – **F**ield, i.e. N → S, or right to left;
Thu**M**b – **M**otion, i.e. upwards;

SEcond finger – **E**.m.f., i.e. **towards the viewer or out of the paper**, as shown in Fig. 32.5(b)

(iii) The polarity of the magnetic system of Fig. 32.4(c) is shown in Fig. 32.5(c) and is obtained using Fleming's right-hand rule.

32.3 Inductance

Inductance is the name given to the property of a circuit whereby there is an e.m.f. induced into the circuit by the change of flux linkages produced by a current change.

When the e.m.f. is induced in the same circuit as that in which the current is changing, the property is called **self inductance**, *L*.

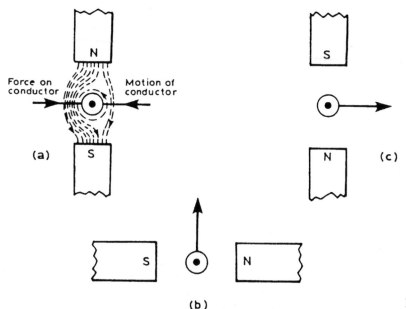

Figure 32.5

When the e.m.f. is induced in a circuit by a change of flux due to current changing in an adjacent circuit, the property is called **mutual inductance**, *M*.

The unit of inductance is the **henry, H**.

A circuit has an inductance of one henry when an e.m.f. of one volt is induced in it by a current changing at the rate of one ampere per second.

Induced e.m.f. in a coil of *N* turns is

$$E = N \frac{\Delta \Phi}{t} \text{ volts}$$

where $\Delta \Phi$ is the change in flux in webers, and *t* is the time taken for the flux to change in seconds.

Induced e.m.f. in a coil of inductance *L* henries is

$$E = L \left(\frac{\Delta I}{t}\right) \text{ volts}$$

where ΔI is the change in current in amperes and *t* is the time taken for the current to change in seconds.

The above equations for e.m.f. *E* are often stated as $E = -N/(\Delta \Phi/t)$ and $E = -L(\Delta I/t)$, the minus sign reminding us of its direction (given by Lenz's law). In the following problems the direction of the e.m.f. is measured.

Problem 6. Determine the e.m.f. induced in a coil of 200 turns when there is a change of flux of 25 mWb linking with it in 50 ms.

Induced e.m.f. *E*

$$= N \left(\frac{\Delta \Phi}{t}\right) = (200) \left(\frac{25 \times 10^{-3}}{50 \times 10^{-3}}\right) = \textbf{100 volts}$$

Problem 7. A flux of 400 Wb passing through a 150-turn coil is reversed in 40 ms. Find the average e.m.f. induced.

Since the flux reverses, the flux changes from +400 μWb to –400 μWb, a total change of flux of 800 μWb. Induced e.m.f. *E*

$$= N \left(\frac{\Delta \Phi}{t}\right) = (150) \left(\frac{800 \times 10^{-6}}{40 \times 10^{-3}}\right)$$

$$= \frac{800 \times 150 \times 10^3}{40 \times 10^6}$$

Hence the average e.m.f. induced *E* = **3 volts**.

Problem 8. Calculate the e.m.f. induced in a coil of inductance 12 H by a current changing at the rate of 4 A/s.

Induced e.m.f. *E*

$$= L \left(\frac{\Delta I}{t}\right) = (12)(4) = \textbf{48 volts}$$

Problem 9. An e.m.f. of 1.5 kV is induced in a coil when a current of 4 A collapses uniformly to zero in 8 ms. Determine the inductance of the coil.

Change in current, $\Delta I = (4 - 0) = 4$ A; $t = 8$ ms = 8×10^{-3} s;

$$\frac{\Delta I}{t} = \frac{4}{8 \times 10^{-3}} = \frac{4000}{8}$$

$$= 500 \text{ A/s}; E = 1.5 \text{ kV} = 1500 \text{ V}$$

Since

$$E = L \left(\frac{\Delta I}{t}\right)$$

$$L = \frac{E}{(\Delta I/t)} = \frac{1500}{500} = \textbf{3 H}$$

Problem 10. An average e.m.f. of 40 V is induced in a coil of inductance 150 mH when a current of 6 A is reversed. Calculate the time taken for the current to reverse.

$E = 40$ V, $L = 150$ mH $= 0.15$ H; change in current, $\Delta I = 6 - (-6) = 12$ A (since the current is reversed). Since $E = L(\Delta I/t)$

$$\text{time } t = \frac{L \Delta I}{E} = \frac{(0.15)(12)}{40} = \textbf{0.045 s or 45 ms}$$

32.4   Inductance of a coil

If a current changing from 0 to *I* amperes, produces a flux change from 0 to Φ webers, then $\Delta I = I$ and $\Delta \Phi = \Phi$. Then, from section 32.3, induced e.m.f.

$$E = \frac{N\Phi}{t} = \frac{LI}{t}$$

from which **inductance of coil**

$$L = \frac{N\Phi}{I} \text{ henrys}$$

Problem 11. Calculate the coil inductance when a current of 4 A in a coil of 800 turns produces a flux of 5 mWb linking with the coil.

For a coil, inductance $L = N\Phi/I$

$$= \frac{(800)(5 \times 10^{-3})}{4} = \textbf{1 H}$$

Problem 12. When a current of 1.5 A flows in a coil the flux linking with the coil is 90 µWb. If the coil inductance is 0.60 H calculate the number of turns of the coil.

For a coil, $L = N\Phi/I$. Thus $N = 2LI/\Phi$

$$= \frac{(0.6)(1.5)}{90 \times 10^{-6}} = \textbf{10 000 turns}$$

Problem 13. When carrying a current of 3 A, a coil of 750 turns has a flux of 12 mWb linking with it. Calculate the coil inductance and the e.m.f. induced in the coil when the current collapses to zero in 18 ms.

Coil inductance, $L = N\Phi/I$

$$= \frac{(750)(12 \times 10^{-3})}{3} = \textbf{3 H}$$

Induced e.m.f. $E = L(\Delta I/t)$

$$= 3\left(\frac{3 - 0}{18 \times 10^{-3}}\right) = \textbf{500 V}$$

(Alternatively, $E = N(\Delta\Phi/t)$

$$= (750)\left(\frac{12 \times 10^{-3}}{18 \times 10^{-3}}\right) = \textbf{500 V})$$

32.5 Mutual inductance

Mutually induced e.m.f. in the second coil is given by

$$\boxed{E_2 = M\left(\frac{\Delta I_1}{t}\right) \text{ volts}}$$

where M is the mutual inductance between two coils in henrys, ΔI_1 is the change in current in the first coil in amperes, and t is the time the current takes to change in the first coil in seconds.

Problem 14. Calculate the mutual inductance between two coils when a current changing at 200 A/s in one coil induces an e.m.f. of 1.5 V in the other.

Induced e.m.f. $E_2 = M(\Delta I_1/t)$, or $1.5 = M(200)$. Thus mutual inductance, $M = (1.5/200)$

$$= \textbf{0.0075 H} \text{ or } \textbf{7.5 mH}$$

Problem 15. The mutual inductance between two coils is 18 mH. Calculate the steady rate of change of current in one coil to induce an e.m.f. of 0.72 V in the other.

Induced e.m.f. $E_2 = M(\Delta I_1/t)$. Hence rate of change of current, $\Delta I_1/t$

$$= \frac{E_2}{M} = \frac{0.72}{0.018} = \textbf{40 A/S}$$

Problem 16. Two coils have a mutual inductance of 0.2 H. If the current in one coil is changed from 10 A to 4 A in 10 ms, calculate (a) the average induced e.m.f. in the second coil (b) the change of flux linked with the second coil if it is wound with 500 turns.

(a) Induced e.m.f. $E_2 = M(\Delta I_1/t)$

$$= (0.2)\left(\frac{10 - 4}{10 \times 10^{-3}}\right) = \textbf{120 V}$$

(b) Induced e.m.f. $E = N(\Delta\Phi/t)$, hence $\Delta\Phi = Et/N$. Thus the change of flux, $\Delta\Phi$

$$= \frac{120 (10 \times 10^{-3})}{500} = \textbf{24 mWb}$$

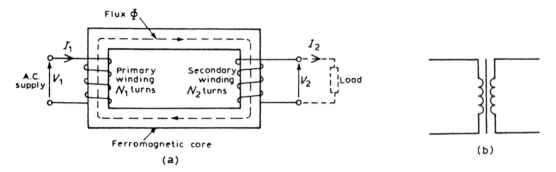

Figure 32.6

32.6 The transformer

A **transformer** is a device which uses the phenomenon of mutual induction to change the values of alternating voltages and currents. In fact, one of the main advantages of a.c. transmission and distribution is the ease with which an alternating voltage can be increased or decreased by transformers.

Losses in transformers are generally low and thus efficiency is high. Being static they have a long life and are very stable.

Transformers range in size from the miniature units used in electronic applications to the large power transformers used in power stations. The principle of operation is the same for each.

A transformer is represented in Fig. 32.6(a) as consisting of two electrical circuits linked by a common ferromagnetic core. One coil is termed the **primary winding** which is connected to the supply of electricity, and the other the **secondary winding**, which may be connected to a load. A circuit diagram symbol for a transformer is shown in Fig. 32.6(b).

Transformer principle of operation

When the secondary is an open-circuit and an alternating voltage V_1 is applied to the primary winding, a small current – called the no-load current I_0 – flows, which sets up a magnetic flux in the core. This alternating flux links with both primary and secondary coils and induces in them e.m.f.s of E_1 and E_2, respectively, by mutual induction. The induced e.m.f. E in a coil of N turns is given by $E = N(\Delta\Phi/t)$ where $\Delta\Phi/t$ is the rate of change of flux. In an ideal transformer,

the rate of change of flux is the same for both primary and secondary and thus $(E_1/N_1) = (E_2/N_2)$, i.e. **the induced e.m.f. per turn is constant**.

Assuming no losses, $E_1 = V_1$ and $E_2 = V_2$, hence

$$\frac{V_1}{N_1} = \frac{V_2}{N_2} \text{ or } \frac{V_1}{V_2} = \frac{N_1}{N_2} \qquad (32.1)$$

V_1/V_2 is called the **voltage ratio** and N_1/N_2 the **turns ratio**, or the '**transformation ratio**' of the transformer. If N_2 is less than N_1 then V_2 is less than V_1 and the device is termed a **step-down transformer**. If N_2 is greater than N_1 then V_2 is greater than V_1 and the device is termed a **step-up transformer**.

When a load is connected across the secondary winding, a current I_2 flows. In an ideal transformer losses are neglected and a transformer is considered to be 100% efficient. Hence input power = output power, or $V_1 I_1 = V_2 I_2$, i.e. in an ideal transformer, the **primary and secondary ampere-turns are equal**.

Thus

$$\frac{V_1}{V_2} = \frac{I_2}{I_1} \qquad (32.2)$$

Combining equations (32.1) and (32.2) gives:

$$\boxed{\frac{V_1}{V_2} = \frac{N_1}{N_2} = \frac{I_2}{I_1}} \qquad (32.3)$$

The **rating** of a transformer is stated in terms of the volt-amperes that it can transform without overheating. With reference to Fig. 32.6(a), the transformer rating is either $V_1 I_1$ or $V_2 I_2$, where I_2 is the full-load secondary current.

Problem 17. A transformer has 500 primary turns and 3000 secondary turns. If the primary voltage is 240 V, determine the secondary voltage, assuming an ideal transformer.

For an ideal transformer, voltage ratio = turns ratio, i.e.

$$\frac{V_1}{V_2} = \frac{N_1}{N_2} \quad \text{hence} \quad \frac{240}{V_2} = \frac{500}{3000}$$

Thus secondary voltage V_2

$$= \frac{(3000)(240)}{(500)} = \textbf{1440 V} \text{ or } \textbf{1.44 kV}$$

Problem 18. An ideal transformer with a turns ratio of $2 : 7$ is fed from a 240 V supply. Determine its output voltage.

A turns ratio of $2 : 7$ means that the transformer has 2 turns on the primary for every 7 turns on the secondary (i.e. a step-up transformer). Thus, $(N_1/N_2) = (2/7)$.

For an ideal transformer, $(N_1/N_2) = (V_1/V_2)$; hence $(2/7) = (240/V_2)$. Thus the secondary voltage $V_2 = (240)(7)/(2) = \textbf{840 V}$.

Problem 19. An ideal transformer has a turns ratio of $8 : 1$ and the primary current is 3 A when it is supplied at 240 V. Calculate the secondary voltage and current.

A turns ratio of $8 : 1$ means $(N_1/N_2) = (8/1)$, i.e. a step-down transformer.
$(N_1/N_2) = (V_1/V_2)$, or secondary voltage, V_2

$$= V_1 \left(\frac{N_2}{N_1}\right) = 240 \left(\frac{1}{8}\right) = \textbf{30 V}$$

Also $(N_1/N_2) = (I_2/I_1)$; hence secondary current $I_2 = I_1(N_1/N_2) = 3(8/1) = \textbf{24 A}$.

Problem 20. An ideal transformer, connected to a 240 V mains, supplies a 12 V, 150 W lamp. Calculate the transformer turns ratio and the current taken from the supply.

$V_1 = 240$ V; $V_2 = 12$ V, $I_2 = (P/V_2) = (150/12) = 12.5$ A. Turns ratio $= (N_1/N_2) = (V_1/V_2) = (240/12) = \textbf{20}$.
$(V_1/V_2) = (I_2/I_1)$, from which

$$I_1 = I_2 \left(\frac{V_2}{V_1}\right) = 12.5 \left(\frac{12}{240}\right)$$

Hence current taken from the supply, $I_1 = (12.5/20) = \textbf{0.625 A}$.

Problem 21. A 12 Ω resistor is connected across the secondary winding of an ideal transformer whose secondary voltage is 120 V. Determine the primary voltage if the supply current is 4 A.

Secondary current $I_2 = (V_2/R_2) = (120/12) = 10$ A.
$(V_1/V_2) = (I_2/I_1)$, from which the primary voltage $V_1 = V_2(I_2/I_1) = 120(10/4) = \textbf{300 V}$.

32.7 Multi-choice questions on electromagnetic induction

(Answers on page 356.)

1. A current changing at a rate of 5 A/s in a coil of inductance 5 H induces an e.m.f. of:
   (a) 25 V in the same direction as the applied voltage
   (b) 1 V in the same direction as the applied voltage
   (c) 25 V in the opposite direction to the applied voltage
   (d) 1 V in the opposite direction to the applied voltage
2. A bar magnet is moved at a steady speed of 1.0 m/s towards a coil of wire which is connected to a centre-zero galvanometer. The magnet is now withdrawn along the same path at 0.5 m/s. The deflection of the galvanometer is in the:
   (a) same direction as previous with the magnitude of the deflection doubled
   (b) opposite direction as previous with the magnitude of the deflection halved
   (c) same direction as previous with the magnitude of the deflection halved
   (d) opposite direction as previous with the magnitude of the deflection doubled
3. When a magnetic flux of 10 Wb links with a circuit of 20 turns in 2 s, the induced e.m.f. is:
   (a) 1 V   (b) 4 V   (c) 100 V   (d) 400 V

4. A current of 10 A in a coil of 1000 turns produces a flux of 10 mWb linking with the coil. The coil inductance is:
   (a) 10^6 H   (b) 1 H   (c) 1 μH   (d) 1 mH
5. An e.m.f. of 1 V is induced in a conductor moving at 10 cm/s in a magnetic field of 0.5 T. The effective length of the conductor in the magnetic field is:
   (a) 20 cm   (b) 5 m   (c) 20 m   (d) 50 m
6. Which of the following statements is false?
   (a) Fleming's left-hand rule or Lenz's law may be used to determine the direction of an induced e.m.f.
   (b) An induced e.m.f. is set up whenever the magnetic field linking that circuit changes.
   (c) The direction of an induced e.m.f. is always such as to oppose the effect producing it.
   (d) The induced e.m.f. in any circuit is proportional to the rate of change of the magnetic flux linking the circuit.
7. The mutual inductance between two coils, when a current changing at 20 A/s in one coil induces an e.m.f. of 10 mV in the other, is:
   (a) 0.5 H   (b) 200 mH   (c) 0.5 mH   (d) 2 H
8. A transformer has 800 primary turns and 100 secondary turns. To obtain 40 V from the secondary winding the voltage applied to the primary winding must be:
   (a) 5 V   (b) 320 V   (c) 2.5 V   (d) 20 V
9. An e.m.f. is induced into a conductor in the direction shown in Fig. 32.7 when the conductor is moved at a uniform speed in the field between the two magnets. The polarity of the system is:
   (a) North pole on right, South pole on left
   (b) North pole on left, South pole on right

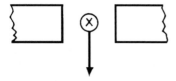

Figure 32.7

10. An ideal transformer has a turns ratio of 1:5 and is supplied at 200 V when the primary current is 3 A. Which of the following statements is false?
    (a) The turns ratio indicates a step-up transformer.

(b) The secondary voltage is 40 V.
(c) The secondary current is 15 A.
(d) The transformer rating is 0.6 kVA.
(e) The secondary voltage is 1 kV.
(f) The secondary current is 0.6 A.

32.8   Short answer questions on electromagnetic induction

1. What is electromagnetic induction?
2. State Faraday's laws of electromagnetic induction.
3. State Lenz's law.
4. Explain briefly the principle of the generator.
5. The direction of an induced e.m.f. in a generator may be determined using Fleming's .... rule.
6. The e.m.f. E induced in a moving conductor may be calculated using the formula $E = Blv$. Name the quantities represented and their units.
7. What is self-inductance? State its symbol.
8. State and define the unit of inductance.
9. When a circuit has an inductance L and the current changes at a rate of $(\Delta I/t)$ then the induced e.m.f. E is given by $E =$ .... volts.
10. If a current of I A flowing in a coil of N turns produces a flux of Φ Wb, the coil inductance L is given by $L =$ .... henrys.
11. What is mutual inductance? State its symbol.
12. The mutual inductance between two coils is M. The e.m.f. E_2 induced in one coil by the current changing at $(\Delta I_1/t)$ in the other is given by $E_2 =$ .... volts.
13. Explain briefly how a voltage is induced in the secondary winding of a transformer.
14. Draw the circuit diagram symbol for a transformer.
15. State the relationship between turns and voltage ratios for a transformer.

32.9   Further questions on electromagnetic induction

1. A conductor of length 15 cm is moved at 750 mm/s at right angles to a uniform flux density of 1.2 T. Determine the e.m.f. induced in the conductor.   [0.135 V]
2. Find the speed that a conductor of length 120 mm must be moved at right angles to a magnetic field of flux density 0.6 T to induce in it an e.m.f. of 1.8 V.   [25 m/s]

3. A 25 cm long conductor moves at a uniform speed of 8 m/s through a uniform magnetic field of flux density 1.2 T. Determine the current flowing in the conductor when (a) its ends are open-circuited (b) its ends are connected to a load of 15 ohms resistance.
[(a) 0 (b) 0.16 A]

4. A straight conductor 500 mm long is moved with constant velocity at right angles both to its length and to a uniform magnetic field. Given that the e.m.f. induced in the conductor is 2.5 V and the velocity is 5 m/s, calculate the flux density of the magnetic field. If the conductor forms part of a closed circuit of total resistance 5 ohms, calculate the force on the conductor. [1 T, 0.25 N]

5. A car is travelling at 80 km/h. Assuming the back axle of the car is 1.76 m in length and the vertical component of the earth's magnetic field is 40 μT, find the e.m.f. generated in the axle due to motion.
[1.56 mV]

6. A conductor moves with a velocity of 20 m/s at an angle of (a) 90° (b) 45° (c) 30° to a magnetic field produced between two square-faced poles of side length 2.5 cm. If the flux on the pole face is 60 mWb, find the magnitude of the induced e.m.f. in each case.
[(a) 48 V (b) 33.9 V (c) 24 V]

7. Find the e.m.f. induced in a coil of 200 turns when there is a change of flux of 30 mWb linking with it in 40 ms. [150 V]

8. An e.m.f. of 25 V is induced in a coil of 300 turns when the flux linking with it changes by 12 mWb. Find the time, in milliseconds, in which the flux makes the change. [144 ms]

9. An ignition coil having 10 000 turns has an e.m.f. of 8 kV induced in it. What rate of change of flux is required for this to happen?
[0.8 Wb/s]

10. A flux of 0.35 mWb passing through a 125-turn coil is reversed in 25 ms. Find the average e.m.f. induced. [3.5 V]

11. Calculate the e.m.f. induced in a coil of inductance 6 H by a current changing at a rate of 15 A/s. [90 V]

12. An e.m.f. of 2 kV is induced in a coil when a current of 5 A collapses uniformly to zero in 10 ms. Determine the inductance of the coil.
[4 H]

13. An average e.m.f. of 50 V is induced in a coil of inductance 160 mH when a current of 7.5 A is reversed. Calculate the time taken for the current to reverse. [40 ms]

14. A coil of 2500 turns has a flux of 10 mWb linking with it when carrying a current of 2 A. Calculate the coil inductance and the e.m.f. induced in the coil when the current collapses to zero in 20 ms.
[12.5 H, 1.25 kV]

15. Calculate the coil inductance when a current of 5 A in a coil of 1000 turns produces a flux of 8 mWb linking with the coil. [1.6 H]

16. A coil is wound with 600 turns and has a self-inductance of 2.5 H. What current must flow to set up a flux of 20 mWb? [4.8 A]

17. When a current of 2 A flows in a coil, the flux linking with the coil is 80 μWb. If the coil inductance is 0.5 H, calculate the number of turns of the coil. [12 500]

18. A coil of 1200 turns has a flux of 15 mWb linking with it when carrying a current of 4 A. Calculate the coil inductance and the e.m.f. induced in the coil when the current collapses to zero in 25 ms.
[4.5 H, 720 V]

19. A coil has 300 turns and an inductance of 4.5 mH. How many turns would be needed to produce a 0.72 mH coil assuming the same core is used? [120 turns]

20. A steady current of 5 A when flowing in a coil of 1000 turns produces a magnetic flux of 500 μWb. Calculate the inductance of the coil. The current of 5 A is then reversed in 12.5 ms. Calculate the e.m.f. induced in the coil. [0.1 H, 80 V]

21. The mutual inductance between two coils is 150 mH. Find the e.m.f. induced in one coil when the current in the other is increasing at a rate of 30 A/s. [4.5 V]

22. Determine the mutual inductance between two coils when a current changing at 50 A/s in one coil induces an e.m.f. of 80 mV in the other. [1.6 mH]

23. Two coils have a mutual inductance of 0.75 H. Calculate the e.m.f. induced in one coil when a current of 2.5 A in the other coil is reversed in 15 ms. [250 V]

24. The mutual inductance between two coils is 240 mH. If the current in one coil changes from 15 A to 6 A in 12 ms, calculate (a) the average e.m.f. induced in the other coil, (b) the change of flux linked with the other coil if it is wound with 400 turns.
[(a) 180 V (b) 5.4 mWb]

25. A mutual inductance of 0.6 H exists between two coils. If a current of 6 A in one coil is reversed in 0.8 s calculate (a) the average

e.m.f. induced in the other coil, (b) the number of turns on the other coil if the flux change linking with the other coil is 5 mWb.
[(a) 0.9 V (b) 144]

26. A transformer has 800 primary turns and 2000 secondary turns. If the primary voltage is 160 V, determine the secondary voltage assuming an ideal transformer. [400 V]

27. An ideal transformer with a turns ratio of 3:8 is fed from a 240 V supply. Determine its output voltage. [640 V]

28. An ideal transformer has a turns ratio of 12:1 and is supplied at 192 V. Calculate the secondary voltage. [16 V]

29. A transformer primary winding connected across a 415 V supply has 750 turns. Determine how many turns must be wound on the secondary side if an output of 1.66 kV is required. [3000 turns]

30. (a) Describe the transformer principle of operation.
    (b) An ideal transformer has a turns ratio of 12 : 1 and is supplied at 180 V when the primary current is 4 A. Calculate the secondary voltage and current.
    [15 V, 48 A]

31. A step-down transformer having a turns ratio of 20 : 1 has a primary voltage of 4 kV and a load of 10 kW. Neglecting losses, calculate the value of the secondary current.
    [50 A]

32. A transformer has a primary-to-secondary turns ratio of 1 : 15. Calculate the primary voltage necessary to supply a 240 V load. If the load current is 3 A, determine the primary current. Neglect any losses.
    [16 V, 45 A]

33

Alternating voltages and currents

At the end of this chapter you should be able to:

- appreciate why a.c. is used in preference to d.c.
- describe the principle of operation of an a.c. generator
- distinguish between unidirectional and alternating waveforms
- define cycle, period or periodic time T and frequency f of a waveform
- perform calculations using $t = 1/f$
- define instantaneous, peak, mean and r.m.s. values, form factor and peak factor, for a sine wave
- calculate mean and r.m.s. values, form factor and peak factor for given waveforms

33.1 Introduction

Electricity is produced by generators at power stations and then distributed by a vast network of transmission lines (called the National Grid system) to industry and for domestic use. It is easier and cheaper to generate **alternating current** (a.c.) than direct current (d.c.) and a.c. is more conveniently distributed than d.c. since its voltage can be readily altered using transformers. Whenever d.c. is needed in preference to a.c., devices called rectifiers are used for conversion.

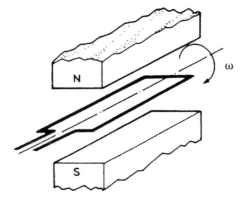

Figure 33.1

33.2 The a.c. generator

Let a single turn coil be free to rotate at constant angular velocity ω symmetrically between the poles of a magnet system as shown in Fig. 33.1. An e.m.f. is generated in the coil (from Faraday's law) which varies in magnitude and reverses its direction at regular intervals. The reason for this is shown in Fig. 33.2.

In positions (a), (e) and (i) the conductors of the loop are effectively moving along the magnetic field, no flux is cut and hence no e.m.f. is induced. In position (c) maximum e.m.f. is induced. In position (g), maximum flux is cut and hence maximum e.m.f. is again induced. However, using Fleming's right-hand rule, the induced e.m.f. is in the opposite direction to that in position (c) and is thus shown as $-E$. In positions (b), (d), (f) and (h) some flux is cut and hence some e.m.f. is induced. If all such positions of the coil are considered, in one revolution of

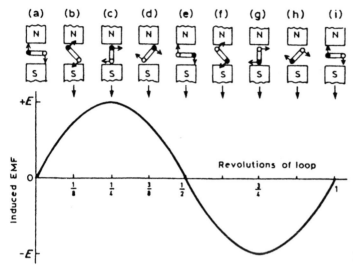

Figure 33.2

the coil, one cycle of alternating e.m.f. is produced as shown. This is the principle of operation of the **a.c. generator** (i.e. the **alternator**).

33.3 Waveforms

If values of quantities which vary with time t are plotted to a base of time, the resulting graph is called a **waveform**. Some typical waveforms are shown in Fig. 33.3. Waveforms (a) and (b) are **unidirectional waveforms**, for, although they vary considerably with time, they flow in one direction only (i.e. they do not cross the time axis and become negative). Waveforms (c) to (g) are called **alternating waveforms** since their quantities are continually changing in direction (i.e. alternately positive and negative).

A waveform of the type shown in Fig. 33.3(g) is called a **sine wave**. It is the shape of the waveform of e.m.f. produced by an alternator and thus the mains electricity supply is of 'sinusoidal' form.

One complete series of values is called a **cycle** (i.e. from O to P in Fig. 33.3(g)). The time taken for an alternating quantity to complete one cycle is called the **period** or the **periodic time**, T, of the waveform. The number of cycles completed in one second is called the **frequency**, f, of the supply and is measured in **hertz, Hz**. The standard frequency of the electricity supply in Great Britain is 50 Hz.

$$T = \frac{1}{f} \text{ or } f = \frac{1}{T}$$

Problem 1. Determine the periodic time for frequencies of (a) 50 Hz and (b) 20 kHz.

(a) Periodic time T

$$= \frac{1}{f} = \frac{1}{50} = \textbf{0.02 s or 20 ms}$$

(b) Periodic time T

$$= \frac{1}{f} = \frac{1}{20\,000} = \textbf{0.000 05 s or 50 μs}$$

Problem 2. Determine the frequencies for periodic times of (a) 4 ms, (b) 4 μs.

(a) Frequency f

$$= \frac{1}{T} = \frac{1}{4 \times 10^{-3}} = \frac{1000}{4} = \textbf{250 Hz}$$

(b) Frequency f

$$= \frac{1}{T} = \frac{1}{4 \times 10^{-6}} = \frac{1\,000\,000}{4}$$

$$= \textbf{250 000 Hz or 250 kHz or 0.25 MHz}$$

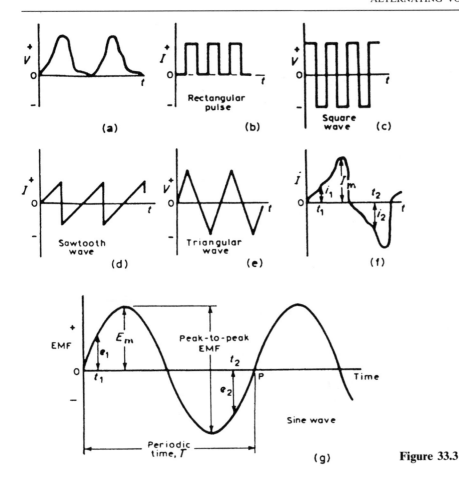

Figure 33.3

Problem 3. An alternating current completes 5 cycles in 8 ms. What is its frequency?

Time for 1 cycle = (8/5) ms = 1.6 ms = periodic time T.
Frequency f

$$= \frac{1}{T} = \frac{1}{1.6 \times 10^{-3}} = \frac{1000}{1.6} = \frac{10\,000}{16} = \textbf{625 Hz}$$

33.4 A.c. values

Instantaneous values are the values of the alternating quantities at any instant of time. They are represented by small letters, i, v, e, etc. (see Figs. 33.3(f) and (g)).

The largest value reached in a half cycle is called the **peak value** or the **maximum value** or the **crest value** or the **amplitude** of the waveform. Such values are represented by V_m, I_m, E_m, etc. (see Fig. 33.3(f) and (g)). A **peak-to-peak** value of e.m.f. is shown in Fig. 33.3(g) and is the difference between the maximum and minimum values in a cycle.

The **average** or **mean value** of a symmetrical alternating quantity (such as a sine wave) is the average value measured over a half cycle (since over a complete cycle the average value is zero).

Average or mean value
$$= \frac{\textbf{area under the curve}}{\textbf{length of base}}$$

The area under the curve is found by approximate methods such as the trapezoidal rule, the mid-ordinate rule or Simpson's rule. Average values are represented by V_{AV}, I_{AV}, etc.

For a sine wave,
average value = 0.637 × maximum value
(i.e. 2/π × maximum value)

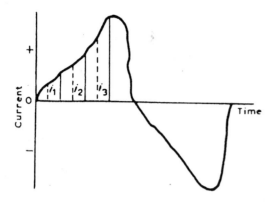

Figure 33.4

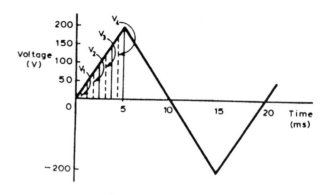

Figure 33.5

The **effective value** of an alternating current is that current which will produce the same heating effect as an equivalent direct current. The effective value is called the **root mean square (r.m.s.) value** and whenever an alternating quantity is given, it is assumed to be the r.m.s. value. For example, the domestic mains supply in Great Britain is 240 V and is assumed to mean '240 V r.m.s.'. The symbols used for r.m.s. values are I, V, E, etc. For a non-sinusoidal waveform as shown in Fig. 33.4 the r.m.s. value is given by:

$$I = \sqrt{\left(\frac{i_1^2 + i_2^2 + \ldots + i_n^2}{n}\right)}$$

where n is the number of intervals used.

> **For a sine wave,**
> **r.m.s. value = 0.707 × maximum value**
>
> **(i.e. $1/\sqrt{2}$ × maximum value)**

> **Form factor =** $\dfrac{\textbf{r.m.s. value}}{\textbf{average value}}$

For a sine wave, form factor = 1.11.

> **Peak factor =** $\dfrac{\textbf{maximum value}}{\textbf{r.m.s. value}}$

For a sine wave, peak factor = 1.41.

The values of form and peak factors give an indication of the shape of waveforms.

> Problem 4.   For the periodic waveforms shown in Fig. 33.5 and 33.6 determine for each: (i) frequency, (ii) average value over half a cycle, (iii) r.m.s. value, (iv) form factor, and (v) peak factor.

(a)   **Triangular waveform** (Fig. 33.5)
  (i)   Time for 1 complete cycle = 20 ms = periodic time T, hence frequency f

$$= \frac{1}{T} = \frac{1}{20 \times 10^{-3}} = \frac{1000}{20} = \textbf{50 Hz}$$

  (ii)   Area under the triangular waveform for a half cycle

$$= \frac{1}{2} \times \text{base} \times \text{height} = \frac{1}{2} \times (10 \times 10^{-3}) \times 200$$

$$= 1 \text{ volt second}$$

   Average value of waveform

$$= \frac{\text{area under curve}}{\text{length of base}} = \frac{1 \text{ volt second}}{10 \times 10^{-3} \text{ second}}$$

$$= \frac{1000}{10} = \textbf{100 V}$$

  (iii)   In Fig. 33.5 the first 1/4 cycle is divided into 4 intervals. Thus

$$\text{r.m.s. value} = \sqrt{\left(\frac{i_1^2 + i_2^2 + i_3^2 + i_4^2}{4}\right)}$$

$$= \sqrt{\left(\frac{25^2 + 75^2 + 125^2 + 175^2}{4}\right)}$$

$$= \textbf{114.6 V}$$

(Note that the greater the number of intervals chosen, the greater the accuracy of the result. For example, if twice the number of ordinates as that chosen above are used, the r.m.s. value is found to be 115.6 V.)

(iv)

$$\text{Form factor} = \frac{\text{r.m.s. value}}{\text{average value}} = \frac{114.6}{100} = \textbf{1.15}$$

(v)

$$\text{Peak factor} = \frac{\text{maximum value}}{\text{r.m.s. value}} = \frac{200}{114.6}$$

$$= 1.75$$

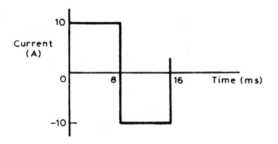

Figure 33.6

(b) **Rectangular waveform** (Fig. 33.6)

(i) Time for 1 complete cycle = 16 ms = periodic time, T, hence frequency, f

$$= \frac{1}{T} = \frac{1}{16 \times 10^{-3}} = \frac{1000}{16}$$

$$= \textbf{62.5 Hz}$$

(ii) Average value over half a cycle

$$= \frac{\text{area under curve}}{\text{length of base}}$$

$$= \frac{10 \times (8 \times 10^{-3})}{8 \times 10^{-3}}$$

$$= \textbf{10 A}$$

(iii) The r.m.s. value is given by

$$\sqrt{\left(\frac{i_1^2 + i_2^2 + \ldots + i_n^2}{n}\right)} = \textbf{10 A}$$

however many intervals are chosen, since the waveform is rectangular.

(iv)

$$\text{Form factor} = \frac{\text{r.m.s. value}}{\text{average value}} = \frac{10}{10} = \textbf{1}$$

(v)

$$\text{Peak factor} = \frac{\text{maximum value}}{\text{r.m.s. value}} = \frac{10}{10} = \textbf{1}$$

Problem 5. The following table gives the corresponding values of current and time for a half cycle of alternating current.

time t (ms)	0	0.5	1.0	1.5	2.0	2.5
current i (A)	0	7	14	23	40	56

time t (ms)	3.0	3.5	4.0	4.5	5.0
current i (A)	68	76	60	5	0

Assuming the negative half cycle is identical in shape to the positive half cycle, plot the waveform and find (a) the frequency of the supply, (b) the instantaneous values of current after 1.25 ms and 3.8 ms, (c) the peak or maximum value, (d) the mean or average value, and (e) the r.m.s. value of the waveform.

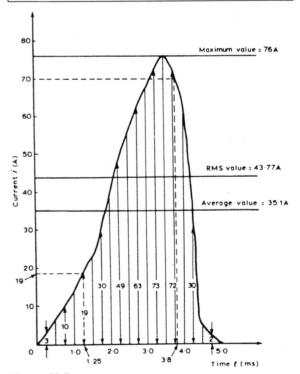

Figure 33.7

The half cycle of alternating current is shown plotted in Fig. 33.7.

(a) Time for a half cycle = 5 ms. Hence the time for 1 cycle, i.e. the periodic time, T = 10 ms or 0.01 s.

Frequency, $f = \dfrac{1}{T} = \dfrac{1}{0.01}$ = **100 Hz**

(b) Instantaneous value of current after 1.25 ms is **19 A**, from Fig. 33.7.
Instantaneous value of current after 3.8 ms is **70 A**, from Fig. 33.7.

(c) Peak or maximum value = **76 A**.

(d) Mean or average value

$$= \dfrac{\text{area under curve}}{\text{length of base}}$$

Using the mid-ordinate rule with 10 intervals, each of width 0.5 ms gives: area under curve

$= (0.5 \times 10^{-3})[3 + 10 + 19 + 30 + 49 + 63 + 73 + 72 + 30 + 2]$ (see Fig. 33.7)

$= (0.5 \times 10^{-3})(351)$

Hence mean or average value

$= \dfrac{(0.5 \times 10^{-3})(351)}{5 \times 10^{-3}}$ = **35.1 A**

(e) r.m.s. value

$$= \sqrt{\left(\dfrac{3^2+10^2+19^2+30^2+49^2+63^2+73^2+72^2+30^2+2^2}{10}\right)}$$

$$= \sqrt{\left(\dfrac{19\,157}{10}\right)} = \textbf{43.8 A}$$

Problem 6. Calculate the r.m.s. value of a sinusoidal current of maximum value 20 A.

For a sine wave,

r.m.s. value = 0.707 × maximum value

= 0.707 × 20 = **14.14 A**

Problem 7. Determine the peak and mean values for a 240 V mains supply.

For a sine wave, r.m.s. value of voltage V = 0.707 × V_m. A 240 V mains supply means that 240 V is the r.m.s. value. Hence

$$V_m = \dfrac{V}{0.707} = \dfrac{240}{0.707} = \textbf{339.5 V} = \text{peak value}$$

Mean value, V_{AV} = 0.637 V_m
= 0.637 × 339.5 = **216.3 V**

Problem 8. A supply voltage has a mean value of 150 V. Determine its maximum value and its r.m.s. value.

For a sine wave, mean value = 0.637 × maximum value, hence

maximum value =

$$\dfrac{\text{mean value}}{0.637} = \dfrac{150}{0.637} = \textbf{235.5 V}$$

r.m.s. value = 0.707 × maximum value = 0.707 × 235.5 = **166.5 V**

33.5 Multi-choice questions on alternating voltages and currents

(Answers on page 356.)

1. The value of an alternating current at any given instant is
   (a) a maximum value
   (b) a peak value
   (c) an instantaneous value
   (d) an r.m.s. value
2. An alternating current completes 100 cycles in 0.1 s. Its frequency is (a) 20 Hz (b) 100 Hz (c) 0.002 Hz (d) 1 kHz
3. In Fig. 33.8, at the instant shown the generated e.m.f. will be
   (a) zero   (b) an r.m.s. value
   (c) an average value   (d) a maximum value

Figure 33.8

4. The supply of electrical energy for a consumer is usually by a.c. because:
   (a) transmission and distribution are more easily affected
   (b) it is most suitable for variable speed motors
   (c) the volt drop in cables is minimal
   (d) cable power losses are negligible
5. Which of the following statements is false?
   (a) It is cheaper to use a.c. than d.c.
   (b) Distribution of a.c. is more convenient than with d.c. since voltages may be readily altered using transformers.
   (c) An alternator is an a.c. generator.
   (d) A rectifier changes d.c. into a.c.
6. An alternating voltage of maximum value 100 V is applied to a lamp. Which of the following direct voltages, if applied to the lamp, would cause the lamp to light with the same brilliance?
   (a) 100 V   (b) 63.7 V   (c) 70.7 V
   (d) 141.4 V
7. The value normally stated when referring to alternating currents and voltages is the:
   (a) instantaneous value   (b) r.m.s. value
   (c) average value   (d) peak value
8. State which of the following is false. For a sine wave:
   (a) the peak factor is 1.414
   (b) the r.m.s. value is 0.707 × peak value
   (c) the average value is 0.637 × r.m.s. value
   (d) the form factor is 1.11
9. An a.c. supply is 70.7 V, 50 Hz. Which of the following statements is false?
   (a) The periodic time is 20 ms.
   (b) The peak value of the voltage is 70.7 V.
   (c) The r.m.s. value of the voltage is 70.7 V.
   (d) The peak value of the voltage is 100 V.

33.6 Short answer questions on alternating voltages and currents

1. Briefly explain the principle of the simple alternator.
2. What is the difference between an alternating and a unidirectional waveform.
3. What is meant by (a) waveform (b) cycle.
4. The time to complete one cycle of a waveform is called the ....
5. What is frequency? Name its unit.
6. The mains supply voltage has a special shape of waveform called a ....

7. Define peak value.
8. What is meant by the r.m.s. value?
9. The domestic mains electricity supply voltage in Great Britain is ....
10. What is the mean value of a sinusoidal alternating e.m.f. which has a maximum value of 100 V?
11. The effective value of a sinusoidal waveform is (......... × maximum value).

33.7 Further questions on alternating voltages and currents

Frequency and periodic time

1. Determine the periodic time for the following frequencies:
   (a) 2.5 Hz (b) 100 Hz (c) 40 kHz.
   [(a) 0.4 s (b) 10 ms (c) 25 µs]
2. Calculate the frequency for the following periodic times:
   (a) 5 ms (b) 50 µs (c) 0.2 s.
   [(a) 0.2 kHz (b) 20 kHz (c) 5 Hz]
3. An alternating current completes 4 cycles in 5 ms. What is the frequency?
   [800 Hz]

a.c. values of non-sinusoidal waveforms

4. An alternating current varies with time over half a cycle as follows:

Current (A)	0	0.7	2.0	4.2	8.4	8.2	2.5	1.0
time (ms)	0	1	2	3	4	5	6	7
Current (A)	0.4	0.2	0					
time (ms)	8	9	10					

The negative half cycle is similar. Plot the curve and determine: (a) the frequency (b) the instantaneous values at 3.4 ms and 5.8 ms (c) its mean value and (d) its r.m.s. value.
   [(a) 50 Hz (b) 5.5 A, 3.4 A (c) 2.8 A
   (d) 4.0 A]
5. For the waveforms shown in Fig. 33.9 determine for each (i) the frequency (ii) the average value over half a cycle (iii) the r.m.s. value (iv) the form factor (v) the peak factor.
   (a)(i) 100 Hz (ii) 2.50 A (iii) 2.88 A (iv) 1.15
   (v) 1.74
   (b)(i) 250 Hz (ii) 20 V (iii) 20 V (iv) 1.0
   (v) 1.0
   (c)(i) 125 Hz (ii) 18 A (iii) 19.56 A (iv) 1.09
   (v) 1.23
   (d)(i) 250 Hz (ii) 25 V (iii) 50 V (iv) 2.0
   (v) 2.0

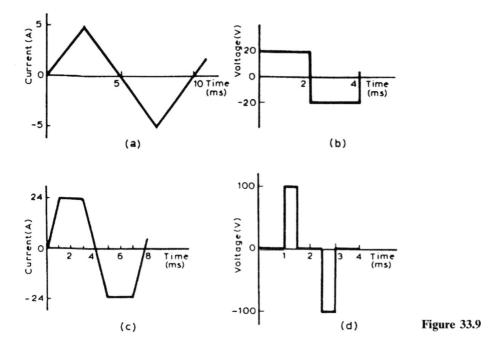

Figure 33.9

6. An alternating voltage is triangular in shape, rising at a constant rate to a maximum of 300 V in 8 ms and then falling to zero at a constant rate in 4 ms. The negative half cycle is identical in shape to the positive half cycle. Calculate (a) the mean voltage over half a cycle, and (b) the r.m.s. voltage.

[(a) 150 V (b) 170 V]

7. An alternating e.m.f. varies with time over half a cycle as follows:

| E.m.f. (V) | 0 | 45 | 80 | 155 | 215 | 320 | 210 | 95 | 0 |
| time (ms) | 0 | 1.5 | 3.0 | 4.5 | 6.0 | 7.5 | 9.0 | 10.5 | 12.0 |

The negative half cycle is identical in shape to the positive half cycle. Plot the waveform and determine (a) the periodic time and frequency, (b) the instantaneous value of voltage at 3.75 ms, (c) the times when the voltage is 125 V, (d) the mean value, and (e) the r.m.s. value.

(a) 24 ms, 41.67 Hz (b) 115 V (c) 4 ms and 10.1 ms (d) 142 V (e) 171 V

a.c. values of sinusoidal waveforms

8. Calculate the r.m.s. value of a sinusoidal curve of maximum value 300 V.

[212.1 V]

9. Find the peak and mean values for a 200 V mains supply.   [282.9 V, 180.2 V]

10. Plot a sine wave of peak value 10.0 A. Show that the average value of the waveform is 6.37 A over half a cycle, and that the r.m.s. value is 7.07 A.

11. A sinusoidal voltage has a maximum value of 120 V. Calculate its r.m.s. and average values.   [84.8 V, 76.4 V]

12. A sinusoidal current has a mean value of 15.0 A. Determine its maximum and r.m.s. values.

[23.55 A, 16.65 A]

34

Capacitors and their effects in electric circuits

At the end of this chapter you should be able to:

- define electrostatics

- explain how a capacitor stores energy

- appreciate $Q = It$, $E = V/d$, $\sigma = Q/A$, $Q = CV$ and $(\sigma/E) = \epsilon_0\epsilon_r$ and perform calculations using these formulae

- state the unit of capacitance

- define a dielectric

- state the value of ϵ_0

- appreciate typical values of ϵ_r

- appreciate that the capacitance C of a parallel plate capacitor is given by $C = (\epsilon_0\epsilon_r A(n - 1))/d$

- perform calculations involving the parallel plate capacitor

- perform calculations for total capacitance and charge when capacitors are connected in parallel and in series

- define dielectric strength

- determine the energy stored in a capacitor and state its unit

- describe practical types of capacitor

- describe the effect on capacitor voltage and current when a d.c. is applied to a CR series circuit

- state that the time constant, $\tau = CR$

- describe the effect in a d.c. circuit when a capacitor is discharged

- perform calculations on transient CR circuits

- define and calculate capacitive reactance in an a.c. circuit

- describe how current leads voltage by 90° in a purely capacitive a.c. circuit

34.1 Electrostatics

Electrostatics is the branch of electricity which is concerned with the study of electrical charges at rest. An electrostatic field accompanies a static charge and this is utilized in the capacitor.

Charged bodies attract or repel each other, depending on the nature of the charge. The rule is: **like charges repel, unlike charges attract**.

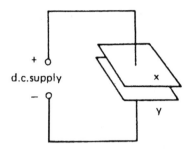

Figure 34.1

34.2 Capacitors and capacitance

A **capacitor** is a device capable of storing electrical energy. Figure 34.1 shows a capacitor consisting of a pair of parallel metal plates X and Y separated by an insulator, which could be air. Since the plates are electrical conductors each will contain a large number of mobile electrons. Because the plates are connected to a d.c. supply the electrons on plate X, which have a small negative charge, will be attracted to the positive pole of the supply and will be repelled from the negative pole of the supply on to plate Y. X will become positively charged due to its shortage of electrons whereas Y will have a negative charge due to its surplus of electrons.

The difference in charge between the plates results in a p.d. existing between them, the flow of electrons dying away and ceasing when the p.d. between the plates equals the supply voltage. The plates are then said to be **charged** and there exists an **electric field** between them. Figure 34. 2 shows a side view of the plates with the field represented by 'lines of electrical flux'. If the plates are disconnected from the supply and connected

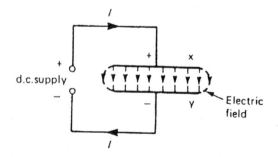

Figure 34.2

together through a resistor the surplus of electrons on the negative plate will flow through the resistor to the positive plate. This is called **discharging**. The current flow decreases to zero as the charges on the plates reduce. The current flowing in the resistor causes it to liberate heat showing that **energy is stored in the electric field**.

Summary of important formulae and definitions

From Chapter 26, charge Q is given by:

$$Q = I \times t \text{ coulombs}$$

where I is the current in amperes, and t the time in seconds.

A **dielectric** is an insulating medium separating charged surfaces. **Electric field strength**, electric force, or voltage gradient,

$$E = \frac{\text{p.d. across dielectric}}{\text{thickness of dielectric}}$$

i.e.

$$E = \frac{V}{d} \text{ volts/m}$$

Charge density

$$\sigma = \frac{\text{charge}}{\text{area of one plate}}$$

i.e.

$$\sigma = \frac{Q}{A} \text{ C/m}^2$$

Charge Q on a capacitor is proportional the applied voltage, V, i.e. $Q \propto V$

$$Q = CV \quad \text{or} \quad C = \frac{Q}{V}$$

where the constant of proportionality, C, is the **capacitance**.

The unit of capacitance is the **farad**, **F** (or more usually $\mu F = 10^{-6}$ F or pF $= 10^{-12}$ F), which is defined as the capacitance of a capacitor when a p.d. of one volt appears across the plates when charged with one coulomb.

Every system of electrical conductors possesses capacitance. For example, there is capacitance between the conductors of overhead transmission lines and also between the wires of a telephone cable. In these examples the capacitance is undesirable but has to be accepted, minimized or compensated for. There are other situations, such as in capacitors, where capacitance is a desirable property.

The ratio of charge density, σ, to electric field strength, E, is called **absolute permittivity**, ϵ, of a dielectric. Thus

$$\frac{\sigma}{E} = \epsilon$$

Permittivity of free space is a constant, given by

$$\epsilon_o = 8.85 \times 10^{-12} \text{ F/m}$$

Relative permittivity, ϵ_r

$$= \frac{\text{flux density of the field in the dielectric}}{\text{flux density of the field in the vacuum}}$$

(ϵ_r has no units). Examples of the values of ϵ_r include: air = 1, polythene = 2.3, mica = 3–7, glass = 5–10, ceramics = 6–1000.

Absolute permittivity, $\epsilon = \epsilon_o \epsilon_r$, thus

$$\boxed{\frac{\sigma}{E} = \epsilon_o \epsilon_r}$$

Problem 1.   Two parallel rectangular plates measuring 20 cm by 40 cm carry an electric charge of 0.2 µC. Calculate the charge density on the plates. If the plates are spaced 5 mm apart and the voltage between them is 0.25 kV, determine the electric field strength.

Charge, $Q = 0.2$ µC $= 0.2 \times 10^{-6}$ C; area A = 20 cm $\times$ 40 cm = 800 cm^2 = 800 $\times$ 10^{-4} m^2.

$$\textbf{Charge density, } \sigma = \frac{Q}{A} = \frac{0.2 \times 10^{-6}}{800 \times 10^{-4}}$$

$$= \frac{0.2 \times 10^4}{800 \times 10^6} = \frac{2000}{800} \times 10^{-6}$$

$$= \textbf{2.5 µC/m}^2$$

Voltage $V = 0.25$ kV = 250 V; plate spacing, $d = 5$ mm = 5 $\times$ 10^{-3} m.

$$\text{Electric field strength, } E = \frac{V}{d}$$

$$= \frac{250}{5 \times 10^{-3}} = 50 \text{ kV/m}$$

Problem 2.   The charge density of two plates separated by mica of relative permittivity 5 is 2 µC/m². Find the voltage gradient between the plates.

Charge density, $\sigma = 2$ µC/m^2 = 2 $\times$ 10^{-6} C/m^2, ϵ_o = 8.85 $\times$ 10^{-12} F/m; ϵ_r = 5; $(\sigma/E) = \epsilon_o\epsilon_r$. Hence

$$\textbf{voltage gradient, } E = \frac{\sigma}{\epsilon_o\epsilon_r}$$

$$= \frac{2 \times 10^{-6}}{8.85 \times 10^{-12} \times 5} \text{ V/m}$$

$$= \textbf{45.2 kV/m}$$

Problem 3.   Two parallel plates having a p.d. of 200 V between them are spaced 0.8 mm apart. What is the electric field strength? Find also the charge density when the dielectric between the plates is (a) air and (b) polythene of relative permittivity 2.3.

Electric field strength,

$$E = \frac{V}{d} = \frac{200}{0.8 \times 10^{-3}} = \textbf{250 kV/m}$$

(a)   For air: ϵ_r = 1. $(\sigma/E) = \epsilon_o\epsilon_r$, hence **charge density,** $\sigma = E\epsilon_o\epsilon_r$

$$= 250 \times 10^3 \times 8.85 \times 10^{-12} \times 1 \text{ C/m}^2$$

$$= \textbf{2.213 µC/m}^2$$

(b)   For polythene, ϵ_r = 2.3. **Charge density,** $\sigma = E\epsilon_o\epsilon_r$

$$= 250 \times 10^3 \times 8.85 \times 10^{-12} \times 2.3 \text{ C/m}^2$$

$$= \textbf{5.089 µC/m}^2$$

Problem 4.   (a) Determine the p.d. across a 4 µF capacitor when charged with 5 mC. (b) Find the charge on a 50 pF capacitor when the voltage applied to it is 2 kV.

(a) $C = 4 \,\mu\text{F} = 4 \times 10^{-6} \,\text{F}; \; Q = 5 \,\text{mC} = 5 \times 10^{-3} \,\text{C}$. Since $C = (Q/V)$ then

$$V = \frac{Q}{C} = \frac{5 \times 10^{-3}}{4 \times 10^{-6}} = \frac{5 \times 10^{6}}{4 \times 10^{3}} = \frac{5000}{4}$$

Hence p.d. = 1250 V or 1.25 kV.

(b) $C = 50 \,\text{pF} = 50 \times 10^{-12} \,\text{F}; \; V = 2 \,\text{kV} = 2000 \,\text{V}; \; Q = CV = 50 \times 10^{-12} \times 2000 = (5 \times 2)/(10^{8}) = 0.1 \times 10^{-6}$.
Hence charge = 0.1 μC.

Problem 5. A direct current of 4 A flows into a previously uncharged 20 μF capacitor for 3 ms. Determine the p.d. between the plates.

$I = 4 \,\text{A}; \; C = 20 \,\mu\text{F} = 20 \times 10^{-6} \,\text{F}; \; t = 3 \,\text{ms} = 3 \times 10^{-3} \,\text{s}; \; Q = It = 4 \times 3 \times 10^{-3} \,\text{C}$.

$$V = \frac{Q}{C} = \frac{4 \times 3 \times 10^{-3}}{20 \times 10^{-6}} = \frac{12 \times 10^{6}}{20 \times 10^{3}}$$

$$= 0.6 \times 10^{3} = 600 \,\text{V}$$

Hence the p.d. between the plates is 600 V

Problem 6. A 5 μF capacitor is charged so that the p.d. between its plates is 800 V. Calculate how long the capacitor can provide an average discharge current of 2 mA.

$C = 5 \,\mu\text{F} = 5 \times 10^{-6} \,\text{F}; \; V = 800 \,\text{V}; \; I = 2 \,\text{mA} = 2 \times 10^{-3} \,\text{A}; \; Q = CV = 5 \times 10^{-6} \times 800 = 4 \times 10^{-3} \,\text{C}$. Also, $Q = It$, thus

$$t = \frac{Q}{I} = \frac{4 \times 10^{-3}}{2 \times 10^{-3}} = 2 \,\text{s}$$

Hence the capacitor can provide an average discharge current of 2 mA for 2 s.

34.3   The parallel plate capacitor

For a parallel plate capacitor, capacitance is proportional to area A, inversely proportional to the plate spacing (or dielectric thickness) d, and depends on the nature of the dielectric and the number of plates n:

$$\text{Capacitance} \quad \boxed{C = \frac{\epsilon_0 \epsilon_r A (n - 1)}{d} \; \text{F}}$$

Problem 7. (a) A ceramic capacitor has an effective plate area of 4 cm² separated by 0.1 mm of ceramic of relative permittivity 100. Calculate the capacitance of the capacitor in picofarads. (b) If the capacitor in part (a) is given a charge of 1.2 μC, what will be the p.d. between the plates?

(a) Area $A = 4 \,\text{cm}^2 = 4 \times 10^{-4} \,\text{m}^2; \; d = 0.1 \,\text{mm} = 0.1 \times 10^{-3} \,\text{m}; \; \epsilon_0 = 8.85 \times 10^{-12} \,\text{F/m}; \; \epsilon_r = 100$.

$$\textbf{Capacitance, } C = \frac{\epsilon_0 \epsilon_r A}{d} \; \text{farads}$$

$$= \frac{8.85 \times 10^{-12} \times 100 \times 4 \times 10^{-4}}{0.1 \times 10^{-3}} \; \text{F}$$

$$= \frac{8.85 \times 4}{10^{10}} \; \text{F} = \frac{8.85 \times 4 \times 10^{-12}}{10^{10}} \; \text{pF}$$

$$= \textbf{3540 pF}$$

(b) $Q = CV$ thus

$$V = \frac{Q}{C} = \frac{1.2 \times 10^{-6}}{3540 \times 10^{-12}} \; \text{volts} = \textbf{339 V}$$

Problem 8. A waxed paper capacitor has two parallel plates, each of effective area 800 cm². If the capacitance of the capacitor is 4425 pF, determine the effective thickness of the paper if its relative permittivity is 2.5.

$A = 800 \,\text{cm}^2 = 800 \times 10^{-4} \,\text{m}^2 = 0.08 \,\text{m}^2; \; C = 4425 \,\text{pF} = 4425 \times 10^{-12} \,\text{F}; \; \epsilon_0 = 8.85 \times 10^{-12} \,\text{F/m}; \; \epsilon_r = 2.5$. Since

$$C = \frac{\epsilon_0 \epsilon_r A}{d} \quad \text{then} \quad d = \frac{\epsilon_0 \epsilon_r A}{C}$$

Hence

$$d = \frac{8.85 \times 10^{-12} \times 2.5 \times 0.08}{4425 \times 10^{-12}} = 0.0004 \,\text{m}$$

The thickness of the paper is therefore 0.4 mm.

Problem 9. A parallel plate capacitor has 19 interleaved plates, each 75 mm by 75 mm and separated by mica sheets 0.2 mm thick. Assuming the relative permittivity of the mica is 5, calculate the capacitance of the capacitor.

$n = 19$; $n - 1 = 18$; $A = 75 \times 75 = 5625$ mm^2 = 5625×10^{-6} m^2; $\epsilon_r = 5$; $\epsilon_o = 8.85 \times 10^{-12}$ F/m; $d = 0.2$ mm $= 0.2 \times 10^{-3}$ m.

Capacitance C

$$= \frac{\epsilon_o \epsilon_r A(n - 1)}{d}$$

$$= \frac{8.85 \times 10^{-12} \times 5 \times 5625 \times 10^{-6} \times 18}{0.2 \times 10^{-3}} \text{ F}$$

$$= \mathbf{0.0224 \ \mu F} \text{ or } \mathbf{22.4 \ nF}$$

34.4 Capacitors in parallel and in series

For n capacitors connected in **parallel**, the equivalent capacitance C_T is given by:

$$\boxed{C_T = C_1 + C_2 + C_3 + \ldots + C_n}$$

(similar to resistors connected in series).
Also total charge

$$\boxed{Q_T = Q_1 + Q_2 + Q_3 \ldots + Q_n}$$

For n capacitors connected in **series**, the equivalent capacitance C_T is given by:

$$\boxed{\frac{1}{C_T} = \frac{1}{C_1} + \frac{1}{C_2} + \frac{1}{C_3} + \ldots + \frac{1}{C_n}}$$

(similar to resistors connected in parallel).
The charge on each capacitor is the same when connected in series.

Problem 10.   Calculate the equivalent capacitance of two capacitors of 6 μF and 4 μF connected (a) in parallel, and (b) in series.

(a)   In parallel, equivalent capacitance

$$C = C_1 + C_2 = 6 \ \mu F + 4 \ \mu F$$

$$= \mathbf{10 \ \mu F}$$

(b)   In series, equivalent capacitance C is given by:

$$\frac{1}{C} = \frac{1}{C_1} + \frac{1}{C_2} = \frac{C_2 + C_1}{C_1 C_2}$$

i.e.

$$C = \frac{C_1 C_2}{C_1 + C_2} \quad \left(\text{i.e. } \frac{\text{product}}{\text{sum}} \right)$$

This formula is used for the special case of two capacitors in series (which is similar to two resistors in parallel). Thus

$$C = \frac{6 \times 4}{6 + 4} = \frac{24}{10} = \mathbf{2.4 \ \mu F}$$

Problem 11.   What capacitance must be connected in series with a 30 μF capacitor for the equivalent capacitance to be 12 μF?

Let $C = 12 \ \mu F$ (the equivalent capacitance), $C_1 = 30 \ \mu F$ and C_2 be the unknown capacitance.
For two capacitors in series

$$\frac{1}{C} = \frac{1}{C_1} + \frac{1}{C_2}$$

Hence

$$\frac{1}{C_2} = \frac{1}{C} - \frac{1}{C_1} = \frac{C_1 - C}{CC_1}$$

$$C_2 = \frac{CC_1}{C_1 - C} = \frac{12 \times 30}{30 - 12} = \frac{360}{18} = \mathbf{20 \ \mu F}$$

Problem 12.   Capacitances of 1 μF, 3 μF, 5 μF and 6 μF are connected in parallel to a direct voltage supply of 100 V. Determine (a) the equivalent circuit capacitance, (b) the total charge, and (c) the charge on each capacitor.

(a)   The equivalent capacitance C for four capacitors in parallel is given by:

$$C = C_1 + C_2 + C_3 + C_4$$

i.e.

$$C = 1 + 3 + 5 + 6 = \mathbf{15 \ \mu F}$$

(b)   Total charge $Q_T = CV$, where C is the equivalent circuit capacitance, i.e.

$$Q_T = 15 \times 10^{-6} \times 100 = 1.5 \times 10^{-3} \text{ C}$$

$$= \mathbf{1.5 \ mC}$$

(c)   The charge on the 1 μF capacitor:

$Q_1 = C_1V = 1 \times 10^{-6} \times 100 = \textbf{0.1 mC}$

The charge on the 3 µF capacitor:

$Q_2 = C_2V = 3 \times 10^{-6} \times 100 = \textbf{0.3 mC}$

The charge on the 5 µF capacitor:

$Q_3 = C_3V = 5 \times 10^{-6} \times 100 = \textbf{0.5 mC}$

The charge on the 6 µF capacitor:

$Q_4 = C_4V = 6 \times 10^{-6} \times 100 = \textbf{0.6 mC}$

(Check: In a parallel circuit $Q_T = Q_1 + Q_2 + Q_3 + Q_4 = 0.1 + 0.3 + 0.5 + 0.6 = 1.5$ mC $= Q_T$.)

Problem 13. Capacitances of 3 µF, 6 µF and 12 µF are connected in series across a 350 V supply. Calculate (a) the equivalent circuit capacitance, (b) the charge on each capacitor, and (c) the p.d. across each capacitor.

The circuit diagram is shown in Fig. 34.3.

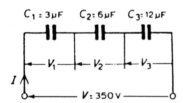

Figure 34.3

(a)   The equivalent circuit capacitance C for three capacitors in series is given by:

$$\frac{1}{C} = \frac{1}{C_1} + \frac{1}{C_2} + \frac{1}{C_3}$$

i.e.

$$\frac{1}{C} = \frac{1}{3} + \frac{1}{6} + \frac{1}{12} = \frac{4 + 2 + 1}{12} = \frac{7}{12}$$

Hence the equivalent circuit capacitance

$$C = \frac{12}{7} = 1\frac{5}{7} \text{ µF}$$

(b)   Total charge $Q_T = CV$. Hence

$$Q_T = \frac{12}{7} \times 10^{-6} \times 350 = 600 \text{ µC or 0.6 mC}$$

Since the capacitors are connected in series 0.6 mC is the charge on each of them.

(c)   The voltage across the 3 µF capacitor:

$$V_1 = \frac{Q}{C_1} = \frac{0.6 \times 10^{-3}}{3 \times 10^{-6}} = \textbf{200 V}$$

The voltage across the 6 µF capacitor:

$$V_2 = \frac{Q}{C_2} = \frac{0.6 \times 10^{-3}}{6 \times 10^{-6}} = \textbf{100 V}$$

The voltage across the 12 µF capacitor:

$$V_3 = \frac{Q}{C_3} = \frac{0.6 \times 10^{-3}}{12 \times 10^{-6}} = \textbf{50 V}$$

(Check: In a series circuit $V = V_1 + V_2 + V_3 = 200 + 100 + 50 = 350$ V = supply voltage.)

In practice capacitors are rarely connected in series unless they are of the same capacitance. The reason for this can be seen from the above problem where the lowest valued capacitor (i.e. 3 µF) has the highest p.d. across it (i.e. 200 V) which means that if all the capacitors have an identical construction they must all be rated at the highest voltage.

Problem 14. For the arrangement shown in Fig. 34.4 find (a) the equivalent capacitance of the circuit, (b) the voltage across QR, and (c) the charge on each capacitor.

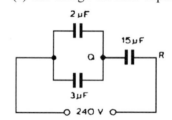

Figure 34.4

(a)   2 µF in parallel with 3 µF gives an equivalent capacitance of

2 µF + 3 µF = 5 µF

The circuit is now as shown in Fig. 34.5. The equivalent capacitance of 5 µF in series with 15 µF is given by

$$\frac{5 \times 15}{5 + 15} \text{ µF, i.e. } \frac{75}{20} \text{ or } \textbf{3.75 µF}$$

(b)   The charge on each of the capacitors shown in Fig. 34.5 will be the same since they are connected in series. Let this charge be Q coulombs. Then

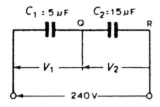

Figure 34.5

$$Q = C_1 V_1 = C_2 V_2$$
i.e.
$$5V_1 = 15V_2$$
$$V_1 = 3V_2$$
Also
$$V_1 + V_2 = 240 \text{ V}$$
Hence
$$3V_2 + V_2 = 240 \text{ V}$$
Thus
$$V_2 = 60 \text{ V and } V_1 = 180 \text{ V}$$
Hence the voltage across QR is 60 V.

(c)   The charge on the 15 µF capacitor is
$$C_2 V_2 = 15 \times 10^{-6} \times 60$$
$$= \textbf{0.9 mC}$$

The charge on the 2 µF capacitor is
$$2 \times 10^{-6} \times 180 = \textbf{0.36 mC}$$

The charge on the 3 µF capacitor is
$$3 \times 10^{-6} \times 180 = \textbf{0.54 mC}$$

34.5   Dielectric strength

The maximum amount of field strength that a dielectric can withstand is called the dielectric strength of the material.

Dielectric strength, $\boxed{E_m = \dfrac{V_m}{d}}$

> **Problem 15.**   A capacitor is to be constructed so that its capacitance is 0.2 µF and to take a p.d. of 1.25 kV across its terminals. The dielectric is to be mica which, after allowing a safety factor of 2, has a dielectric strength of 50 MV/m. Find (a) the thickness of the mica needed, and (b) the area of a plate assuming a two-plate construction. (Assume ϵ_r for mica to be 6.)

(a)   **Dielectric strength, $E = V/d$,** i.e.

$$d = \frac{V}{E} = \frac{1.25 \times 10^3}{50 \times 10^6} \text{ m}$$

$$= \textbf{0.025 mm}$$

(b)   **Capacitance, $C = \epsilon_0 \epsilon_r A/d$,** hence

$$\text{area } A = \frac{Cd}{\epsilon_0 \epsilon_r}$$

$$= \frac{0.2 \times 10^{-6} \times 0.025 \times 10^{-3}}{8.85 \times 10^{-12} \times 6} \text{ m}^2$$

$$= 0.094 \, 16 \text{ m}^2 = \textbf{941.6 cm}^2$$

34.6   Energy stored

The **energy, W,** stored by a capacitor is given by

$$\boxed{W = \frac{1}{2} CV^2 \text{ joules}}$$

> **Problem 16.**   (a) Determine the energy stored in a 3 µF capacitor when charged to 400 V. (b) Find also the average power developed if this energy is dissipated in a time of 10 µs.

(a)

$$\text{Energy stored } W = \frac{1}{2} CV^2 \text{ joules}$$

$$= \frac{1}{2} \times 3 \times 10^{-6} \times 400^2 = \frac{3}{2} \times 16 \times 10^{-2}$$

$$= \textbf{0.24 J}$$

(b)

$$\text{Power} = \frac{\text{energy}}{\text{time}} = \frac{0.24}{10 \times 10^{-6}} \text{ watts} = \textbf{24 kW}$$

> **Problem 17.**   A 12 µF capacitor is required to store 4 J of energy. Find the p.d. to which the capacitor must be charged.

Energy stored $W = \frac{1}{2} CV^2$ hence $V^2 = 2W/C$ and

$$V = \sqrt{\left(\frac{2W}{C}\right)} = \sqrt{\left(\frac{2 \times 4}{12 \times 10^{-6}}\right)} = \sqrt{\left(\frac{2 \times 10^{6}}{3}\right)}$$

$$= \textbf{816.5 volts}$$

Problem 18.   A capacitor is charged with 10 mC. If the energy stored is 1.2 J find (a) the voltage, and (b) the capacitance.

Energy stored $W = \frac{1}{2} CV^2$ and $C = Q/V$ hence

$$W = \frac{1}{2}\left(\frac{Q}{V}\right)V^2 = \frac{1}{2}QV$$

from which

$$V = \frac{2W}{Q}$$

$Q = 10\,\text{mC} = 10 \times 10^{-3}\,\text{C}$ and $W = 1.2\,\text{J}$.

(a)   **Voltage, V**

$$= \frac{2W}{Q} = \frac{2 \times 1.2}{10 \times 10^{-3}} = \textbf{0.24 kV} \text{ or } \textbf{240 volts}$$

(b)   **Capacitance, C**

$$= \frac{Q}{V} = \frac{10 \times 10^{-3}}{240}\,\text{F} = \frac{10 \times 10^{6}}{240 \times 10^{3}}\,\mu\text{F}$$

$$= \textbf{41.67 } \mu\text{F}$$

34.7   Practical types of capacitor

Practical types of capacitor are characterized by the material used for their dielectric. The main types include: variable air, mica, paper, ceramics, plastic and electrolytic.

1.   **Variable air capacitors**. Usually consists of two sets of metal plates (such as aluminium), one fixed, the other variable. The set of moving plates rotates on a spindle as shown by the end view in Fig. 34.6. As the moving

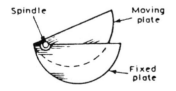

Figure 34.6

plates are rotated through half a revolution, the meshing, and therefore the capacitance, varies from a minimum to a maximum value. Variable air capacitors are used in radio and electronic circuits where very low losses are required, or where a variable capacitance is needed. The maximum value of such capacitors is between 500 pF and 1000 pF.

2.   **Mica capacitors**. A typical older type construction is shown in Fig. 34.7. Usually the whole capacitor is impregnated with wax and placed in a bakelite case. Mica is easily obtained in thin sheets and is a good insulator. However, mica is expensive and is not used in capacitors above about 0.1 μF. A modified form of mica capacitor is the silvered mica type. The mica is coated on both sides with a thin layer of silver which forms the plates. Capacitance is stable and less likely to change with age. Such capacitors have a constant capacitance with change of temperature, a high working voltage rating and a long service life and are used in high frequency circuits with fixed values of capacitance up to about 1000 pF.

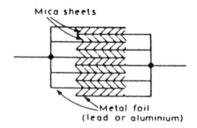

Figure 34.7

3.   **Paper capacitors**. A typical paper capacitor is shown in Fig. 34.8 where the length of the roll corresponds to the capacitance required. The whole is usually impregnated with oil or wax to exclude moisture, and then placed in a plastic or aluminium container for protection. Paper capacitors up to about 1 μF are made in various working voltages. Disadvantages of paper capacitors include variation in capacitance with temperature change and a shorter service life than most other types of capacitor.

4.   **Ceramic capacitors**. These are made in various forms, each type of construction depending on the value of capacitance

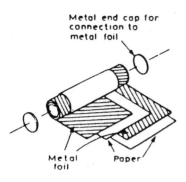

Figure 34.8

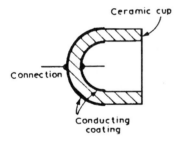

Figure 34.9

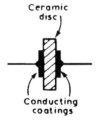

Figure 34.10

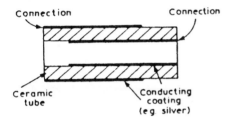

Figure 34.11

required. For high values, a tube of ceramic material is used as shown in the cross-section of Fig. 34.9. For smaller values the cup construction is used as shown in Fig. 34.10,

and for still smaller values the disc construction shown in Fig. 34.11 is used. Certain ceramic materials have a very high permittivity and this enables capacitors of high capacitance to be made which are of small physical size with a high working voltage rating. Ceramic capacitors are available in the range 1 pF to 0.1 μF and may be used in high frequency electronic circuits subject to a wide range of temperature.

5. **Plastic capacitors**. Some plastic materials, such as polystyrene and Teflon, can be used as dielectrics. Construction is similar to the paper capacitor but using a plastic film instead of paper. Plastic capacitors operate well under conditions of high temperature, provide a precise value of capacitance, a very long service life and high reliability.

6. **Electrolytic capacitors**. Construction is similar to the paper capacitor with aluminium foil used for the plates and with a thick absorbent material, such as paper, impregnated with an electrolyte (ammonium borate), separating the plates. The finished capacitor is usually assembled in an aluminium container and hermetically sealed. Its operation depends on the formation of a thin aluminium oxide layer on the positive plate by electrolytic action when a suitable direct potential is maintained between the plates. This oxide layer is very thin and forms the dielectric. (The absorbent paper between the plates is a conductor and does not act as a dielectric.) Such capacitors must only be used on d.c. and must be connected with the correct polarity; if this is not done the capacitor will be destroyed since the oxide layer will be destroyed. Electrolytic capacitors are manufactured with working voltages from 6 V to 500 V, although accuracy is generally not very high. These capacitors possess a much larger capacitance than other types of capacitors of similar dimensions due to the oxide film being only a few microns thick. The fact that they can be used only on d.c. supplies limits their usefulness.

34.8 Effects of capacitance in d.c. circuits

When a d.c. voltage is applied to a capacitor C and resistor R connected in series, there is a short period of time immediately after the voltage is connected during which the current flowing in the

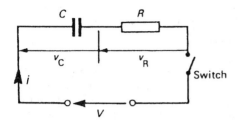

Figure 34.12

circuit and the voltage across C and R are changing. These changing values are called **transients**.

Charging a capacitor

Figure 34.12 shows a capacitor, initially having no charge on its plates, connected in series with resistor R across a d.c. supply. When the switch is closed:

(i)   the initial current flowing, i, is given by $i = V/R$, and the capacitor acts as if it is a short-circuit when $v_C = 0$,

(ii) the current then begins to charge the capacitor so that v_C builds up rapidly across the plates,

(iii) v_R falls to $V - v_C$ (since $V = v_C + v_R$ at all times) and the charging current i reduces to $(V - v_C)/R$,

(iv) eventually the capacitor is charged to the full supply voltage, V, current $i = 0$ and the capacitor acts as an open-circuit. Curves of v_C and i against time are shown in Fig. 34.13 and are natural or exponential curves of growth and decay.

(v)   The **time constant** τ of the CR circuit shown in Fig. 34.12 is defined as: *the time taken for*

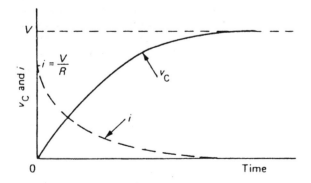

Figure 34.13

a transient to reach its final state if the initial rate of change is maintained.

The time constant τ for any series-connected C–R circuit (as in Fig. 34.12) is given by:

time constant $\boxed{\tau = CR \text{ seconds}}$

(vi) In the time $\tau = CR$ seconds, v_C rises to 63.2% of its final value V and, in practical situations, v_C rises to within 1% of its final value V in a time equal to 5τ seconds.

Problem 19.   A 20 µF capacitor is to be charged through a 100 kΩ, resistor by a constant d.c. supply. Determine (a) the time constant for the circuit, and (b) the additional resistance required to increase the time constant to 5 s.

(a)   **Time constant, $\tau = CR$**
$= 20 \times 10^{-6} \times 100 \times 10^{3} = $ **2 s**.

(b)   When the time constant $\tau = 5$ s, then $5 = CR$, i.e. $5 = 20 \times 10^{-6} \times R$, from which

resistance, $R = \dfrac{5}{20 \times 10^{-6}} = \dfrac{5 \times 10^{6}}{20}$

$= 250\,000$ Ω or 250 kΩ

Thus the additional resistance needed is

$(250 \text{ kΩ} - 100 \text{ kΩ}) = $ **150 kΩ**

Discharging a capacitor

Figure 34.14 shows a capacitor C fully charged to voltage V, as described above, connected in series with a resistor R. When the switch is closed the initial discharge current is given by $i = V/R$. As the capacitor loses its charge, v_C falls and hence i falls. The result is the natural decay curves

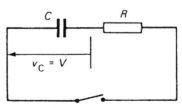

Figure 34.14

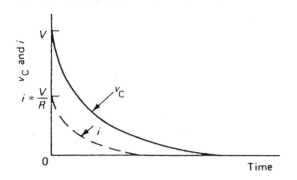

Figure 34.15

shown in Fig. 34.15. In the time $\tau = CR$ seconds, v_C falls to 36.8% of its initial voltage V and, in practical situations, v_C falls to less than 1% of its initial value in a time 5τ seconds.

When a capacitor has been disconnected from the supply it may still be charged and it may retain this charge for some considerable time. Thus precautions must be taken to ensure that the capacitor is automatically discharged after the supply is switched off. This is done by connecting a high value resistor across the capacitor terminals.

Problem 20.   A 0.1 µF capacitor is charged to 200 V before being connected across a 4 kΩ resistor. Determine (a) the initial discharge current, (b) the time constant of the circuit, and (c) the minimum time required for the voltage across the capacitor to fall to less than 2 V.

(a)   Initial discharge current, i

$$= \frac{V}{R} = \frac{200}{4 \times 10^3}$$

$$= \textbf{0.05 A or 50 mA}$$

(b)   Time constant $\tau = CR = 0.1 \times 10^{-6} \times 4 \times 10^3$

$$= \textbf{0.0004 or 0.4 ms}$$

(c)   The minimum time for the capacitor voltage to fall to less than 2 V, i.e. less than 2/200 or 1% of the initial value is given by 5τ.

$$5\tau = 5 \times 0.4 = \textbf{2 ms}$$

In a **d.c. circuit**, a capacitor blocks the current except during the times there are changes in the supply voltage.

34.9   Effects of capacitance in a.c. circuits

In an **a.c. circuit**, a capacitor provides opposition to current flow which results in the voltage and current waveforms being 90° out of phase. This opposition is called **capacitive reactance X_C** and is given by:

$$\boxed{X_C = \frac{1}{2\pi f C} \text{ ohms}}$$

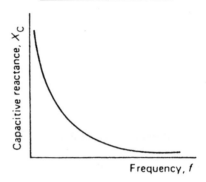

Figure 34.16

A typical graph showing the variation of capacitive reactance with frequency is shown in Fig. 34.16. In a purely capacitive a.c. circuit current I is given by:

$$\boxed{I = \frac{I}{X_C} \text{ amperes}}$$

When a capacitor is connected to an a.c. supply its effects are present at all times since the voltage is continually changing.

Let the instantaneous values of voltage and charge be v volts and q coulombs, respectively. If the voltage increases from zero to v in t seconds and the increase in charge is q coulombs, then $q = Cv$ and $q = it$. Hence

$$\frac{dq}{dt} = C \frac{dv}{dt}$$

where

$$\frac{dq}{dt} \text{ and } \frac{dv}{dt}$$

are the rates of change of charge and voltage, respectively. Therefore

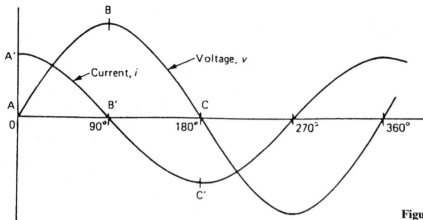

Figure 34.17

$$i = C\frac{dv}{dt}$$

(since current $i = dq/dt$).

Let voltage v be sinusoidal as shown in Fig. 34.17. At point A, the voltage waveform is at its steepest, i.e. dv/dt is a maximum. Hence $i = C(dv/dt)$ is a maximum value and is shown by the point A′. As v increases, the gradient of the curve decreases (i.e. dv/dt decreases) until at point B the gradient is zero. Hence the current is zero, shown by point B′. Between B and C the gradient of the curve (i.e. dv/dt) is negative and hence the current value is negative. At point C the gradient is a maximum negative and thus i is shown by the point C′. The remaining half of the cycle varies in a similar manner to the first half cycle and the resulting current waveform is seen to be sinusoidal but out of phase. The peak of the current waveform A′ occurs 90° before the peak of the voltage waveform B and **the current is said to lead the voltage by 90°**.

Problem 21. Determine the capacitive reactance of a capacitor of 10 μF when connected to a circuit of frequency (a) 50 Hz, (b) 20 kHz.

(a) Capacitive reactance X_C

$$= \frac{1}{2\pi fC} = \frac{1}{2\pi(50)(10 \times 10^{-6})}$$

$$= \frac{10^6}{2\pi(50)(10)}$$

$$= \textbf{318.3 } \Omega$$

(b)

$$X_C = \frac{1}{2\pi fC} = \frac{1}{2\pi(20 \times 10^3)(10 \times 10^{-6})}$$

$$= \frac{10^6}{2\pi(20 \times 10^3)(10)}$$

$$= \textbf{0.796 } \Omega$$

Hence as the frequency is increased from 50 Hz to 20 kHz, X_C decreases from 318.3 Ω to 0.796 Ω (see Fig. 34.16).

Problem 22.   A capacitor has a reactance of 40 Ω when operated on a 50 Hz supply. Determine the value of its capacitance.

Since capacitive reactance $X_C = 1/2\pi fC$, capacitance, C

$$= \frac{1}{2\pi fX_C} = \frac{1}{2\pi(50)(40)} \text{ F} = \frac{10^6}{2\pi(50)(40)} \text{ μF}$$

$$= \textbf{79.58 μF}$$

Problem 23.   Calculate the current taken by a 23 μF capacitor when connected to a 240 V, 50 Hz supply.

$$\text{Current, } I = \frac{V}{X_C} = \frac{V}{\dfrac{1}{2\pi fC}} = 2\pi fCV$$

$$= 2\pi(50)(23 \times 20^{-6})(240)$$

$$= \textbf{1.73 A}$$

34.10 Multi-choice questions on capacitors and their effects in electric circuits

(Answers on page 356.)

1. Electrostatics is a branch of electricity concerned with:
   (a) energy flowing across a gap between conductors
   (b) charges at rest
   (c) charges in motion
   (d) energy in the form of charges
2. The capacitance of a capacitor is the ratio:
   (a) charge to p.d. between plates
   (b) p.d. between plates to plate spacing
   (c) p.d. between plates to thickness of dielectric
   (d) p.d. between plates to charge
3. The p.d. across a 10 μF capacitor to charge it with 10 mC is:
   (a) 100 V   (b) 1 kV   (c) 1 V   (d) 10 V
4. The charge on a 10 pF capacitor when the voltage applied to it is 10 kV is:
   (a) 100 μC   (b) 0.1 C   (c) 0.1 μC   (d) 0.01 μC
5. Four 2 μF capacitors are connected in parallel. The equivalent capacitance is:
   (a) 8 μF   (b) 0.5 μF
6. Four 2 μF capacitors are connected in series. The equivalent capacitance is:
   (a) 8 μF   (b) 0.5 μF
7. State which of the following is false. The capacitance of a capacitor:
   (a) is proportional to the cross-sectional area of the plates
   (b) is proportional to the distance between the plates
   (c) depends on the number of plates
   (d) is proportional to the relative permittivity of the dielectric
8. Which of the following statements is false?
   (a) An air capacitor is normally a variable type.
   (b) A paper capacitor generally has a shorter service life than most other types of capacitor
   (c) An electrolytic capacitor must be used only on a.c. supplies.
   (d) Plastic capacitors generally operate satisfactorily under conditions of high temperature.
9. The energy stored in a 10 μF capacitor when charged to 500 V is:

(a) 1.25 mJ   (b) 0.025 μJ   (c) 1.25 J
(d) 1.25 C

10. The capacitance of a variable air capacitor is a maximum when:
    (a) the movable plates half overlap the fixed plates
    (b) the movable plates are most widely separated from the fixed plates
    (c) both sets of plates are exactly meshed
    (d) the movable plates are closer to one side of the fixed plate than to the other
11. A capacitor of 1 μF is connected to a 50 Hz supply. The capacitive reactance is:
    (a) 50 MΩ   (b) $(10/\pi)$ kΩ   (c) $(\pi/10^4)$ Ω
    (d) $(10/\pi)$ Ω
12. When a capacitor is connected to an a.c. supply the current:
    (a) leads the voltage by 180°
    (b) is in phase with the voltage
    (c) leads the voltage by $\dfrac{\pi}{2}$ rad
    (d) lags the voltage by 90°
13. In a circuit consisting of a 1 μF capacitor and a 100 kΩ resistor connected in series the time constant of the circuit is:
    (a) 10 ms   (b) 100 ms   (c) 1 s   (d) 10 s
14. The current taken by a 10 μF capacitor when connected to a 1 kV, $(500/\pi)$ Hz supply is:
    (a) 5/π A   (b) 10^5 A   (c) 10 A   (d) 10 μA

34.11 Short answer questions on capacitors and their effects in electric circuits

1. Explain the term 'electrostatics'.
2. Complete the statements: Like charges ....; unlike charges ....
3. How can an 'electric field' be established between two parallel metal plates?
4. What is capacitance?
5. State the unit of capacitance.
6. Complete the statement

   Capacitance = $\dfrac{\text{................}}{\text{................}}$
7. Complete the statements:
   (a) 1 μF = ... F
   (b) 1 pF = ... F
8. Complete the statement:

   Electric field strength, $E = \dfrac{\text{................}}{\text{................}}$
9. Complete the statement

   Charge density, $\sigma = \dfrac{\text{................}}{\text{................}}$

10. Draw the electrical circuit diagram symbol for a capacitor.
11. Name two practical examples where capacitance is present, although undesirable.
12. The insulating material separating the plates of a capacitor is called the ....
13. 10 volts applied to a capacitor results in a charge of 5 coulombs. What is the capacitance of the capacitor?
14. Three 3 µF capacitors are connected in parallel. The equivalent capacitance is ....
15. Three 3 µF capacitors are connected in series. The equivalent capacitance is ....
16. State an advantage of series connected capacitors.
17. Name three factors upon which capacitance depends.
18. What does 'relative permittivity' mean?
19. Define 'permittivity of free space'.
20. Name five types of capacitor commonly used.
21. Sketch a typical rolled paper capacitor.
22. Explain briefly the construction of a variable air capacitor.
23. State three advantages and one disadvantage of mica capacitors.
24. Name two disadvantages of paper capacitors.
25. Between what values of capacitance are ceramic capacitors normally available?
26. What main advantages do plastic capacitors possess?
27. Explain briefly the construction of an electrolytic capacitor.
28. What is the main disadvantage of electrolytic capacitors?
29. Name an important advantage of electrolytic capacitors.
30. What safety precautions should be taken when a capacitor is disconnected from a supply?
31. What is meant by the 'dielectric strength' of a material?
32. State the formula used to determine the energy stored by a capacitor.
33. State the effects of a capacitor on the current in a d.c. circuit.
34. State the effects of a capacitor on the current in an a.c. circuit.
35. Sketch typical graphs of capacitor voltage against time for the charge and discharge of a capacitor.

34.12 Further questions on capacitors and their effects in electric currents

(Where appropriate take ϵ_0 as 8.85×10^{-12} F/m.)

Charge density and electric field strength

1. A capacitor uses a dielectric 0.04 mm thick and operates at 30 V. What is the electric field strength across the dielectric at this voltage? [750 kV/m]
2. A two-plate capacitor has a charge of 25 C. If the effective area of each plate is 5 cm² find the charge density of the electric field. [50 kC/m²]
3. A charge of 1.5 µC is carried on two parallel rectangular plates each measuring 60 mm by 80 mm. Calculate the charge density on the plates. If the plates are spaced 10 mm apart and the voltage between them is 0.5 kV, determine the electric field strength. [312.5 µC/m², 50 kV/m]
4. Two parallel plates are separated by a dielectric and charged with 10 µC. Given that the area of each plate is 50 cm², calculate the charge density in the dielectric separating the plates. [2 mC/m²]
5. The charge density between two plates separated by polystyrene of relative permittivity 2.5 is 5 µC/m². Find the voltage gradient between the plates. [226 kV/m]
6. Two parallel plates having a p.d. of 250 V between them are spaced 1 mm apart. Determine the electric field strength. Find also the charge density when the dielectric between the plates is (a) air, and (b) mica of relative permittivity 5.
[250 kV/m (a) 2.213 µC/m²
(b) 11.063 µC/m²]

$Q = CV$ problems

7. Find the charge on a 10 µF capacitor when the applied voltage is 250 V. [2.5 mC]
8. Determine the voltage across a 1000 pF capacitor to charge it with 2 µC. [2 kV]
9. The charge on the plates of a capacitor is 6 mC when the potential between them is 2.4 kV. Determine the capacitance of the capacitor. [2.5 µF]

10. For how long must a charging current of 2 A be fed to a 5 µF capacitor to raise the p.d. between its plates by 500 V. [1.25 ms]

11. A direct current of 10 A flows into a previously uncharged 5 µF capacitor for 1 ms. Determine the p.d. between the plates.
[2 kV]

12. A 16 µF capacitor is charged at the constant current of 4 µA for 2 minutes. Calculate the final p.d. across the capacitor and the corresponding charge in coulombs.
[30 V, 480 µC]

13. A steady current of 10 A flows into a previously uncharged capacitor for 1.5 ms when the p.d. between the plates is 2 kV. Find the capacitance of the capacitor. [7.5 µF]

Parallel plate capacitor

14. A capacitor consists of two parallel plates each of area 0.1 m², spaced 0.01 mm in air. Calculate the capacitance in picofarads.
[885 pF]

15. A waxed paper capacitor has two parallel plates, each of effective area 0.2 m². If the capacitance is 4000 pF, determine the effective thickness of the paper if its relative permittivity is 2. [0.885 mm]

16. Calculate the capacitance of a parallel plate capacitor having five plates, each 30 mm by 20 mm and separated by a dielectric 0.75 mm thick having a relative permittivity of 2.3.
[65.14 pF]

17. How many plates has a parallel plate capacitor having a capacitance of 0.005 µF if each plate is 40 mm square and each dielectric is 0.102 mm thick with a relative permittivity of 6. [7]

18. A parallel plate capacitor is made from 25 plates, each 70 mm by 120 mm interleaved with mica of relative permittivity 5. If the capacitance of the capacitor is 3000 pF determine the thickness of the mica sheet.
[2.97 mm]

19. A capacitor is constructed with parallel plates and has a value of 50 pF. What would be the capacitance of the capacitor if the plate area is doubled and the plate spacing is halved?
[200 µF]

20. The capacitance of a parallel plate capacitor is 1000 pF. It has 19 plates, each 50 mm by 30 mm separated by a dielectric of thickness 0.40 mm. Determine the relative permittivity of the dielectric. [1.67]

21. The charge on the square plates of a multiplate capacitor is 80 µC when the potential between them is 5 kV. If the capacitor has 25 plates separated by a dielectric of thickness 0.102 mm and relative permittivity 4.8, determine the width of a plate. [40 mm]

22. A capacitor is to be constructed so that its capacitance is 4250 pF and to operate at a p.d. of 100 V across its terminals. The dielectric is to be polythene (ϵ_r = 2.3) which, after allowing a safety factor, has a dielectric strength of 20 MV/m. Find (a) the thickness of the polythene needed, and (b) the area of a plate. [(a) 0.005 mm (b) 10.44 cm²]

Capacitors connected in parallel and in series

23. Capacitors of 2 µF and 6 µF are connected (a) in parallel, and (b) in series. Determine the equivalent capacitance in each case.
[(a) 8 µF (b) 1.5 µF]

24. Find the capacitance to be connected in series with a 10 µF capacitor for the equivalent capacitance to be 6 µF.
[15 µF]

25. What value of capacitance would be obtained if capacitors of 0.15 µF and 0.1 µF are connected (a) in series, and (b) in parallel.
[(a) 0.06 µF (b) 0.25 µF]

26. Two 6 µF capacitors are connected in series with one having a capacitance of 1.2 µF. Find the total equivalent circuit capacitance. What capacitance must be added in series to obtain a capacitance of 1.2 µF?
[2.4 µF, 2.4 µF]

27. Determine the equivalent capacitance when the following capacitors are connected (a) in parallel, and (b) in series:
(i) 2 µF, 4 µF and 8 µF,
(ii) 0.02 µF, 0.05 µF and 0.1 µF,
(iii) 50 pF and 450 pF,
(iv) 0.01 µF and 200 pF.
[(a)(i) 14 µF, (ii) 0.17 µF, (iii) 500 pF, (iv) 0.0102 µF
(b)(i) $1\frac{1}{7}$ µF, (ii) 0.0125 µF, (iii) 45 pF, (iv) 196.1 pF]

28. For the arrangement shown in Fig. 34.18 find (a) the equivalent circuit capacitance, and (b) the voltage across a 4.5 µF capacitor.
[(a) 1.2 µF (b) 100 V]

29. Three 12 µF capacitors are connected in series across a 750 V supply. Calculate (a) the equivalent capacitance, (b) the charge on

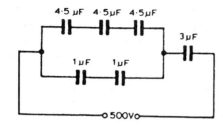

Figure 34.18

each capacitor, and (c) the p.d. across each capacitor.

[(a) 4 µF (b) 3 mC (c) 250 V]

30. If two capacitors having capacitances of 3 µF and 5 µF respectively are connected in series across a 240 V supply determine (a) the p.d. across each capacitor, and (b) the charge on each capacitor.

[(a) 150 V, 90 V (b) 0.45 mC on each]

31. In Fig. 34.19 capacitors P, Q and R are identical and the total equivalent capacitance of the circuit is 3 µF. Determine the values of P, Q and R.   [4.2 µF each]

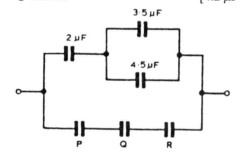

Figure 34.19

32. Capacitances of 4 µF, 8 µF and 16 µF are connected in parallel across a 200 V supply. Determine (a) the equivalent capacitance, (b) the total charge, and (c) the charge on each capacitor.

[(a) 28 µF (b) 5.6 mC (c) 0.8 mC, 1.6 mC, 3.2 mC]

33. A circuit consists of two capacitors P and Q in parallel, connected in series with another capacitor R. The capacitances of P, Q and R are 4 µF, 12 µF and 8 µF, respectively. When the circuit is connected across a 300 V d.c. supply find (a) the total capacitance of the circuit, (b) the p.d. across each capacitor, and (c) the charge on each capacitor.

[(a) $5\frac{1}{3}$ µF (b) 100 V across P, 100 V across Q, 200 V across R (c) 0.4 mC on P, 1.2 mC on Q, 1.6 mC on R]

Energy stored in capacitors

34. When a capacitor is connected across a 200 V supply the charge is 4 µC. Find (a) the capacitance, and (b) the energy stored.

[(a) 0.02 µF (b) 0.4 mJ]

35. Find the energy stored in a 10 µF capacitor when charged to 2 kV.   [20 J]

36. A 3300 pF capacitor is required to store 0.5 mJ of energy. Find the p.d. to which the capacitor must be charged.   [550 V]

37. A capacitor is charged with 8 mC. If the energy stored is 0.4 J find (a) the voltage, and (b) the capacitance.   [(a) 100 V (b) 80 µF]

38. A capacitor, consisting of two metal plates each of area 50 cm² and spaced 0.2 mm apart in air, is connected across a 120 V supply. Calculate (a) the energy stored, (b) the charge density, and (c) the potential gradient.

[(a) 1.593 µJ (b) 5.31 µC/m² (c) 600 kV/m]

39. A bakelite capacitor is to be constructed to have a capacitance of 0.04 F and to have a steady working potential of 1 kV maximum. Allowing a safe value of field stress of 25 MV/m find (a) the thickness of bakelite required, (b) the area of plate required if the relative permittivity of bakelite is 5, (c) the maximum energy stored by the capacitor, and (d) the average power developed if this energy is dissipated in a time of 20 µs.

[(a) 0.04 mm (b) 361.6 cm² (c) 0.02 J (d) 1 kW]

Charging and discharging of capacitors

40. A 15 µF capacitor is to be charged through a 300 kΩ resistor by a constant d.c. supply. Determine (a) the time constant of the circuit, and (b) the additional resistance needed to increase the time constant to 7.5 s.

[(a) 4.5 s (b) 200 kΩ]

41. Explain the charging and discharging of a capacitor through a resistor with reference to a graph of capacitor voltage against time.

42. An electrical circuit is controlled by a time switch which depends on the charging of an 8 µF capacitor. If the circuit time constant is

to be variable between 0.4 s and 2.4 s determine the limits of the value of the resistance required in series with the capacitor.

[50 kΩ to 300 kΩ]

43. A 6 nF capacitor is charged to 600 V before being connected across a 2 kΩ resistor. Determine (a) the initial discharge current, (b) the time constant of the circuit, and (c) the minimum probable time required for the voltage to fall to less than 1% of its initial value. [(a) 0.3 A (b) 12 μs (c) 60 μs]

Capacitors in a.c. circuits

44. Calculate the capacitive reactance of a 20 μF capacitor when connected to an a.c. circuit of frequency (a) 20 Hz, (b) 500 Hz, (c) 4 kHz.

[(a) 397.9 Ω (b) 15.92 Ω (c) 1.989 Ω]

45. A capacitor has a reactance of 80 Ω when connected to a 50 Hz supply. Calculate the value of its capacitance. [39.79 μF]

46. Calculate the current taken by a 10 μF capacitor when connected to a 200 V, 100 Hz supply. [1.257 A]

47. A capacitor has a capacitive reactance of 400 Ω when connected to a 100 V, 25 Hz supply. Determine its capacitance and the current taken from the supply.

[15.92 μF, 0.25 A]

48. Two similar capacitors are connected in parallel to a 200 V, 1 kHz supply. Find the value of each capacitor if the supply current is 0.628 A. [0.25 μF]

35

Inductors and their effects in electric circuits

At the end of this chapter you should be able to:

- define inductance and state its unit

- state the factors which affect the inductance of an inductor

- describe examples of practical inductors and their circuit diagram symbols

- calculate the energy stored in the magnetic field of an inductor and state its unit

- describe the effect on inductor voltage and current when a d.c. is applied to a series *LR* circuit

- state that the time constant $\tau = L/R$

- describe current decay in an *LR* circuit when the supply is removed

- explain the difficulties experienced when switching inductive circuits

- perform calculations on transient *LR* circuits

- define and calculate inductive reactance in an a.c. circuit

- understand that current and voltage are in-phase for a purely resistive a.c. circuit

- describe how current lags voltage by 90° in a purely inductive a.c. circuit

35.1 Inductance

As stated in Chapter 32, **inductance** is the property of a circuit whereby there is an e.m.f. induced into the circuit by the change of flux linkages produced by a current change. When the e.m.f. is induced in the same circuit as that in which the current is changing, the property is called **self-inductance** *L*.

The **unit of inductance** is the **henry, H**:

A circuit has an inductance of one henry when an e.m.f. of one volt is induced in it by a current changing at the rate of one ampere per second.

A component called an **inductor** is used when the property of inductance is required in a circuit. The basic form of an inductor is simply a coil of wire.

Factors which affect the inductance of an inductor include:

(i) the number of turns of wire – the more turns the higher the inductance;

(ii) the cross-sectional area of the coil of wire – the greater the cross-sectional area the higher the inductance;

(iii) the presence of a magnetic core – when the coil is wound on an iron core the same current sets up a more concentrated

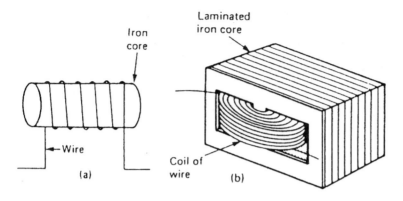

Figure 35.1

magnetic field and the inductance is increased;

(iv) the way the turns are arranged – a short thick coil of wire has a higher inductance than a long thin one.

35.2 Practical inductors

Two examples of practical inductors are shown in Fig. 35.1, and the standard electrical circuit diagram symbols for air-cored and iron-cored inductors are shown in Fig. 35.2.

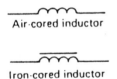
Air-cored inductor

Iron-cored inductor

Figure 35.2

An iron-cored inductor is often called a **choke** since, when used in a.c. circuits, it has a choking effect, limiting the current flowing through it.

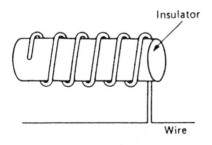
Insulator

Wire

Figure 35.3

Inductance is often undesirable in a circuit. To reduce inductance to a minimum the wire may be bent back on itself, as shown in Fig. 35.3, so that the magnetizing effect of one conductor is neutralized by that of the adjacent conductor. The wire may be coiled around an insulator, as shown, without increasing the inductance. Standard resistors may be non-inductively wound in this manner.

35.3 Energy stored

An inductor possesses an ability to store energy. The energy stored, W, in the magnetic field of an inductor is given by:

$$W = \frac{1}{2} L I^2 \text{ joules}$$

Problem 1.   An 8 H inductor has a current of 3 A flowing through it. How much energy is stored in the magnetic field of the inductor?

Energy stored, $W = \dfrac{1}{2} L I^2 = \dfrac{1}{2} (8)(3)^2$

$$= \textbf{36 joules}$$

Problem 2.   A flux of 25 mWb links with a 1500 turn coil when a current of 3 A passes through the coil. Calculate (a) the inductance of the coil, (b) the energy stored in the magnetic field, and (c) the average e.m.f. induced if the current falls to zero in 150 ms.

(a)   **Inductance, $L = \dfrac{N\Phi}{I} = \dfrac{(1500)(25 \times 10^{-3})}{3}$**

$\quad\quad = \textbf{12.5 H}$

(b)   **Energy stored in field, $W = \dfrac{1}{2}LI^2$**

$\quad\quad = \dfrac{1}{2}(12.5)(3)^2 = \textbf{56.25 J}$

(c)   **Induced e.m.f., $E = N\left(\dfrac{\Phi}{t}\right)$**

$\quad\quad = (1500)\left(\dfrac{25 \times 10^{-3}}{150 \times 10^{-3}}\right) = \textbf{250 V}$

35.4   Effects of inductance in a d.c. circuit

A coil of wire possesses both inductance and resistance, each turn of the coil contributing to both its self-inductance and its resistance. It is not possible to obtain pure inductance. An inductive circuit is usually represented as resistance and inductance connected in series.

When a d.c. voltage is connected to a circuit having inductance L and resistance R there is a short period of time immediately after the voltage is connected, during which the current flowing in the circuit and the voltages across L and R are changing. These changing values are called **transients**.

Current growth

(a)   When the switch S shown in Fig. 35.4 is closed then

$$V = v_L + v_R \tag{35.1}$$

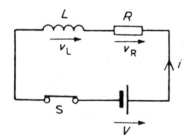

Figure 35.4

(b)   The battery voltage V is constant,

$$v_L = L \times \frac{\text{change of current}}{\text{change of time}}$$

i.e.

$$v_L = L\frac{\Delta I}{t}, \text{ and } v_R = iR$$

Hence at all times

$$V = L\left(\frac{\Delta I}{t}\right) + iR \tag{35.2}$$

(c)   At the instant of closing the switch, the rate of change of current is such that it induces an e.m.f. in the inductance which is equal and opposite to V.
Hence $V = v_L + 0$, i.e. $v_L = V$.
From equation (35.1), $v_R = 0$ and $i = 0$.

(d)   A short time later at time t_1 seconds after closing S, current i_1 is flowing, since there is a rate of change of current initially, resulting in a voltage drop of $i_1 R$ across the resistor. Since V, which is constant, is given by $V = v_L + v_R$, the induced e.m.f. v_L is reduced and equation (35.2) becomes

$$V = L\frac{\Delta I_1}{t_1} + i_1 R$$

(e)   A short time later still, say at time t_2 seconds after closing the switch, the current flowing is i_2, and the voltage drop across the resistor increases to $i_2 R$. Since v_R increases v_L decreases.

(f)   Ultimately, some time after closing S, the current flow is entirely limited by R, the rate of change of current is zero and hence v_L is zero. Thus $V = iR$. Under these conditions, steady state current flows, usually signified by I. Thus $I = V/R$, $v_R = IR$ and $v_L = 0$ at steady-state conditions.

(g)   Curves showing the changes in v_L, v_R and i with time are shown in Fig. 35.5 and show that v_L is a maximum value initially (i.e. equal to V), decaying exponentially to zero, whereas v_R and i grow from zero to their steady-state values of V and $I = (V/R)$, respectively.

The time taken for the current in an inductive circuit to reach its final value depends on the values of L and R. The ratio L/R is called the **time constant** τ of the circuit, i.e.

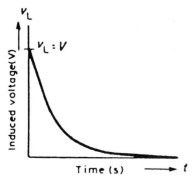

(a) Induced voltage transient

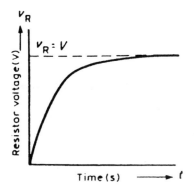

(b) Resistor voltage transient

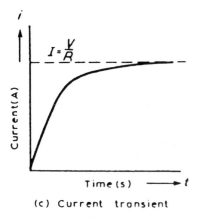

(c) Current transient

Figure 35.5

$$\tau = \frac{L}{R} \text{ seconds}$$

In the time τ seconds the current rises to 63.2% of its final value I and in practical situations the

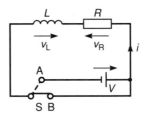

Figure 35.6

current rises to within 1% of its final value in a time equal to 5τ seconds.

Current decay

When a series L–R circuit is connected to a d.c. supply, as shown in Fig. 35.6, with S in position A, a current $I = V/R$ flows after a short time, creating a magnetic field ($\Phi \propto I$) associated with the inductor. When S is moved to position B, the current decreases, causing a decrease in the strength of the magnetic field. Flux linkages occur generating a voltage v_L, equal to $L(\Delta I/t)$. By Lenz's law, this voltage keeps current i flowing in the circuit, its value being limited by R. Since $V = v_L + v_L$, $0 = v_L + v_R$ and $v_L = -v_R$, i.e. v_L and v_R are equal in magnitude but opposite in direction. The current decays exponentially to zero and since v_R is proportional to the current flowing, v_R decays exponentially to zero. Since $v_L = v_R$, v_L also decays exponentially to zero. The curves representing these transients are shown in Fig. 35.7. Summarizing, in a d.c. circuit, inductance has no effect on the current except during the time when there are **changes** in the supply current (i.e. immediately following switching on or switching off).

> Problem 3.   What difficulties can be experienced when switching inductive circuits?

Energy stored in the magnetic field of an inductor exists because a current provides the magnetic field. When the d.c. supply is switched off the current falls rapidly, the magnetic field collapses causing a large induced e.m.f. which will either cause an arc across the switch contacts or will break down the insulation between adjacent turns of the coil. The high induced e.m.f. acts in a direction which tends to keep the current

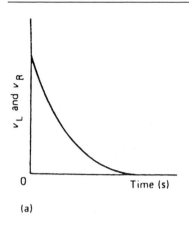

(a)

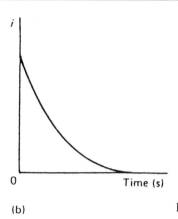

(b)

Figure 35.7

flowing, i.e. in the same direction as the applied voltage. The energy from the magnetic field will thus be aided by the supply voltage in maintaining an arc, which could cause severe damage to the switch. To reduce the induced e.m.f. when the supply switch is opened, a discharge resistor R_D is connected in parallel with the inductor as shown in Fig. 35.8. The magnetic field energy is dissipated as heat in R_D and R and arcing at the switch contacts is avoided.

Problem 4.   A coil of inductance 0.04 H and resistance 10 Ω is connected to a 120 V d.c. supply. Determine (a) the final value of current, (b) the time constant of the circuit, (c) the value of current after a time equal to the time constant from the instant the supply voltage is connected, (d) the expected time for the current to rise to within 1% of its final value.

(a)   Final steady current

$$I = \frac{V}{R} = \frac{120}{10} = \textbf{12 A}$$

(b)   Time constant of the circuit

$$\tau = \frac{L}{R} = \frac{0.04}{10} = \textbf{0.004 s or 4 ms}$$

(c)   In the time τ s the current rises to 63.2% of its final value of 12 A, i.e. in 4 ms the current rises to $0.632 \times 12 = \textbf{7.58 A}$.

(d)   The expected time for the current to rise to within 1% of its final value is given by 5τ, i.e. $5 \times 4 = \textbf{20 ms}$.

Problem 5.   A coil of inductance 0.5 H and resistance 6 Ω is connected in parallel with a resistance of 15 Ω to a 120 V d.c. supply. Determine (a) the current in the 15 Ω resistance, (b) the steady-state current in the coil, and (c) the current in the 15 Ω resistance at the instant the supply is switched off.

The circuit diagram is shown in Fig. 35.9.

(a)   Current flowing in 15 Ω resistor

$$I_R = \frac{V}{R_2} = \frac{120}{15} = \textbf{8 A}$$

(b)   The steady-state current flowing in coil

$$I_{COIL} = \frac{V}{R_1} = \frac{120}{6} = \textbf{20 A}$$

(c)   At the instant the supply is switched off the current flowing in R_2 is the same as the current flowing in the coil, i.e. **20 A** (see Fig. 35.10).

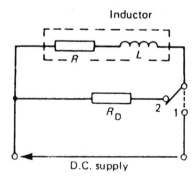

Figure 35.8

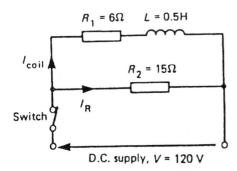

Figure 35.9

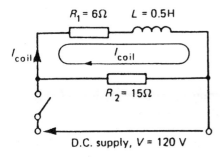

Figure 35.10

Problem 6.   A coil of resistance 20 Ω and inductance 0.8 H is to be discharged from a steady current of 3 A through a parallel resistance of 40 Ω. Determine (a) the induced e.m.f. at the instant the current begins to fall, and (b) the initial rate of change of current.

(a)   The instant when current begins to fall is when the supply is switched off. The circuit is shown in Fig. 35.11.

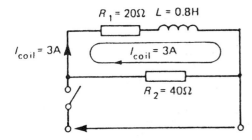

Figure 35.11

Total resistance of the closed loop,

$$R = R_1 + R_2 = 20 + 40 = 60\ \Omega$$

e.m.f. induced

$$e = I_{COIL}R = (3)(60) = \mathbf{180\ V}$$

(b)   Induced e.m.f. $e = L(\Delta I/t)$. Hence initial rate of change of current,

$$\left(\frac{\Delta I}{t}\right) = \frac{e}{L} = \frac{180}{0.8} = \mathbf{225\ A/s}$$

35.5   Effects of inductance in an a.c. circuit

In an a.c. circuit containing inductance, induced e.m.f.

$$\left[e = L\left(\frac{\Delta I}{t}\right)\right]$$

is present at nearly all times. The higher the frequency the greater the induced e.m.f.. The opposition offered by the induced e.m.f. to an applied voltage tends to limit the current in an a.c. circuit. This opposition is called **inductive reactance**, X_L, and is given by:

$$\boxed{X_L = 2\pi fL\ \mathbf{ohms}}$$

where f = frequency in hertz and L = inductance in henrys.

A graph of inductive resistance against frequency is shown in Fig. 35.12 and shows that a linear relationship exists between X_L and f.

Problem 7.   (a) Calculate the reactance of a coil of inductance 0.32 H when it is connected to a 50 Hz supply. (b) A coil has a reactance of 124 Ω in a circuit with a supply of frequency 5 kHz. Determine the inductance of the coil.

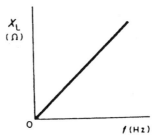

Figure 35.12

(a)   Inductive reactance $X_L = 2\pi fL = 2\pi(50)(0.32) = \mathbf{100.5\ \Omega}$.

(b)   Since $X_L = 2\pi fL$, inductance

$$L = \frac{X_L}{2\pi f} = \frac{124}{2\pi(5000)}\ \text{H}$$

$$= \mathbf{3.95\ mH}$$

Problem 8.   A coil of inductance 0.2 H has an inductive reactance of 754 Ω when connected to an a.c. supply. Calculate the frequency of the supply.

$L = 0.2$ H; $X_L = 754\ \Omega$. Since $X_L = 2\pi fL$,

$$\mathbf{frequency},\ f = \frac{X_L}{2\pi L} = \frac{754}{2\pi(0.2)} = \mathbf{600\ Hz}$$

Problem 9.   A coil has an inductance of 40 mH and negligible resistance. Calculate its inductive reactance and the resulting current if connected to (a) a 240 V, 50 Hz supply, and (b) a 100 V, 1 kHz supply.

(a)   Inductive reactance, X_L
$= 2\pi fL = 2\pi(50)(40 \times 10^{-3}) = \mathbf{12.57\ \Omega}$.

$$\text{Current},\ I = \frac{V}{X_L} = \frac{240}{12.57} = \mathbf{19.09\ A}$$

(b)   Inductive reactance, X_L
$= 2\pi(1000)(40 \times 10^{-3}) = \mathbf{251.3\ \Omega}$.

$$\text{Current},\ I = \frac{V}{X_L} = \frac{100}{251.3} = \mathbf{0.398\ A}$$

35.6   Purely resistive a.c. circuit

If a sinusoidal voltage is applied to a purely resistive circuit the resulting current is also sinusoidal and 'in-phase' with the voltage, i.e. voltage and current waveforms pass through their zero values at the same instant and attain their maximum values at the same instant, as shown in Fig. 35.13.

35.7   Purely inductive a.c. circuit

Let an a.c. circuit be purely inductive, i.e. the resistance is zero (which is a theoretical condition), and the applied current be sinusoidal, as shown in Fig. 35.14. The magnitude of the induced e.m.f.

$$\left[e = L\left(\frac{\Delta I}{t}\right)\right]$$

is directly proportional to the rate of change of current ($\Delta I/t$), which is given by the gradient of the tangent to the current curve.

At point A, in Fig. 35.14(b), the rate of change of current is a maximum positive value and thus e is a maximum value but acts in opposition to the applied voltage from Lenz's law. This is shown as point A'.

At point B, the rate of change of current, i.e. the gradient of the curve, is zero and hence $e = 0$, shown as point B'.

At point C, the rate of change of current is a maximum negative value and thus e is a maximum positive value shown as point C'.

With similar reasoning applied to the second half of the cycle the induced e.m.f. e is seen to be

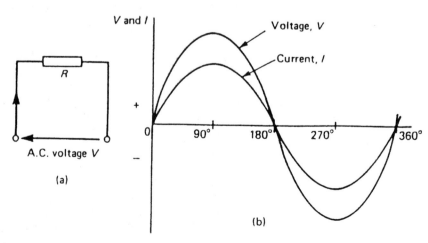

(a)

A.C. voltage V

(b)

Figure 35.13

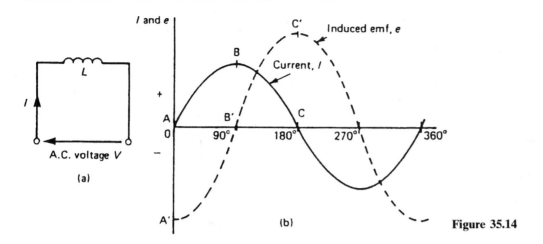

Figure 35.14

a sine wave. The induced e.m.f., e, has a polarity that is always opposite to the applied voltage V, i.e. e and V are 180° out of phase. The current I and applied voltage V are shown in Fig. 35.15 and are seen to be 90° out of phase. The maximum value of I, shown as point P, occurs 90° after the maximum value of V, shown as point Q, and thus the current is said to lag the voltage by 90°.

In practice, both resistance and inductance are present in an inductive circuit which results in current lagging voltage by an angle which is greater than 0° and less than 90°.

Summarizing, in an a.c. circuit, pure inductance introduces opposition to current flow, called inductive reactance X_L ($= 2\pi fL$), and also causes the current to lag the voltage by 90°.

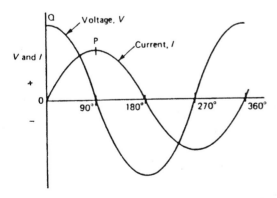

Figure 35.15

35.8   Multi-choice questions on inductors and their effects in electric circuits

(Answers on page 356.)

1. The effect of inductance occurs in an electrical circuit when:
   (a) the resistance is changing
   (b) the flux is changing
   (c) the current is changing
2. Which of the following statements is false? The inductance of an inductor increases:
   (a) with a short, thick, coil
   (b) when wound on an iron core
   (c) as the number of turns increases
   (d) as the cross-sectional area of the coil decreases
3. The time constant for a d.c. circuit containing resistance R and inductance L is given by:
   (a) $\dfrac{R}{L}$ s   (b) $\dfrac{L}{R}$ s   (c) LR s

An inductor of inductance 0.1 H and negligible resistance is connected in series with a 50 Ω resistor to a 24 V d.c. supply. In questions 4 to 8, use this data to determine the value required, selecting your answers from those given below.
(a) 5 ms   (b) 12.6 V   (c) 0.4 A   (d) 500 ms
(e) 7.4 V   (f) 2.5 A   (g) 2 ms   (h) 0 V
(i) 0 A   (j) 20 V

4. The value of the time constant of the circuit
5. The approximate value of the voltage across the resistor at a time equal to the time constant after being connected to the supply

6. The final value of current flowing in the circuit
7. The initial value of voltage across the inductor
8. The final value of the steady-state voltage across the inductor
9. A circuit comprising a 12 Ω resistor and a 4 H inductance connected across a 60 V d.c. supply has a steady current flowing of:
    (a) 5 A   (b) 15 A   (c) $3\frac{3}{4}$ A   (d) 3 A
10. An inductance of 10 mH connected across a 100 V, 50 Hz supply has an inductive reactance of:
    (a) 10π Ω   (b) 1000π Ω   (c) π Ω   (d) π H
11. When the frequency of an a.c. circuit containing resistance and inductance connected in series is increased (the voltage remaining constant), the current:
    (a) decreases
    (b) increases
    (c) stays the same
12. Pure inductance in an a.c. circuit results in a current that:
    (a) leads the voltage by 90°
    (b) is in phase with the voltage
    (c) leads the voltage by π rad
    (d) lags the voltage by $(\pi/2)$ rad

35.9   Short answer questions on inductors and their effects in electrical circuits

1. Define inductance and name its unit.
2. What factors affect the inductance of an inductor?
3. What is an inductor? Sketch a typical practical inductor.
4. Explain how a standard resistor may be non-inductively wound.
5. Energy W stored in the magnetic field of an inductor is given by $W = $ ....
6. What is a transient?
7. State briefly the effects of inductance on the current in a d.c. circuit.
8. The opposition to current in a purely inductive a.c. circuit is called .... and is calculated using the formula ....
9. In an a.c. circuit, what effect does pure inductance have on the phase angle between current and voltage?

35.10   Further questions on inductors and their effects in electrical circuits

1. Describe the basic form of an inductor and state the factors which affect inductance.
2. Describe the effect of an inductor on the current in a d.c. circuit. Explain how difficulties encountered in switching inductive circuits can be overcome.
3. An inductor of 20 H has a current of 2.5 A flowing in it. Find the energy stored in the magnetic field of the inductor.   [62.5 J]
4. Calculate the value of the energy stored when a current of 30 mA is flowing in a coil of inductance 400 mH.   [0.18 mJ]
5. The energy stored in the magnetic field of an inductor is 80 J when the current flowing in the inductor is 2 A. Calculate the inductance of the coil.   [40 H]
6. A flux of 30 mWb links with a 1200 turn coil when a current of 5 A is passing through the coil. Calculate (a) the inductance of the coil, (b) the energy stored in the magnetic field, and (c) the average e.m.f. induced if the current is reduced to zero in 0.20 seconds.
    [(a) 7.2 H (b) 90 J (c) 180 V]
7. A coil of resistance 20 Ω and inductance 500 mH is connected to a d.c. supply of 160 V. Determine (a) the final value of current, (b) the time constant, (c) the current after a time equal to the time constant from the instant the supply voltage is connected, and (d) the expected time for the current to rise to within 1% of its final value.
    [(a) 8 A (b) 25 ms (c) 5.06 A (b) 0.125 s]
8. An inductive circuit has a time constant of 50 ms. If the steady value of current flowing through the circuit is 2 A when connected to a 200 V d.c. supply, calculate the value of resistance and the value of inductance.
    [$R = 100$ Ω, $L = 5$ H]
9. A coil of inductance 0.3 H and resistance 10 Ω is connected in parallel with a resistance of 25 Ω to a 100 V d.c. supply. Determine (a) the current in the 25 Ω resistor, (b) the steady-state current in the coil, and (c) the current in the 25 Ω resistance at the instant the supply is switched off.
    [(a) 4 A (b) 10 A (c) 10 A]
10. A coil of resistance 15 Ω and inductance 0.3 H is to be discharged from a steady current of 5 A through a parallel resistance

of 30 Ω. Determine (a) the induced e.m.f. at the instant the current begins to fall, and (b) the initial rate of change of current.

[(a) 225 V (b) 750 A/s]

11. Calculate the inductive reactance of a coil of inductance 0.2 H when it is connected to (a) a 50 Hz, (b) a 600 Hz, and (c) a 4 kHz supply.

[(a) 62.83 Ω (b) 754 Ω (c) 5.027 kΩ]

12. A coil has an inductive reactance of 120 Ω in a circuit with a supply frequency of 4 kHz. Calculate the inductance of the coil.

[4.77 mH]

13. A supply of 240 V, 50 Hz is connected across a pure inductance and the resulting current is 1.2 A. Calculate the inductance of the coil.

[0.637 H]

14. An e.m.f. of 200 V at a frequency of 2 kHz is applied to a coil of pure inductance 50 mH. Determine (a) the reactance of the coil, and (b) the current flowing in the coil.

[(a) 628 Ω (b) 0.318 A]

15. A 120 mH inductor has a 50 mA, 1 kHz alternating current flowing through it. Find the p.d. across the inductor.

[37.7 V]

16. Describe the effects of an inductor on the current in an a.c. circuit.

Part 4

Measurements

36

Electrical measuring instruments and measurements

At the end of this chapter you should be able to:

- recognize the importance of testing and measurements in electric circuits
- appreciate the essential devices comprising an analogue instrument
- explain the operation of an attraction and a repulsion type of moving-iron instrument
- explain the operation of a moving-coil rectifier instrument
- compare moving-coil, moving-iron and moving-coil rectifier instruments
- calculate values of shunts for ammeters and multipliers for voltmeters
- understand the operation of an ohmmeter/megger
- appreciate the operation of multimeters/Avometers
- understand the operation of a voltmeter
- understand the operation of a C.R.O. for d.c. and a.c. measurements
- calculate periodic time, frequency, peak-to-peak values from waveforms on a C.R.O.
- understand null methods of measurement for a Wheatstone bridge and d.c. potentiometer
- appreciate the most likely source of errors in measurements

36.1 Introduction

Tests and measurements are important in designing, evaluating, maintaining and servicing electrical circuits and equipment. In order to detect electrical quantities such as current, voltage, resistance or power, it is necessary to transform an electrical quantity or condition into a visible indication. This is done with the aid of instruments (or meters) that indicate the magnitude of quantities either by the position of a pointer moving over a graduated scale (called an analogue instrument) or in the form of a decimal number (called a digital instrument).

36.2 Analogue instruments

All analogue electrical indicating instruments require three essential devices:

(a) **A deflecting or operating device**. A mechanical force is produced by the current or voltage which causes the pointer to deflect from its zero position.

(b) **A controlling device**. The controlling force acts in opposition to the deflecting force and ensures that the deflection shown on the meter is always the same for a given measured quantity. It also prevents the

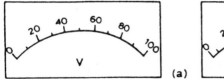
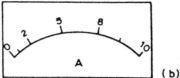

Figure 36.1

pointer always going to the maximum deflection. There are two main types of controlling device – spring control and gravity control.

(c) **A damping device.** The damping force ensures that the pointer comes to rest in its final position quickly and without undue oscillation. There are three main types of damping used – eddy-current damping, air-friction damping and fluid-friction damping.

There are basically two types of scale – linear and non-linear. A **linear scale** is shown in Fig. 36.1(a), where the divisions or graduations are evenly spaced. The voltmeter shown has a range 0–100 V, i.e. a full-scale deflection (f.s.d.) of 100 V. A **non-linear scale** is shown in Fig. 36.1(b). The scale is cramped at the beginning and the graduations are uneven throughout the range. The ammeter shown has a f.s.d. of 10 A.

36.3 Moving-iron instrument

(a) An **attraction type** of moving-iron instrument is shown diagrammatically in Fig. 36.2(a). When current flows in the solenoid, a pivoted soft-iron disc is attracted towards the solenoid and the movement causes a pointer to move across a scale.

(b) In the **repulsion type** moving-iron instrument shown diagrammatically in Fig. 36.2(b), two pieces of iron are placed inside the solenoid, one being fixed, and the other attached to the spindle carrying the pointer. When current passes through the solenoid, the two pieces of iron are magnetized in the same direction and therefore repel each other. The pointer thus moves across the scale. The force moving the pointer is, in each type, proportional to I^2. Because of this the direction of current does not matter and the moving-iron instrument can be used on d.c. or a.c. The scale, however, is non-linear.

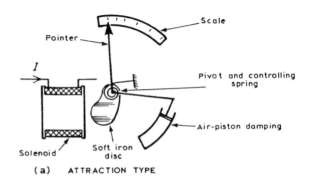

(a)   ATTRACTION TYPE

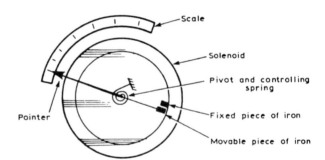

(b)   REPULSION TYPE

Figure 36.2

36.4 The moving-coil rectifier instrument

A moving-coil instrument, which measures only d.c., may be used in conjunction with a bridge rectifier circuit as shown in Fig. 36.3 to provide an indication of alternating currents and voltages (see Chapter 33). The average value of the full-wave rectified current is $0.637 I_m$. However, a meter being used to measure a.c. is usually calibrated in r.m.s. values. For sinusoidal quantities the indication is

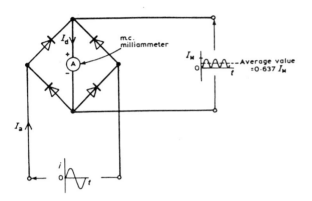

Figure 36.3

$$\frac{0.707I_\mathrm{m}}{0.637I_\mathrm{m}}$$

i.e. 1.11 times the mean value. Rectifier instruments have scales calibrated in r.m.s. quantities and it is assumed by the manufacturer that the a.c. is sinusoidal.

36.5 Comparison of moving-coil, moving-iron and moving-coil rectifier instruments

See Table 36.1.

36.6 Shunts and multipliers

An **ammeter**, which measures current, has a low resistance (ideally zero) and must be connected in series with the circuit.

A **voltmeter**, which measures p.d., has a high resistance (ideally infinite) and must be connected in parallel with the part of the circuit whose p.d. is required.

Table 36.1

Type of instrument	Moving-coil	Moving-iron	Moving-coil rectifier
Suitable for measuring	Direct current and voltage	Direct and alternating currents and voltage (reading in r.m.s. value)	Alternating current and voltage (reads average value but scale is adjusted to give r.m.s. value for sinusoidal waveforms)
Scale	Linear	Non-linear	Linear
Method of control	Hairsprings	Hairsprings	Hairsprings
Method of damping	Eddy current	Air	Eddy current
Frequency limits	–	20–200 Hz	20–100 kHz
Advantages	1 Linear scale 2 High sensitivity 3 Well shielded from stray magnetic fields 4 Lower power consumption	1 Robust construction 2 Relatively cheap 3 Measures d.c. and a.c. 4 In frequency range 20–100 Hz reads r.m.s. correctly regardless of supply waveform	1 Linear scale 2 High sensitivity 3 Well shielded from stray magnetic fields 4 Low power consumption 5 Good frequency range
Disadvantages	1 Only suitable for d.c. 2 More expensive than moving-iron type 3 Easily damaged	1 Non-linear scale 2 Affected by stray magnetic fields 3 Hysteresis errors in d.c. circuits 4 Liable to temperature errors 5 Due to the inductance of the solenoid, readings can be affected by variation of frequency	1 More expensive than moving-iron type 2 Errors caused when supply is non-sinusoidal

(For the principle of operation of a moving-coil instrument, see Chapter 31, page 254.)

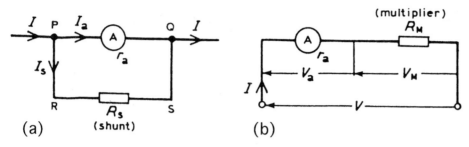

Figure 36.4

There is no difference between the basic instrument used to measure current and voltage since both use a milliammeter as their basic part. This is a sensitive instrument which gives f.s.d. for currents of only a few milliamperes. When an ammeter is required to measure currents of larger magnitude, a proportion of the current is diverted through a low-value resistance connected in parallel with the meter. Such a diverting resistor is called a **shunt**.

From Fig. 36.4(a), $V_{PQ} = V_{RS}$. Hence $I_a r_a = I_s R_S$. Thus the value of the shunt

$$R_S = \frac{I_a r_a}{I_s} \text{ ohms}$$

The milliammeter is converted into a voltmeter by connecting a high resistance (called a **multiplier**) in series with it as shown in Fig. 36.4(b). From Fig. 36.4(b), $V = V_a + V_M = I r_a + I R_M$. Thus the value of the multiplier

$$R_M = \frac{V - I r_a}{I} \text{ ohms}$$

Problem 1. A moving-coil instrument gives a f.s.d. when the current is 40 mA and its resistance is 25 Ω. Calculate the value of the shunt to be connected in parallel with the meter to enable it to be used as an ammeter for measuring currents up to 50 A.

The circuit diagram is shown in Fig. 36.5, where r_a = resistance of instrument = 25 Ω
r_s = resistance of shunt
I_a = maximum permissible current flowing in instrument = 40 mA = 0.04 A

I_s = current flowing in shunt
I = total circuit current required to give f.s.d. = 50 A.

Since $I = I_a + I_s$ then $I_s = I - I_a = 50 - 0.04 = 49.96$ A. $V = I_a r_a = I_s R_S$, hence

$$R_S = \frac{I_a r_a}{I_s} = \frac{(0.04)(25)}{49.96} = 0.020\ 02\ \Omega = \textbf{20.02 m}\Omega$$

Thus for the moving-coil instrument to be used as an ammeter with a range 0–50 A, a resistance of value 20.02 mΩ needs to be connected in parallel with the instrument.

Problem 2. A moving-coil instrument having a resistance of 10 Ω gives a f.s.d. when the current is 8 mA. Calculate the value of the multiplier to be connected in series with the instrument so that it can be used as a voltmeter for measuring p.d.s. up to 100 V.

The circuit diagram is shown in Fig. 36.6, where r_a = resistance of instrument = 10 Ω
R_M = resistance of multiplier
I = total permissible instrument current = 8 mA = 0.008 A,
V = total p.d. required to give f.s.d. = 100 V.

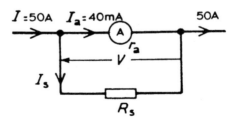

Figure 36.5

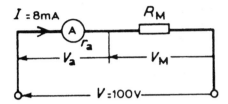

Figure 36.6

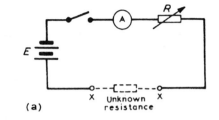

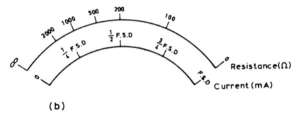

Figure 36.7

$V = V_a + V_M = Ir_a + IR_M$, i.e. $100 = (0.008)(10) + (0.008)R_M$, or $100 - 0.08 = 0.008\,R_M$, thus

$$R_M = \frac{99.92}{0.008} = 12\,490\ \Omega = \mathbf{12.49\ k\Omega}$$

Hence for the moving-coil instrument to be used as a voltmeter with a range 0–100 V, a resistance of value 12.49 kΩ needs to be connected in series with the instrument.

An **electronic voltmeter** can be used to measure with accuracy e.m.f. or p.d. from millivolts to kilovolts by incorporating in its design amplifiers and attenuators. The loading effect of an electronic voltmeter is minimal.

A **digital voltmeter** (**DVM**) has, like the electronic voltmeter, a high input resistance. For power frequencies and d.c. measurements a DVM will normally be preferable to an analogue instrument.

36.7 The ohmmeter

An ohmmeter is an instrument for measuring electrical resistance. A simple ohmmeter circuit is shown in Fig. 36.7(a). Unlike the ammeter or voltmeter, the ohmmeter circuit does not receive the energy necessary for its operation from the circuit under test. In the ohmmeter this energy is supplied by a self-contained source of voltage, such as a battery. Initially, terminals XX are short-circuited and R adjusted to give f.s.d. on the milliammeter. If current I is at a maximum value and voltage E is constant, then resistance $R = E/I$ is at a minimum value. Thus f.s.d. on the milliammeter is made zero on the resistance scale. When terminals XX are open-circuited no current flows and R (= E/O) is infinity, ∞.

The milliammeter can thus be calibrated directly in ohms. A cramped (non-linear) scale results and is 'back to front', as shown in Fig. 36.7(b). When calibrated, an unknown resistance

is placed between terminals XX and its value determined from the position of the pointer on the scale. An ohmmeter designed for measuring low values of resistance is called a **continuity tester**. An ohmmeter designed for measuring high values of resistance (i.e. megohms) is called an **insulation resistance tester** (e.g. **megger**).

36.8 Multimeters

Instruments are manufactured that combine a moving-coil meter with a number of shunts and series multipliers, to provide a range of readings on a single scale graduated to read current and voltage. If a battery is incorporated then resistance can also be measured. Such instruments are called **multimeters** or **universal instruments** or **multirange instruments**. An 'Avometer' is a typical example. A particular range may be selected either by the use of separate terminals or by a selector switch. Only one measurement can be performed at one time. Often such instruments can be used in a.c. as well as d.c. circuits when a rectifier is incorporated in the instrument.

36.9 Wattmeters

A **wattmeter** is an instrument for measuring electrical power in a circuit. Figure 36.8 shows typical connections of a wattmeter used for measuring power supplied to a load. The instrument has two coils:

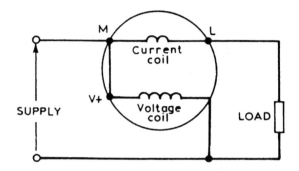

Figure 36.8

(i) a current coil, which is connected in series with the load, like an ammeter, and
(ii) a voltage coil, which is connected in parallel with the load, like a voltmeter.

36.10   Instrument 'loading' effect

Some measuring instruments depend for their operation on power taken from the circuit in which measurements are being made. Depending on the 'loading' effect of the instrument (i.e. the current taken to enable it to operate), the prevailing circuit conditions may change.

The resistance of voltmeters may be calculated since each has a stated sensitivity (or 'figure of merit'), often stated in 'kΩ per volt' of FSD. A voltmeter should have as high a resistance as possible (ideally infinite). In a.c. circuits the impedance of the instrument varies with frequency and thus the loading effect of the instrument can change.

Problem 3.   Calculate the power dissipated by the voltmeter and by resistor R in Fig. 36.9 when (a) $R = 250\,\Omega$, (b) $R = 2\,M\Omega$. Assume that the voltmeter sensitivity (sometimes called figure of merit) is $10\,k\Omega/V$.

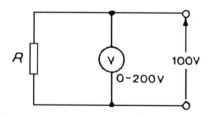

Figure 36.9

(a) Resistance of voltmeter, R_v = sensitivity $\times$ f.s.d. Hence, $R_v = 10\,k\Omega/V \times 200\,V = 2000\,k\Omega = 2\,M\Omega$.
Current flowing in voltmeter
$$I_v = \frac{V}{R_v} = \frac{100}{2 \times 10^6} = 50 \times 10^{-6}\,A$$
Power dissipated by voltmeter $= VI_v = (100)(50 \times 10^{-6}) = \textbf{5 mW}$.
When $R = 250\,\Omega$, current in resistor
$$I_R = \frac{V}{R} = \frac{100}{250}\quad \textbf{0.4 A}$$
Power dissipated in load resistor $R = VI_R = (100)(0.4) = \textbf{40 W}$. Thus the power dissipated in the voltmeter is insignificant in comparison with the power dissipated in the load.

(b) When $R = 2\,M\Omega$, current in resistor
$$I_R = \frac{V}{R} = \frac{100}{2 \times 10^6}$$
$$= 50 \times 10^{-6}\,A$$
Power dissipated in load resistor $R = VI_R$
$$= 100 \times 50 \times 10^{-6}$$
$$= \textbf{5 mW}$$

In this case the higher load resistance reduced the power dissipated such that the voltmeter is using as much power as the load.

Problem 4.   An ammeter has a f.s.d. of 100 mA and a resistance of 50 Ω. The ammeter is used to measure the current in a load of resistance 500 Ω when the supply voltage is 10 V. Calculate (a) the ammeter reading expected (neglecting its resistance), (b) the actual current in the circuit, (c) the power dissipated in the ammeter, (d) the power dissipated in the load.

From Fig. 36.10,

(a) expected ammeter reading
$$= \frac{V}{R} = \frac{10}{500} = \textbf{20 mA}$$

(b) Actual ammeter reading
$$= \frac{V}{R + r_a} = \frac{10}{500 + 50} = \textbf{18.18 mA}$$

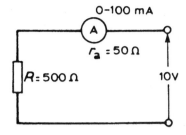

Figure 36.10

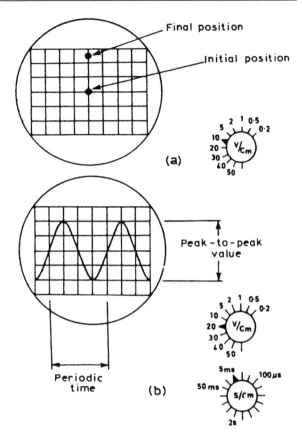

Figure 36.11

Thus the ammeter itself has caused the circuit conditions to change from 20 mA to 18.18 mA.

(c) Power dissipated in the ammeter = I^2R

$$= (18.18 \times 10^{-3})^2(50)$$

$$= \textbf{16.53 mW}$$

(d) Power dissipated in the load resistor = I^2R

$$= (18.18 \times 10^{-3})^2(500) = \textbf{165.3 mW}$$

36.11 Cathode ray oscilloscope

The cathode ray oscilloscope (c.r.o.) may be used in the observation of waveforms and for the measurement of voltage, current, frequency, phase and periodic time. For examining periodic waveforms the electron beam is deflected horizontally (i.e. in the X direction) by a sawtooth generator acting as a timebase. The signal to be examined is applied to the vertical deflection system (Y direction) usually after amplification.

Oscilloscopes normally have a transparent grid of 10 mm by 10 mm squares in front of the screen, called a graticule. Among the timebase controls is a 'variable' switch which gives the sweep speed as time per centimetre. This may be in s/cm, ms/cm or μs/cm, a large number of switch positions being available. Also on the front panel of a c.r.o. is a Y amplifier switch marked in volts per centimetre, with a large number of available switch positions.

(i) With **direct voltage measurements**, only the Y amplifier 'volts/cm' switch on the c.r.o. is used. With no voltage applied to the Y plates the position of the spot trace on the screen is noted. When a direct voltage is applied to the Y plates the new position of the spot trace is an indication of the magnitude of the voltage.

For example, in Fig. 36.11(a), with no voltage applied to the Y plates, the spot trace is in the centre of the screen (initial position) and then the spot trace moves 2.5 cm to the final position shown, on application of a d.c. voltage. With the 'volts/cm' switch on 10 volts/cm the magnitude of the direct voltage is 2.5 cm × 10 volts/cm, i.e. 25 volts.

(ii) With **alternating voltage measurement**, let a sinusoidal waveform be displayed on a c.r.o. screen as shown in Fig. 36.11(b). If the s/cm switch is on, say, 5 ms/cm then the **periodic time** T of the sinewave is 5 ms/cm × 4 cm, i.e. **20 ms or 0.02 s**.

Since

$$\text{frequency } f = \frac{1}{T}, \quad \textbf{frequency} = \frac{1}{0.02} = \textbf{50 Hz}$$

If the 'volts/cm' switch is on, say, 20 volts/cm then the **amplitude** or **peak value** of the sinewave shown is 20 volts/cm × 2 cm, i.e. **40 V**.

Since

$$\text{r.m.s. voltage} = \frac{\text{peak voltage}}{\sqrt{2}} \quad \text{(see Chapter 33)}$$

$$\textbf{r.m.s. voltage} = \frac{40}{\sqrt{2}} = \textbf{28.28 V}$$

Double beam oscilloscopes are useful whenever two signals are to be compared simultaneously. The c.r.o. demands reasonable skill in adjustment and use. However its greatest advantage is in observing the shape of a waveform – a feature not possessed by other measuring instruments.

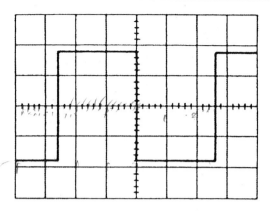

Figure 36.12

Problem 5.   Describe how a simple c.r.o. is adjusted to give (a) a spot trace, (b) a continuous horizontal trace on the screen, explaining the functions of the various controls.

(a) To obtain a spot trace on a typical c.r.o. screen:
  (i)   Switch on the c.r.o.
  (ii) Switch the timebase control to off. This control is calibrated in time per centimetres – for example, 5 ms/cm or 100 μs/cm. Turning it to zero ensures no signal is applied to the X-plates. The Y-plate input is left open-circuited.
  (iii) Set the intensity, X-shift and Y-shift controls to about the mid-range positions.
  (iv) A spot trace should now be observed on the screen. If not, adjust either or both of the X and Y-shift controls. The X-shift control varies the position of the spot trace in a horizontal direction while the Y-shift control varies its vertical position.
  (v)   Use the X and Y-shift controls to bring the spot to the centre of the screen and use the focus control to focus the electron beam into a small circular spot.
(b) To obtain a continuous horizontal trace on the screen the same procedure as in (a) is initially adopted. Then the timebase control is switched to a suitable position, initially the millisecond timebase range, to ensure that the repetition rate of the sawtooth is sufficient for the persistence of the vision time of the screen phosphor to hold a given trace.

Problem 6.   For the c.r.o. square voltage waveform shown in Fig. 36.12 determine (a) the periodic time, (b) the frequency and (c) the peak-to-peak voltage. The 'time/cm' (or timebase control) switch is on 100 μs/cm and the 'volts/cm' (or signal amplitude control) switch is on 20 V/cm.

(In Figs. 36.12 to 36.15 assume that the squares shown are 1 cm by 1 cm.)

(a)   The width of one complete cycle is 5.2 cm, hence the periodic time, $T = 5.2 \text{ cm} \times 100 \times 10^{-6} \text{ s/cm} = \textbf{0.52 ms}$.

(b)   Frequency, $f = \dfrac{1}{T} = \dfrac{1}{0.52 \times 10^{-3}} = 1.92 \text{ kHz}$

(c)   The peak-to-peak height of the display is 3.6 cm, hence the peak-to-peak voltage = $3.6 \text{ cm} \times 20 \text{ V/cm} = \textbf{72 V}$.

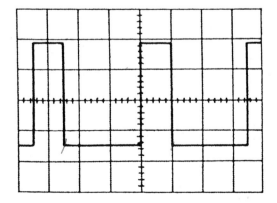

Figure 36.13

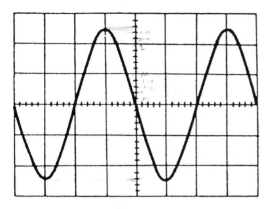

Figure 36.14

Problem 7. For the c.r.o. display of a pulse waveform shown in Fig. 36.13 the 'time/cm' switch is on 50 ms/cm and the 'volts/cm' switch is on 0.2 V/cm. Determine (a) the periodic time, (b) the frequency, (c) the magnitude of the pulse voltage.

(a) The width of one complete cycle is 3.5 cm, hence the periodic time, T = 3.5 cm × 50 ms/cm = **175 ms**.

(b) Frequency, $f = \dfrac{1}{T} = \dfrac{1}{0.175}$ = **5.71 Hz**

(c) The height of a pulse is 3.4 cm, hence the magnitude of the pulse voltage = 3.4 cm × 0.2 V/cm = **0.68 V.**

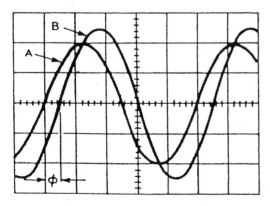

Figure 36.15

Problem 8. A sinusoidal voltage trace displayed by a c.r.o. is shown in Fig. 36.14. If the 'time/cm' switch is on 500 μs/cm and the 'volts/cm' switch is on 5 V/cm, find, for the waveform, (a) the frequency, (b) the peak-to-peak voltage, (c) the amplitude, (d) the r.m.s. value.

(a) The width of one complete cycle is 4 cm. Hence the periodic time, T is 4 cm × 500 μs/cm, i.e. 2 ms.

Frequency, $f = \dfrac{1}{T} = \dfrac{1}{2 \times 10^{-3}}$ = **500 Hz**

(b) The peak-to-peak height of the waveform is 5 cm. Hence the peak-to-peak voltage = 5 cm × 5 V/cm = **25 V.**

(c) Amplitude = $\frac{1}{2}$ × 25 V = 12.5 V.

(d) The peak value of voltage is the amplitude, i.e. 12.5 V

r.m.s. voltage = $\dfrac{\text{peak voltage}}{\sqrt{2}} = \dfrac{12.5}{\sqrt{2}}$ = **8.84 V**

Problem 9. For the double-beam oscilloscope displays shown in Fig. 36.15 determine (a) their frequency, (b) their r.m.s. values, (c) their phase difference. The 'time/cm' switch is on 100 μs/cm and the 'volts/cm' switch on 2 V/cm.

(a) The width of each complete cycle is 5 cm for both waveforms. Hence the periodic time, T, of each waveform is 5 cm × 100 μs/cm, i.e. 0.5 ms.

Frequency of each waveform

$$f = \dfrac{1}{T} = \dfrac{1}{0.5 \times 10^{-3}} = \textbf{2 kHz}$$

(b) The peak value of waveform A is 2 cm × 2 V/cm = **4 V.**
Hence the r.m.s. value of waveform A = $(4/\sqrt{2})$ = **2.83 V.**
The peak value of waveform B is 2.5 cm × 2 V/cm = **5 V.**
Hence the r.m.s. value of waveform B = $(5/\sqrt{2})$ = **3.54 V.**

(c) Since 5 cm represents 1 cycle, then 5 cm represents 360°,
i.e. 1 cm represents (360/5) = 72°.
The phase angle ϕ = 0.5 cm = 0.5 cm × 72°/cm = 36°.
Hence waveform A leads waveform B by 36°.

36.12 Null method of measurement

A **null method of measurement** is a simple, accurate and widely used method which depends on an instrument reading being adjusted to read zero current only. The method assumes:

(i) if there is any deflection at all, then some current is flowing;
(ii) if there is no deflection, then no current flows (i.e. a null condition).

Hence it is unnecessary for a meter sensing current flow to be calibrated when used in this way. A sensitive milliammeter or microammeter with centre-zero position setting is called a **galvanometer**. Two examples where the method is used are in the Wheatstone bridge and in the d.c. potentiometer.

Figure 36.16 shows a **Wheatstone bridge** circuit which compares an unknown resistance R_x with others of known values, i.e. R_1 and R_2 which have fixed values, and R_3 which is variable. R_3 is varied until zero deflection is obtained on the galvanometer G. No current then flows through the meter, $V_A = V_B$, and the bridge is said to be 'balanced'. At balance, $R_1 R_x = R_2 R_3$, i.e.

$$R_x = \frac{R_2 R_3}{R_1} \text{ ohms}$$

> Problem 10. In a Wheatstone bridge ABCD, a galvanometer is connected between A and C, and a battery between B and D. A resistor of unknown value is connected between A and B. When the bridge is balanced, the resistance between B and C is 100 Ω, that between C and D is 10 Ω and that between D and A is 400 Ω. Calculate the value of the unknown resistance.

The Wheatstone bridge is shown in Fig. 36.17 where R_x is the unknown resistance. At balance, equating the products of opposite ratio arms gives: $(R_x)(10) = (100)(400)$, hence

$$R_x = \frac{(100)(400)}{10} = 4000 \ \Omega$$

Hence the unknown resistance, R_x = 4 kΩ.

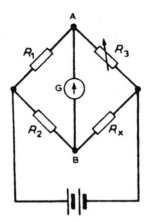

Figure 36.16

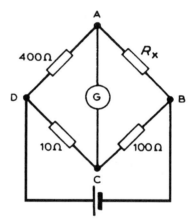

Figure 36.17

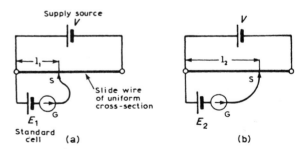

Figure 36.18

The **d.c. potentiometer** is a null-balance instrument used for determining values of e.m.f.s and p.d.s by comparison with a known e.m.f. or p.d.

In Fig. 36.18(a), using a standard cell of known e.m.f. E_1, the slider S is moved along the slide wire until balance is obtained (i.e. the galvanometer deflection is zero), shown as length l_1.

The standard cell is now replaced by a cell of unknown e.m.f. E_2 (see Fig. 36.18(b)) and again balance is obtained (shown as l_2).
Since $E_1 \propto l_1$ and $E_2 \propto l_2$ then

$$\frac{E_1}{E_2} = \frac{l_1}{l_2} \text{ and } \boxed{E_2 = E_1\left(\frac{l_2}{l_1}\right) \text{ volts}}$$

A potentiometer may be arranged as a resistive two-element potential divider in which the division ratio is adjustable to give a simple variable d.c. supply. Such devices may be constructed in the form of a resistive element carrying a sliding contact which is adjusted by a rotary or linear movement of the control knob.

Problem 11.   In a d.c. potentiometer balance is obtained at a length of 400 mm when using a standard cell of 1.0186 volts. Determine the e.m.f. of a dry cell if balance is obtained with a length of 650 mm.

$E_1 = 1.0186$ V, $l_1 = 400$ mm, $l_2 = 650$ mm. With reference to Fig. 36.18

$$\frac{E_1}{E_2} = \frac{l_1}{l_2}$$

from which

$$E_2 = E_1\left(\frac{l_2}{l_1}\right) = (1.0186)\left(\frac{650}{400}\right) = \mathbf{1.655 \ V}$$

36.13   Measurement errors

The errors most likely to occur in measurements are those due to:

(i)   the limitations of the instrument;
(ii)   the operator;
(iii) the instrument disturbing the circuit.

Errors in the limitations of the instrument

The calibration accuracy of an instrument depends on the precision with which it is constructed. Every instrument has a margin of error which is expressed as a percentage of the indication. For example, industrial grade instruments have an accuracy of ± 2% of f.s.d. Thus if a voltmeter has a f.s.d. of 100 V and it indicates 40 V say, then the actual voltage may be anywhere between 40 ± (2% of 100), or 40 ± 2, i.e. between 38 V and 42 V.

When an instrument is calibrated, it is compared against a standard instrument and a graph is drawn of 'error' against 'meter deflection'. A typical graph is shown in Fig. 36.19 where it is seen that the accuracy varies over the scale length. Thus a meter with a ± 2% f.s.d. accuracy would tend to have an accuracy which is much better than ± 2% f.s.d. over much of the range.

Errors by the operator

It is easy for an operator to misread an instrument reading. With linear scales the values of the sub-divisions are reasonably easy to determine; non-linear scale graduations are more difficult to estimate. Also, scales differ from instrument to instrument and some meters have more than one scale (as with multimeters) and mistakes in reading indications are easily made. When

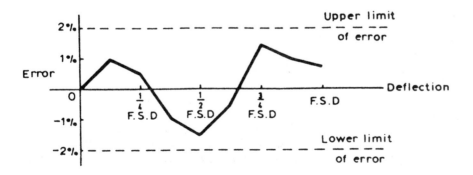

Figure 36.19

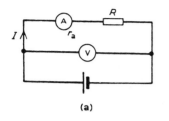 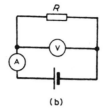

Figure 36.20

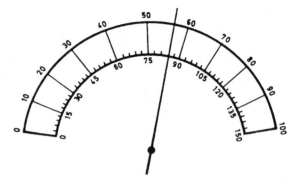

Figure 36.21

reading a meter scale it should be viewed from an angle perpendicular to the surface of the scale at the location of the pointer; a meter scale should not be viewed 'at an angle'.

Errors due to the instrument disturbing the circuit

Any instrument connected into a circuit will affect that circuit to some extent. Meters require some power to operate, but provided this power is small compared with the power in the measured circuit, then little error will result. Incorrect positioning of instruments in a circuit can be a source of errors. For example, let a resistance be measured by the voltmeter–ammeter method as shown in Fig. 36.20. Assuming 'perfect' instruments, the resistance should be given by the voltmeter reading divided by the ammeter reading (i.e. $R = V/I$). However, in Fig. 36.20(a), $(V/I) = R + r_a$ and in Fig. 36.20(b) the current through the ammeter is that through the resistor plus that through the voltmeter. Hence the voltmeter reading divided by the ammeter reading will not give the true value of the resistance R for either methods of connection.

36.14   Multi-choice questions on electrical measuring instruments and measurements

(Answers on page 356.)

1. Which of the following would apply to a moving-coil instrument?
   (a) An uneven scale, measuring d.c.
   (b) An even scale, measuring a.c.
   (c) An uneven scale, measuring a.c.
   (d) An even scale, measuring d.c.
2. In question 1, which would refer to a moving-iron instrument?

3. In question 1, which would refer to a moving-coil rectifier instrument?
4. Which of the following is needed to extend the range of a milliammeter to read voltages of the order of 100 V?
   (a) A parallel high-value resistance
   (b) A series high-value resistance
   (c) A parallel low-value resistance
   (d) A series low-value resistance
5. Figure 36.21 shows a scale of a multi-range ammeter. What is the current indicated when switched to a 25 A scale?
   (a) 84 A   (b) 5.6 A   (c) 14 A   (d) 8.4 A

A sinusoidal waveform is displayed on a c.r.o. screen. The peak-to-peak distance is 5 cm and the distance between cycles is 4 cm. The 'variable' switch is on 100 μs/cm and the 'volts/cm' switch is on 10 V/cm. In questions 6 to 10, select the correct answer from the following:
(a) 25 V   (b) 5 V   (c) 0.4 ms
(d) 35.4 V   (e) 4 ms   (f) 50 V
(g) 250 Hz   (h) 2.5 V   (i) 2.5 kHz
(j) 17.7 V

6. Determine the peak-to-peak voltage.
7. Determine the periodic time of the waveform.
8. Determine the maximum value of the voltage.
9. Determine the frequency of the waveform.
10. Determine the r.m.s. value of the waveform.

Figure 36.22 shows double-beam c.r.o. waveform traces. For the quantities stated in questions 11 to 17, select the correct answer from the following:
(a) 30 V   (b) 0.2 s   (c) 50 V   (d) $(15/\sqrt{2})$ V
(e) 54° leading   (f) $(250/\sqrt{2})$ V   (g) 15 V
(h) 100 μs   (i) $(50/\sqrt{2})$ V   (j) 250 V   (k) 10 kHz

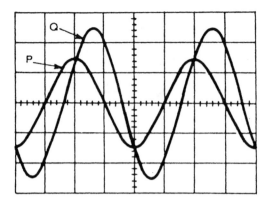

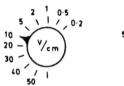

Figure 36.22

(l) 75 V   (m) 40 μs   (n) (3π/10) rad lagging
(o) (25/√2) V   (p) 5 kHz   (q) (30/√2) V
(r) 25 kHz   (s) (75/√2) V   (t) (3π/10) rad leading

11. Amplitude of waveform P
12. Peak-to-peak value of waveform Q
13. Periodic time of both waveforms
14. Frequency of both waveforms
15. R.m.s. value of waveform P
16. R.m.s. value of waveform Q
17. Phase displacement of waveform Q relative to waveform P

36.15   Short answer questions on electrical measuring instruments and measurements

1. What is the main difference between an analogue and a digital type measuring instrument?
2. Name the three essential devices for all analogue electrical indicating instruments.
3. Complete the following statements:
   (a) An ammeter has a .... resistance and is connected .... with the circuit.
   (b) A voltmeter has a .... resistance and is connected .... with the circuit.
4. State two advantages and two disadvantages of a moving-coil instrument.
5. What effect does the connection of (a) a shunt (b) a multiplier have on a milliammeter?
6. State two advantages and two disadvantages of a moving-iron instrument.
7. Briefly explain the principle of operation of an ohmmeter.

8. Name a type of ohmmeter used for measuring (a) low resistance values (b) high resistance values.
9. What is a multimeter?
10. When may a rectifier instrument be used in preference to either the moving-coil or moving-iron instrument?
11. What is the principle of the Wheatstone bridge?
12. How may a d.c. potentiometer be used to measure p.d.s?
13. What is meant by a null method of measurement?
14. Define 'calibration accuracy' as applied to a measuring instrument.
15. State three main areas where errors are most likely to occur in measurements.
16. Name five quantities that a c.r.o. is capable of measuring.

36.16   Further questions on electrical measuring instruments and measurements

1. A moving-coil instrument gives f.s.d. for a current of 10 mA. Neglecting the resistance of the instrument, calculate the approximate value of series resistance needed to enable the instrument to measure up to (a) 20 V (b) 100 V (c) 250 V.
   [(a) 2 kΩ (b) 10 kΩ (c) 25 kΩ]
2. A meter of resistance 50 Ω has a f.s.d. of 4 mA. Determine the value of shunt resistance required in order that f.s.d. should be (a) 15 mA (b) 20 A (c) 100 A.
   [(a) 18.18 Ω (b) 10.00 mΩ (c) 2.00 mΩ]

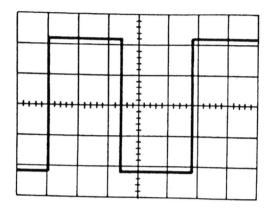

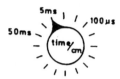

Figure 36.23

3. A moving-coil instrument having a resistance of 20 Ω gives a f.s.d. when the current is 5 mA. Calculate the value of the multiplier to be connected in series with the instrument so that it can be used as a voltmeter for measuring p.d.s up to 200 V.   [39.98 kΩ]

4. A moving-coil instrument has a f.s.d. current of 20 mA and a resistance of 25 Ω. Calculate the values of resistance required to enable the instrument to be used (a) as a 0–10 A ammeter, and (b) as a 0–100 V voltmeter. State the mode of resistance connection in each case.
   [(a) 50.10 mΩ in parallel (b) 4.975 kΩ in series]

5. A meter has a resistance of 40 Ω and registers a maximum deflection when a current of 15 mA flows. Calculate the value of resistance that converts the movement into (a) an ammeter with a maximum deflection of 50 A (b) a voltmeter with a range 0–250 V.
   [(a) 12.00 mΩ in parallel (b) 16.63 kΩ in series]

6. (a) Describe, with the aid of diagrams, the principle of operation of a moving-iron instrument.
   (b) Draw a circuit diagram showing how a moving-coil instrument may be used to measure alternating current.
   (c) Discuss the advantages and disadvantages of moving-coil rectifier instruments when compared with moving-iron instruments.

7. (a) Describe, with the aid of a diagram, the principle of the Wheatstone bridge and hence deduce the balance condition giving the unknown resistance in terms of known values of resistance.
   (b) In a Wheatstone bridge PQRS, a galvanometer is connected between Q and S and a voltage source between P and R. An unknown resistor R_x is connected between P and Q. When the bridge is balanced, the resistance between Q and R is 200 Ω, that between

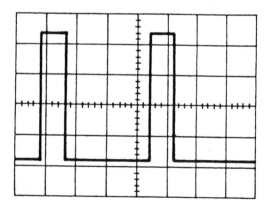

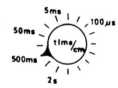

Figure 36.24

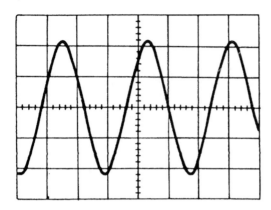

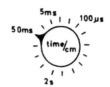

Figure 36.25

R and S is 10 Ω and that between S and P is 150 Ω. Calculate the value of R_x.

[3 kΩ]

8. (a) Describe, with the aid of a diagram, how a d.c. potentiometer can be used to measure the e.m.f. of a cell.

(b) Balance is obtained in a d.c. potentiometer at a length of 31.2 cm when using a standard cell of 1.0186 volts. Calculate the e.m.f. of a dry cell if balance is obtained with a length of 46.7 cm.

[1.525 V]

9. List the errors most likely to occur in the measurements of electrical quantities. A 240 V supply is connected across a load resistance R. Also connected across R is a voltmeter having a f.s.d. of 300 V and a figure or merit (i.e. sensitivity) of 8 kΩ/V. Calculate the power dissipated by the voltmeter and by the load resistance if (a) $R = 100$ Ω (b) $R = 1$ MΩ. Comment on the results obtained.

[(a) 24 mW, 576 W (b) 24 mW, 57.6 mW]

10. A 0–1 A ammeter having a resistance of 50Ω is used to measure the current flowing in a 1 kΩ resistor when the supply voltage is 250 V. Calculate: (a) the approximate value of current (neglecting the ammeter resistance), (b) the actual current in the circuit, (c) the power dissipated in the ammeter, (d) the power dissipated in the 1 kΩ resistor.

[(a) 0.250 A (b) 0.238 A (c) 2.832 W (d) 56.64 W]

11. For the square voltage waveform displayed on a c.r.o. shown in Fig. 36.23, find (a) its frequency, (b) its peak-to-peak voltage.

[(a) 41.7 Hz (b) 176 V]

12. For the pulse waveform shown in Fig. 36.24, find (a) its frequency, (b) the magnitude of the pulse voltage.

[(a) 0.56 Hz (b) 8.4 V]

13. For the sinusoidal waveform shown in Fig. 36.25, determine (a) its frequency, (b) the peak-to-peak voltage, (c) the r.m.s. voltage.

[(a) 7.14 Hz (b) 220 V (c) 77.78 V]

37

The measurement of temperature

At the end of this chapter you should be able to:

- describe the construction, principle of operation and practical applications of the following temperature measuring devices:

    (a) liquid-in-glass thermometer (including advantages of mercury, and sources of error)

    (b) thermocouples (including advantages and sources of errors)

    (c) resistance thermometer (including limitations and advantages of platinum coil)

    (d) thermistors

    (e) pyrometers (total radiation and optical types, including advantages and disadvantages)

- describe the principles of operation of

    (a) temperature indicating paints and crayons

    (b) bimetallic thermometers

    (c) mercury-in-steel thermometer

    (d) gas thermometer

- select the appropriate temperature measuring device for a particular application

37.1 Introduction

A change in temperature of a substance can often result in a change in one or more of its physical properties. Thus, although temperature cannot be measured directly, its effects can be measured. Some properties of substances used to determine changes in temperature include changes in dimensions, electrical resistance, state, type and volume of radiation and colour.

Temperature measuring devices available are many and varied. Those described in sections 37.2 to 37.10 are those most often used in science and industry.

37.2 Liquid-in-glass thermometer

A **liquid-in-glass thermometer** uses the expansion of a liquid with increase in temperature as its principle of operation.

Construction

A typical liquid-in-glass thermometer is shown in Fig. 37.1 and consists of a sealed stem of uniform small-bore tubing, called a capillary tube, made of glass, with a cylindrical glass bulb formed at one end. The bulb and part of the stem are filled

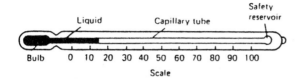

Figure 37.1

with a liquid such as mercury or alcohol and the remaining part of the tube is evacuated. A temperature scale is formed by etching graduations on the stem. A safety reservoir is usually provided, into which the liquid can expand without bursting the glass if the temperature is raised beyond the upper limit of the scale.

Principle of operation

The operation of a liquid-in-glass thermometer depends on the liquid expanding with increase in temperature and contracting with decrease in temperature. The position of the end of the column of liquid in the tube is a measure of the temperature of the liquid in the bulb – shown as 15°C in Fig. 37.1, which is about room temperature. Two fixed points are needed to calibrate the thermometer, with the interval between these points being divided into 'degrees'. In the first thermometer, made by Celsius, the fixed points chosen were the temperature of melting ice (0°C) and that of boiling water at standard atmospheric pressure (100°C), in each case the blank stem being marked at the liquid level. The distance between these two points, called the fundamental interval, was divided into 100 equal parts, each equivalent to 1°C, thus forming the scale.

The **clinical thermometer**, with a limited scale around body temperature, the **maximum and/or minimum thermometer**, recording the maximum day temperature and minimum night temperature, and the **Beckman** thermometer, which is used only in accurate measurement of temperature change and has no fixed points, are particular types of liquid-in-glass thermometer which all operate on the same principle.

Advantages

The liquid-in-glass thermometer is simple in construction, relatively inexpensive, easy to use

and portable, and is the most widely used method of temperature measurement having industrial, chemical, clinical and meteorological applications.

Disadvantages

Liquid-in-glass thermometers tend to be fragile and hence easily broken, can only be used where the liquid column is visible, cannot be used for surface temperature measurements, cannot be read from a distance and are unsuitable for high temperature measurements.

Advantages of mercury

The use of mercury in a thermometer has many advantages, for mercury:

(i)   is clearly visible,
(ii) has a fairly uniform rate of expansion,
(iii) is readily obtainable in the pure state,
(iv) does not 'wet' the glass,
(v)   is a good conductor of heat.

Mercury has a freezing point of –39°C and cannot be used in a thermometer below this temperature. Its boiling point is 357°C but before this temperature is reached some distillation of the mercury occurs if the space above the mercury is a vacuum. To prevent this, and to extend the upper temperature limits to over 500°C, an inert gas such as nitrogen under pressure is used to fill the remainder of the capillary tube. Alcohol, often dyed red to be seen in the capillary tube, is considerably cheaper than mercury and has a freezing point of –113°C, which is considerably lower than for mercury. However it has a low boiling point at about 79°C.

Errors

Typical errors in liquid-in-glass thermometers may occur due to:

(i)   the slow cooling rate of glass,
(ii) incorrect positioning of the thermometer,
(iii) a delay in the thermometer becoming steady (i.e. slow response time),
(iv) non-uniformity of the bore of the capillary tube, which means that equal intervals marked on the stem do not correspond to equal temperature intervals.

37.3 Thermocouples

Thermocouples use the e.m.f. set up when the junction of two dissimilar metals is heated.

Principle of operation

At the junction between two different metals, say, copper and constantan, there exists a difference in electrical potential, which varies with the temperature of the junction. This is known as the 'thermo-electric effect'. If the circuit is completed with a second junction at a different temperature, a current will flow round the circuit. This principle is used in the thermocouple. Two different metal conductors having their ends twisted together are shown in Fig. 37.2. If the two junctions are at different temperatures, a current I flows round the circuit.

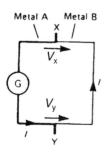

Figure 37.2

The deflection on the galvanometer G depends on the difference in temperature between junctions X and Y and is caused by the difference between voltages V_X and V_Y. The higher temperature junction is usually called the 'hot junction' and the lower temperature junction the 'cold junction'. If the cold junction is kept at a constant known temperature, the galvanometer can be calibrated to indicate the temperature of the hot junction directly. The cold junction is then known as the reference junction.

In many instrumentation situations, the measuring instrument needs to be located far from the point at which the measurements are to be made. Extension leads are then used, usually made of the same material as the thermocouple but of smaller gauge. The reference junction is then effectively moved to their ends. The thermo-couple is used by positioning the hot junction where the temperature is required. The meter will indicate the temperature of the hot junction only if the reference junction is at 0°C for:

(temperature of hot junction) =
(temperature of the cold junction)
+ (temperature difference)

In a laboratory the reference junction is often placed in melting ice, but in industry it is often positioned in a thermostatically controlled oven or buried underground where the temperature is constant.

Construction

Thermocouple junctions are made by twisting together two wires of dissimilar metals before welding them. The construction of a typical copper–constantan thermocouple for industrial use is shown in Fig. 37.3. Apart from the actual junction the two conductors used must be insulated electrically from each other with appropriate insulation and is shown in Fig. 37.3 as twin-holed tubing. The wires and insulation are usually inserted into a sheath for protection from environments in which they might be damaged or corroded. A copper–constantan thermocouple can measure temperature from –250°C up to about 400°C, and is used typically with boiler flue gases, food processing and with sub-zero temperature measurement.

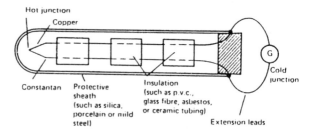

Figure 37.3

Applications

An iron–constantan thermocouple can measure temperature from –200°C to about 850°C, and is used typically in paper and pulp mills, re-heat and annealing furnaces and in chemical reactors. A chromel–alumel thermocouple can measure temperatures from –200°C to about 1100°C and is used typically with blast furnace gases, brick kilns and in glass manufacture.

For the measurement of temperatures above 1100°C radiation pyrometers are normally used. However, thermocouples are available made of platinum–platinum/rhodium, capable of measuring temperatures up to 1400°C, or tungsten–molybdenum which can measure up to 2600°C.

Advantages

A thermocouple:

(i) has a very simple, relatively inexpensive construction,
(ii) can be made very small and compact,
(iii) is robust,
(iv) is easily replaced if damaged,
(v) has a small response time,
(vi) can be used at a distance from the actual measuring instrument and is thus ideal for use with automatic and remote-control systems.

Sources of error

Sources of error in the thermocouple which are difficult to overcome include:

(i) voltage drops in leads and junctions,
(ii) possible variations in the temperature of the cold junction,
(iii) stray thermoelectric effects, which are caused by the addition of further metals into the 'ideal' two-metal thermocouple circuit.

Additional leads are frequently necessary for extension leads or voltmeter terminal connections.

A thermocouple may be used with a battery- or mains-operated electronic thermometer instead of a millivoltmeter. These devices amplify the small e.m.f.s from the thermocouple before feeding them to a multi-range voltmeter calibrated directly with temperature scales. These devices have great accuracy and are almost unaffected by voltage drops in the leads and junctions.

Problem 1.   A chromel–alumel thermocouple generates an e.m.f. of 5 mV. Determine the temperature of the hot junction if the cold junction is at a temperature of 15°C and the sensitivity of the thermocouple is 0.04 mV/°C.

Temperature difference for 5 mV

$$= \frac{5\,\text{mV}}{0.04\,\text{mV/°C}} = 125°C$$

Temperature at hot junction

= temperature of cold junction + temperature difference

= 15°C + 125°C = **140°C**

37.4   Resistance thermometers

Resistance thermometers use the change in electrical resistance caused by temperature change.

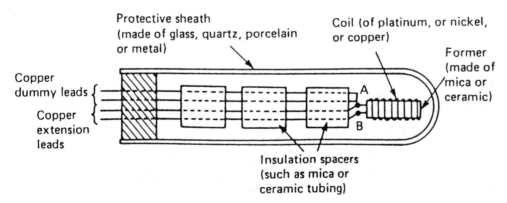

Figure 37.4

Construction

Resistance thermometers are made in a variety of sizes, shapes and forms depending on the application for which they are designed. A typical resistance thermometer is shown diagrammatically in Fig. 37.4. The most common metal used for the coil in such thermometers is platinum even though its sensitivity is not as high as other metals such as copper and nickel. However, platinum is a very stable metal and provides reproducible results in a resistance thermometer. A platinum resistance thermometer is often used as a calibrating device. Since platinum is expensive, connecting leads of another metal, usually copper, are used with the thermometer to connect it to a measuring circuit.

The platinum and the connecting leads are shown joined at A and B in Fig. 37.4, although sometimes this junction may be made outside of the sheath. However, these leads often come into close contact with the heat source which can introduce errors into the measurements. These may be eliminated by including a pair of identical leads, called dummy leads, which experience the same temperature change as the extension leads.

Principle of operation

With most metals a rise in temperature causes an increase in electrical resistance, and since resistance can be measured accurately this property can be used to measure temperature. If the resistance of a length of wire at $0°C$ is R_o, and its resistance at $\theta°C$ is R_θ, then $R_\theta = R_o (1 + \alpha\theta)$, where α is the temperature coefficient of resistance of the material (see Chapter 27).

Rearranging gives:

$$\text{temperature} \quad \boxed{\theta = \frac{R_\theta - R_o}{\alpha R_o}}$$

Values of R_o and α may be determined experimentally or obtained from existing data. Thus, if R_θ can be measured, temperature θ can be calculated. This is the principle of operation of a resistance thermometer. Although a sensitive ohmmeter can be used to measure R_θ, for more accurate determinations a Wheatstone bridge circuit is used as shown in Fig. 37.5. This circuit compares an unknown resistance R_θ with others

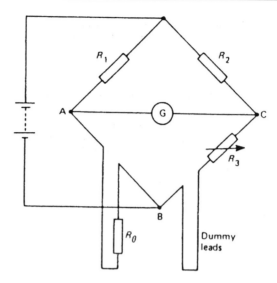

Figure 37.5

of known values, R_1 and R_2 being fixed values and R_3 being variable. Galvanometer G is a sensitive centre-zero microammeter. R_3 is varied until zero deflection is obtained on the galvanometer, i.e. no current flows through G and the bridge is said to be 'balanced'.

At balance

$$R_2 R_\theta = R_1 R_3$$

from which

$$\boxed{R_\theta = \frac{R_1 R_3}{R_2}}$$

and if R_1 and R_2 are of equal value, then $R_\theta = R_3$.

A resistance thermometer may be connected between points A and B in Fig. 37.5 and its resistance R_θ at any temperature θ accurately measured. Dummy leads included in arm BC help to eliminate errors caused by the extension leads which are normally necessary in such a thermometer.

Limitations

Resistance thermometers using a nickel coil are used mainly in the range $-100°C$ to $300°C$, whereas platinum resistance thermometers are capable of measuring with greater accuracy temperatures in the range $-200°C$ to about $800°C$.

This upper range may be extended to about 1500°C if high melting point materials are used for the sheath and coil construction.

Advantages and disadvantages of a platinum coil

Platinum is commonly used in resistance thermometers since it is chemically inert, i.e. unreactive, resists corrosion and oxidation and has a high melting point of 1769°C. A disadvantage of platinum is its slow response to temperature variation.

Applications

Platinum resistance thermometers may be used as calibrating devices or in applications such as heat treating and annealing processes and can be adapted easily for use with automatic recording or control systems. Resistance thermometers tend to be fragile and easily damaged especially when subjected to excessive vibration or shock.

Problem 2.   A platinum resistance thermometer has a resistance of 25 Ω at 0°C. When measuring the temperature of an annealing process a resistance value of 60 Ω is recorded. To what temperature does this correspond? Take the temperature coefficient of resistance of platinum as 0.0038/°C.

$R_\theta = R_0(1 + \alpha\theta)$, where $R_0 = 25\ \Omega$, $R_\theta = 60\ \Omega$ and $\alpha = 0.0038/°C$. Rearranging gives

$$\textbf{temperature } \theta = \frac{R_\theta - R_0}{\alpha R_0} = \frac{60 - 25}{(0.0038)(25)}$$

$$= \textbf{368.4°C}$$

37.5   Thermistors

A thermistor is a semiconducting material – such as mixtures of oxides of copper, manganese, cobalt, etc. – in the form of a fused bead connected to two leads. As its temperature is increased its resistance rapidly decreases. Typical resistance/temperature curves for a thermistor and common metals are shown in Fig. 37.6. The resistance of a typical thermistor can vary from 400 Ω at 0°C to 100 Ω at 140°C.

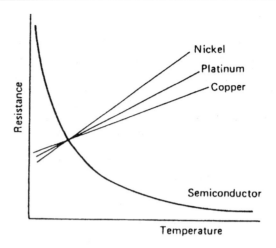

Figure 37.6

Advantages

The main advantages of a thermistor are its high sensitivity and small size. It provides an inexpensive method of measuring and detecting small changes in temperature.

37.6   Pyrometers

A pyrometer is a device for measuring very high temperatures and uses the principle that all substances emit radiant energy when hot, the rate of emission depending on their temperature. The measurement of thermal radiation is therefore a convenient method of determining the temperature of hot sources and is particularly useful in industrial processes. There are two main types of pyrometer, namely the total radiation pyrometer and the optical pyrometer.

Pyrometers are very convenient instruments since they can be used at a safe and comfortable distance from the hot source. Thus applications of pyrometers are found in measuring the temperature of molten metals, the interiors of furnaces or the interiors of volcanoes. Total radiation pyrometers can also be used in conjunction with devices which record and control temperature continuously.

Total radiation pyrometer

A typical arrangement of a **total radiation pyrometer** is shown in Fig. 37.7. Radiant energy from a hot source, such as a furnace, is focused

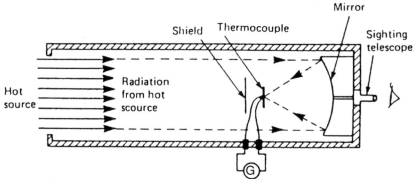

Figure 37.7

on to the hot junction of a thermocouple after reflection from a concave mirror. The temperature rise recorded by the thermocouple depends on the amount of radiant energy received, which in turn depends on the temperature of the hot source. The galvanometer G shown connected to the thermocouple records the current which results from the e.m.f. developed and may be calibrated to give a direct reading of the temperature of the hot source. The thermocouple is protected from direct radiation by a shield as shown and the hot source may be viewed through the sighting telescope. For greater sensitivity, a thermopile may be used, a thermopile being a number of thermocouples connected in series. Total radiation pyrometers are used to measure temperature in the range 700°C to 2000°C.

Optical pyrometers

When the temperature of an object is raised sufficiently two visual effects occur; the object appears brighter and there is a change in colour of the light emitted. These effects are used in the optical pyrometer where a comparison or matching is made between the brightness of the glowing hot source and the light from a filament of known temperature.

The most frequently used optical pyrometer is the disappearing filament pyrometer and a typical arrangement is shown in Fig. 37.8. A filament lamp is built into a telescope arrangement which receives radiation from a hot source, an image of which is seen through an eyepiece. A red filter is incorporated as a protection to the eye.

The current flowing through the lamp is controlled by a variable resistor. As the current is increased the temperature of the filament increases and its colour changes. When viewed through the eyepiece the filament of the lamp appears superimposed on the image of the radiant energy from the hot source. The current is varied until the filament glows as brightly as the background. It will then merge into the

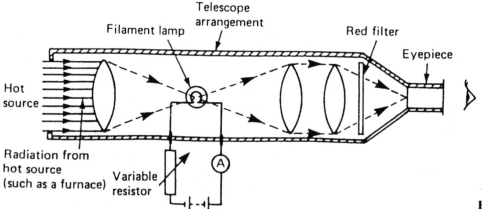

Figure 37.8

background and seem to disappear. The current required to achieve this is a measure of the temperature of the hot source and the ammeter can be calibrated to read the temperature directly. Optical pyrometers may be used to measure temperatures up to, and even in excess of, 3000°C.

Advantages of pyrometers

(i)   There is no practical limit to the temperature that a pyrometer can measure.
(ii) A pyrometer need not be brought directly into the hot zone and so is free from the effects of heat and chemical attack that can often cause other measuring devices to deteriorate in use.
(iii) Very fast rates of change of temperature can be followed by a pyrometer.
(iv) The temperature of moving bodies can be measured.
(v)   The lens system makes the pyrometer virtually independent of its distance from the source.

Disadvantages of pyrometers

(i)   A pyrometer is often more expensive than other temperature measuring devices.
(ii) A direct view of the heat process is necessary.
(iii) Manual adjustment is necessary.
(iv) A reasonable amount of skill and care is required in calibrating and using a pyrometer. For each new measuring situation the pyrometer must be re-calibrated.
(v)   The temperature of the surroundings may affect the reading of the pyrometer and such errors are difficult to eliminate.

37.7   Temperature indicating paints and crayons

Temperature indicating paints contain substances which change their colour when heated to certain temperatures. This change is usually due to chemical decomposition, such as loss of water, in which the change in colour of the paint after having reached the particular temperature will be a permanent one. However, in some types the original colour returns after cooling.

Temperature indicating paints are used where the temperature of inaccessible parts of apparatus and machines is required. They are particularly useful in heat-treatment processes where the temperature of the component needs to be known before a quenching operation. There are several such paints available and most have only a small temperature range so that different paints have to be used for different temperatures. The usual range of temperatures covered by these paints is from about 30°C to 700°C.

Temperature sensitive crayons consist of fusible solids compressed into the form of a stick. The melting point of such crayons is used to determine when a given temperature has been reached. The crayons are simple to use but indicate a single temperature only, i.e. its melting point temperature. There are over 100 different crayons available, each covering a particular range of temperature. Crayons are available for temperatures within the range of 50°C to 1400°C. Such crayons are used in metallurgical applications such as preheating before welding, hardening, annealing or tempering, or in monitoring the temperature of critical parts of machines or for checking mould temperatures in the rubber and plastics industries.

37.8   Bimetallic thermometers

Bimetallic thermometers depend on the expansion of metal strips which operate an indicating pointer. Two thin metal strips of differing thermal expansion are welded or riveted together and the curvature of the bimetallic strip changes with temperature change. For greater sensitivity the strips may be coiled into a flat spiral or helix, one end being fixed and the other being made to rotate a pointer over a scale. Bimetallic thermometers are useful for alarm and overtemperature applications where extreme accuracy is not essential. If the whole is placed in a sheath, protection from corrosive environments is achieved but with a reduction in response characteristics. The normal upper limit of temperature measurement by this thermometer is about 200°C, although with special metals the range can be extended to about 400°C.

37.9   Mercury-in-steel thermometer

The **mercury-in-steel thermometer** is an extension of the principle of the mercury-in-glass thermometer. Mercury in a steel bulb expands via a small bore capillary tube into a pressure indicating device, say a Bourdon gauge, the position of the pointer indicating the amount of expansion and thus the temperature. The advantages of this instrument are that it is robust and, by increasing the length of the capillary tube, the gauge can be placed some distance from the bulb and can thus be used to monitor temperatures in positions which are inaccessible to the liquid-in-glass thermometer. Such thermometers may be used to measure temperatures up to 600°C.

37.10 Gas thermometers

The **gas thermometer** consists of a flexible U-tube of mercury connected by a capillary tube to a vessel containing gas. The change in the volume of a fixed mass of gas at constant pressure, or the change in pressure of a fixed mass of gas at constant volume, may be used to measure temperature. This thermometer is cumbersome and rarely used to measure temperature directly, but it is often used as a standard with which to calibrate other types of thermometer. With pure hydrogen the range of the instrument extends from –240°C to 1500°C and measurements can be made with extreme accuracy.

37.11 Choice of measuring device

Problem 3. State which device would be most suitable to measure the following:

(a) metal in a furnace, in the range 50°C to 1600°C
(b) the air in an office in the range 0°C to 40°C
(c) boiler flue gas in the range 15°C to 300°C
(d) a metal surface, where a visual indication is required when it reaches 425°C
(e) materials in a high-temperature furnace in the range 2000°C to 2800°C
(f) to calibrate a thermocouple in the range –100°C to 500°C
(g) brick in a kiln up to 900°C
(h) an inexpensive method for food processing applications in the range –25°C to –75°C

(a) Radiation pyrometer
(b) Mercury-in-glass thermometer
(c) Copper–constantan thermocouple
(d) Temperature sensitive crayon
(e) Optical pyrometer
(f) Platinum resistance thermometer or gas thermometer
(g) Chromel–alumel thermocouple
(h) Alcohol-in-glass thermometer

37.12 Multi-choice questions on the measurement of temperature

(Answers on page 356.)

1. The most suitable device for measuring very small temperature changes is a (a) thermopile (b) thermocouple (c) thermistor
2. When two wires of different metals are twisted together and heat applied to the junction, an e.m.f. is produced. This effect is used in a thermocouple to measure:
   (a) e.m.f. (b) temperature (c) expansion (d) heat
3. A cold junction of a thermocouple is at room temperature of 15°C. A voltmeter connected to the thermocouple circuit indicates 10 mV. If the voltmeter is calibrated as 20°C/mV, the temperature of the hot source is:
   (a) 185°C (b) 200°C (c) 35°C (d) 215°C
4. The e.m.f. generated by a copper–constantan thermometer is 15 mV. If the cold junction is at a temperature of 20°C, the temperature of the hot junction when the sensitivity of the thermocouple is 0.03 mV/°C is:
   (a) 480°C (b) 520°C (c) 20.45°C (d) 500°C

In questions 5 to 12, select the most appropriate temperature measuring device from this list.
(a) copper–constantan thermocouple
(b) thermistor
(c) mercury-in-glass thermometer
(d) total radiation pyrometer
(e) platinum resistance thermometer
(f) gas thermometer
(g) temperature sensitive crayon
(h) alcohol-in-glass thermometer
(i) bimetallic thermometer
(j) mercury-in-steel thermometer
(k) optical pyrometer

5. Overtemperature alarm at about 180°C
6. Food processing plant in the range –250°C to +250°C

7. Automatic recording system for a heat treating process in the range 90°C to 250°C
8. Surface of molten metals in the range 1000°C to 1800°C
9. To calibrate accurately a mercury-in-glass thermometer
10. Furnace up to 3000°C
11. Inexpensive method of measuring very small changes in temperature
12. Metal surface where a visual indication is required when the temperature reaches 520°C

37.13 Short answer questions on the measurement of temperature

For each of the temperature measuring devices listed in 1 to 10, state very briefly its principle of operation and the range of temperatures that it is capable of measuring.

1. Mercury-in-glass thermometer
2. Alcohol-in-glass thermometer
3. Thermocouple
4. Platinum resistance thermometer
5. Total radiation pyrometer
6. Optical pyrometer
7. Temperature sensitive crayons
8. Bimetallic thermometer
9. Mercury-in-steel thermometer
10. Gas thermometer

37.14 Further questions on the measurement of temperature

For each of the temperature measuring devices listed in questions 1 to 10, (a) describe, with appropriate sketches, their construction and state how they operate, (b) state their characteristics and range, (c) discuss their limitations, advantages and disadvantages, (d) state typical applications where they may be used.

1. Thermocouple
2. Mercury-in-glass thermometer
3. Optical pyrometer
4. Platinum resistance thermometer
5. Alcohol-in-glass thermometer
6. Gas thermometer
7. Total radiation pyrometer
8. Temperature indicating paints and temperature sensitive crayons
9. Mercury-in-steel thermometer
10. Bimetallic thermometer
11. A platinum–platinum/rhodium thermocouple generates an e.m.f. of 7.5 mV. If the cold junction is at a temperature of 20°C, determine the temperature of the hot junction. Assume the sensitivity of the thermocouple to be 6 μV/°C.   [1270°C]
12. Explain how temperature is measured using a resistance thermometer incorporating a Wheatstone bridge circuit. Explain why dummy leads are used in such a circuit.
    Why is platinum most often used in a resistance thermometer?
13. A platinum resistance thermometer has a resistance of 100 Ω at 0°C. When measuring the temperature of a heat process a resistance value of 177 Ω is measured using a Wheatstone bridge. Given that the temperature coefficient of resistance of platinum is 0.0038/°C, determine the temperature of the heat process correct to the nearest degree.

[203°C]

38

The measurement of fluid flow

At the end of this chapter you should be able to:

- appreciate the importance of the measurement of fluid flow

- understand the construction, principles of operation and practical applications of the following differential pressure flowmeters:

  (a) orifice plate (including advantages and disadvantages)

  (b) Venturi tube (including advantages and disadvantages)

  (c) flow nozzles

  (d) Pitot-static tube (including advantages and disadvantages)

- understand the construction, principle of operation and practical applications of the following mechanical flowmeters:

  (a) deflecting vane flowmeter

  (b) turbine type meters (cup anemometer, rotary vane positive displacement meter and turbine flowmeter)

- understand the construction, principle of operation and practical applications of:

  (a) float and tapered-tube meter

  (b) electromagnetic flowmeter

  (c) hot wire anemometer

- select the appropriate flowmeter for a particular application

38.1 Introduction

The measurement of fluid flow is of great importance in many industrial processes, some examples including air flow in the ventilating ducts of a coal mine, the flow rate of water in a condenser at a power station, the flow rate of liquids in chemical processes, the control and monitoring of the fuel, lubricating and cooling fluids of ships and aircraft engines, and so on. Fluid flow is one of the most difficult of industrial measurements to carry out, since flow behaviour depends on a great many variables concerning the physical properties of a fluid.

There are available a large number of fluid flow measuring instruments generally called **flowmeters**, which can measure the flow rate of liquids (in m^3/s) or the mass flow rate of gaseous fluids (in kg/s). The two main categories of flowmeters are differential pressure flowmeters and mechanical flowmeters.

38.2 Differential pressure flowmeters

When certain flowmeters are installed in pipelines they often cause an obstruction to the fluid flowing in the pipe by reducing the cross-

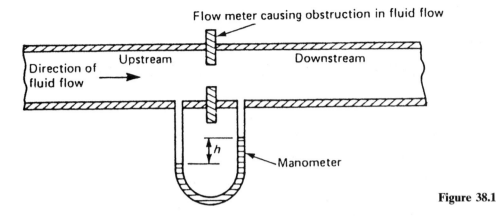

Figure 38.1

sectional area of the pipeline. This causes a change in the velocity of the fluid, with a related change in pressure. Fig. 38.1 shows a section through a pipeline into which a flowmeter has been inserted. The flow rate of the fluid may be determined from a measurement of the difference between the pressures on the walls of the pipe at specified distances upstream and downstream of the flowmeter. Such devices are known as **differential pressure flowmeters**.

The pressure difference in Fig. 38.1 is measured using a manometer connected to appropriate pressure tapping points. The pressure is seen to be greater upstream of the flowmeter than downstream, the pressure difference being shown as h.

Calibration of the manometer depends on the shape of the obstruction, the positions of the pressure tapping points and the physical properties of the fluid.

In industrial applications the pressure difference is detected by a differential pressure cell, the output from which is either an amplified pressure signal or an electrical signal.

Examples of differential pressure flowmeters commonly used include:

(a) Orifice plate (see section 38.3)
(b) Venturi tube (see section 38.4)
(c) Flow nozzles (see section 38.5)
(d) Pitot-static tube (see section 38.6)

British Standard reference BS 1042: Part 1: 1964 and Part 2A: 1973 'Methods for the measurement of fluid flow in pipes' gives specifications for measurement, manufacture, tolerances, accuracy, sizes, choice, and so on, of differential flowmeters.

38.3 Orifice plate

Construction

An orifice plate consists of a circular, thin, flat plate with a hole (or orifice) machined through its centre to fine limits of accuracy. The orifice has a diameter less than the pipeline into which the plate is installed and a typical section of an installation is shown in Fig. 38.2(a). Orifice plates are manufactured in stainless steel, monel metal, polyester glass fibre, and for large pipes, such as sewers or hot gas mains, in brick and concrete.

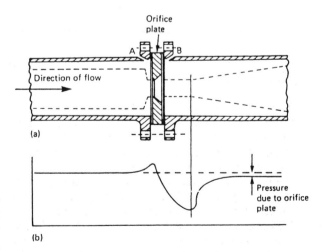

Figure 38.2

Principles of operation

When a fluid moves through a restriction in a pipe, the fluid accelerates and a reduction in pressure occurs, the magnitude of which is related to the flow rate of the fluid. The variation of pressure near an orifice plate is shown in Fig. 38.2(b). The position of minimum pressure is located downstream from the orifice plate where the flow stream is narrowest. This point of minimum cross-sectional area of the jet is called the 'vena contracta'. Beyond this point the pressure rises but does not return to the original upstream value and there is a permanent pressure loss. This loss depends on the size and type of orifice plate, the positions of the upstream and downstream pressure tappings and the change in fluid velocity between the pressure tappings which depends on the flow rate and the dimensions of the orifice plate.

In Fig. 38.2(a) corner pressure tappings are shown at A and B. Alternatively, with an orifice plate inserted into a pipeline of diameter d, pressure tappings are often located at distances of d and $d/2$ from the plate respectively upstream and downstream. At distance d upstream the flow pattern is not influenced by the presence of the orifice plate and distance $d/2$ coincides with the vena contracta.

Advantages of orifice plates

(i)   They are relatively inexpensive.
(ii) They are usually thin enough to fit between an existing pair of pipe flanges.

Disadvantages of orifice plates

(i)   The sharpness of the edge of the orifice can become worn with use, causing calibration errors.
(ii) The possible build-up of matter against the plate.
(iii) A considerable loss in the pumping efficiency due to the pressure loss downstream of the plate.

Applications

Orifice plates are usually used in medium and large pipes and are best suited to the indication

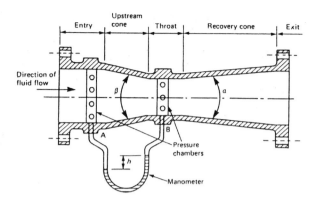

Figure 38.3

and control of essentially constant flow rates. Several applications are found in the general process industries.

38.4   Venturi tube

Construction

The Venturi tube or venturimeter is an instrument for measuring with accuracy the flow rate of fluids in pipes. A typical arrangement of a section through such a device is shown in Fig. 38.3, and consists of a short converging conical tube called the inlet or upstream cone, leading to a cylindrical portion called the throat. This is followed by a diverging section called the outlet or recovery cone. The entrance and exit diameter is the same as that of the pipeline into which it is installed. Angle β is usually a maximum of 21°, giving a taper of $\beta/2$ of $10\frac{1}{2}°$. The length of the throat is made equal to the diameter of the throat. Angle α is about 5° to 7° to ensure a minimum loss of energy but where this is unimportant α can be as large as 14° to 15°.

Pressure tappings are made at the entry (at A) and at the throat (at B) and the pressure difference h which is measured using a manometer, a differential pressure cell or similar gauge, is dependent on the flow rate through the meter. Usually pressure chambers are fitted around the entrance pipe and the throat circumference with a series of tapping holes made in the chamber to which the manometer is connected. This ensures that an average pressure is recorded. The loss of energy due to turbulence which occurs just downstream with an orifice plate is largely avoided in the venturimeter due to the gradual divergence beyond the throat.

Venturimeters are usually made a permanent installation in a pipeline and are manufactured usually from stainless steel, cast iron, monel metal or polyester glass fibre.

Advantages of venturimeters

(i)   High accuracy results are possible.
(ii) There is a low pressure loss in the tube (typically only 2% to 3% in a well proportioned tube).
(iii) Venturimeters are unlikely to trap any matter from the fluid being metered.

Disadvantages of venturimeters

(i) High manufacturing costs.
(ii) The installation tends to be rather long (typically 120 mm for a pipe of internal diameter 50 mm).

38.5 Flow nozzle

The flow nozzle lies between an orifice plate and the venturimeter both in performance and cost. A typical section through a flow nozzle is shown in Fig. 38.4 where pressure tappings are located immediately adjacent to the upstream and downstream faces of the nozzle (i.e. at points A and B). The fluid flow does not contract any further as it leaves the nozzle and the pressure loss created is considerably less than that occurring with orifice plates. Flow nozzles are suitable for use with high velocity flows for they do not suffer the wear that occurs in orifice plate edges during such flows.

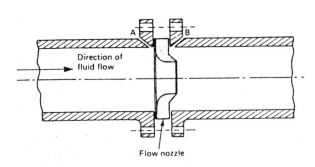

Figure 38.4

38.6 Pitot-static tube

A Pitot-static tube is a device for measuring the velocity of moving fluids or of the velocity of bodies moving through fluids. It consists of one tube, called the Pitot tube, with an open end facing the direction of the fluid motion, shown as pipe R in Fig. 38.5, and a second tube, called the piezometer tube, with the opening at 90° to the fluid flow, shown as T in Fig. 38.5. Pressure recorded by a pressure gauge moving with the flow, i.e. static or stationary relative to the fluid, is called free stream pressure and connecting a pressure gauge to a small hole in the wall of a pipe, such as point T in Fig. 38.5, is the easiest method of recording this pressure. The difference in pressure $(p_R - p_T)$, shown as h in the manometer of Fig. 38.5, is an indication of the speed of the fluid in the pipe.

Figure 38.6 shows a practical Pitot-static tube consisting of a pair of concentric tubes. The centre tube is the impact probe which has an open end which faces 'head-on' into the flow. The

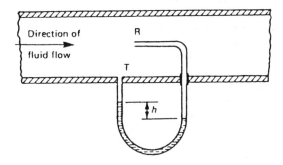

Figure 38.5

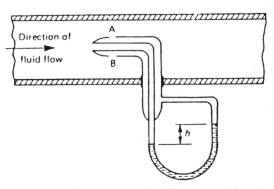

Figure 38.6

outer tube has a series of holes around its circumference located at right angles to the flow, as shown by A and B in Fig. 38.6. The manometer, showing a pressure difference of h, may be calibrated to indicate the velocity of flow directly.

Applications

A Pitot-static tube may be used for both turbulent and non-turbulent flow. The tubes can be made very small compared with the size of the pipeline and the monitoring of flow velocity at particular points in the cross-section of a duct can be achieved. The device is generally unsuitable for routine measurements and in industry is often used for making preliminary tests of flow rate in order to specify permanent flow measuring equipment for a pipeline. The main use of Pitot tubes is to measure the velocity of solid bodies moving through fluids, such as the velocity of ships. In these cases, the tube is connected to a Bourdon pressure gauge which can be calibrated to read velocity directly. A development of the Pitot tube, a pitometer, tests the flow of water in water mains and detects leakages.

Advantages of Pitot-static tubes

(i) They are inexpensive devices.
(ii) They are easy to install.
(iii) They produce only a small pressure loss in the tube.
(iv) They do not interrupt the flow.

Disadvantages of Pitot-static tubes

(i) Due to the small pressure difference, they are only suitable for high velocity fluids.
(ii) They can measure the flow rate only at a particular position in the cross-section of the pipe.
(iii) They easily become blocked when used with fluids carrying particles.

38.7   Mechanical flowmeters

With mechanical flowmeters a sensing element situated in a pipeline is displaced by the fluid flowing past it.

Examples of mechanical flowmeters commonly used include:

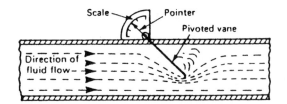

Figure 38.7

(a) Deflecting vane flowmeter (see section 38.8)
(b) Turbine type meters (see section 38.9)

38.8   Deflecting vane flowmeter

The deflecting vane flowmeter consists basically of a pivoted vane suspended in the fluid flow stream as shown in Fig. 38.7.

When a jet of fluid impinges on the vane it deflects from its normal position by an amount proportional to the flow rate. The movement of the vane is indicated on a scale which may be calibrated in flow units. This type of meter is normally used for measuring liquid flow rates in open channels or for measuring the velocity of air in ventilation ducts. The main disadvantages of this device is that it restricts the flow rate and it needs to be recalibrated for fluids of differing densities.

38.9   Turbine type meters

Turbine type flowmeters are those which use some form of multi-vane rotor and are driven by the fluid being investigated. Three such devices are the cup anemometer, the rotary vane positive displacement meter and the turbine flowmeter.

(a) **Cup anemometer**. An anemometer is an instrument which measures the velocity of moving gases and is most often used for the measurement of wind speed. The cup anemometer has three or four cups of hemispherical shape mounted at the end of arms radiating horizontally from a fixed point. The cup system spins round the vertical axis with a speed approximately proportional to the velocity of the wind. With the aid of a mechanical and/or electrical counter the wind speed can be determined and the device is easily adapted for automatic recording.

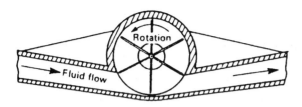

Figure 38.8

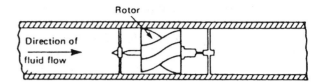

Figure 38.9

(b) **Rotary vane positive displacement meters** measure the flow rate by indicating the quantity of liquid flowing through the meter in a given time. A typical such device is shown in section in Fig. 38.8 and consists of a cylindrical chamber into which is placed a rotor containing a number of vanes (six in this case). Liquid entering the chamber turns the rotor and a known amount of liquid is trapped and carried round to the outlet. If x is the volume displaced by one blade then for each revolution of the rotor in Fig. 38.8 the total volume displaced is $6x$. The rotor shaft may be coupled to a mechanical counter and electrical devices which may be calibrated to give flow volume. This type of meter in its various forms is used widely for the measurement of domestic and industrial water consumption, for the accurate measurement of petrol in petrol pumps and for the consumption and batch control measurements in the general process and food industries for measuring flows as varied as solvents, tar and molasses (i.e. thickish treacle).

(c) A **turbine flowmeter** contains in its construction a rotor to which blades are attached which spin at a velocity proportional to the velocity of the fluid which flows through the meter. A typical section through such a meter is shown in Fig. 38.9. The number of revolutions made by the turbine blades may be determined by a mechanical or electrical device enabling the flow rate or total flow to be determined. Advantages of turbine flowmeters include a compact durable form, high accuracy, wide temperature and pressure capability and good response characteristics. Applications include the volumetric measurement of both crude and refined petroleum products in pipelines up to 600 mm bore, and in the water, power, aerospace, process and food industries, and

with modification may be used for natural, industrial and liquid gas measurements. Turbine flowmeters require periodic inspection and cleaning of the working parts.

38.10 Float and tapered-tube meter

Principle of operation

With orifice plates and venturimeters the area of the opening in the obstruction is fixed and any change in the flow rate produces a corresponding change in pressure. With the float and tapered-tube meter the area of the restriction may be varied so as to maintain a steady pressure differential. A typical meter of this type is shown diagrammatically in Fig. 38.10 where a vertical tapered tube contains a 'float' which has a density greater than the fluid.

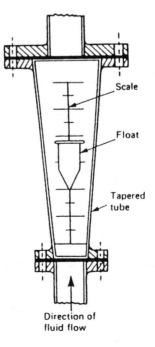

Figure 38.10

The float in the tapered tube produces a restriction to the fluid flow. The fluid can only pass in the annular area between the float and the walls of the tube. This reduction in area produces an increase in velocity and hence a pressure difference, which causes the float to rise. The greater the flow rate, the greater is the rise in the float position, and vice versa. The position of the float is a measure of the flow rate of the fluid and this is shown on a vertical scale engraved on a transparent tube of plastic or glass. For air, a small sphere is used for the float but for liquids there is a tendency to instability and the float is then designed with vanes which cause it to spin and thus stabilize itself as the liquid flows past. Such meters are often called 'rotameters'. Calibration of float and tapered tube flowmeters can be achieved using a Pitot-static tube or, more often, by using a weighing meter in an instrument repair workshop.

Advantages of float and tapered-tube flowmeters

(i) They have a very simple design.
(ii) They can be made direct reading.
(iii) They can measure very low flow rates.

Disadvantages of float and tapered-tube flowmeters

(i) They are prone to errors, such as those caused by temperature fluctuations.
(ii) They can only be installed vertically in a pipeline.
(iii) They cannot be used with liquids containing large amounts of solids in suspension.
(iv) They need to be recalibrated for fluids of different densities.

Practical applications of float and tapered-tube meters are found in the medical field, in instrument purging, in mechanical engineering test rigs and in simple process applications, in particular for very low flow rates. Many corrosive fluids can be handled with this device without complications.

38.11   Electromagnetic flowmeter

The flow rate of fluids which conduct electricity, such as water or molten metal, can be measured

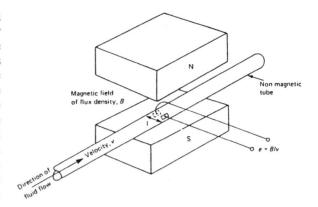

Figure 38.11

using an electromagnetic flowmeter whose principle of operation is based on the laws of electromagnetic induction. When a conductor of length l moves at right angles to a magnetic field of density B at a velocity v, an induced e.m.f. e is generated, given by $e = Blv$ (see Chapter 32, page 259).

With the electromagnetic flowmeter arrangement shown in Fig. 38.11, the fluid is the conductor and the e.m.f. is detected by two electrodes placed across the diameter of the non-magnetic tube.

Rearranging $e = Blv$ gives

$$\text{velocity} \quad \boxed{v = \frac{e}{Bl}}$$

Thus with B and l known, when e is measured, the velocity of the fluid can be calculated.

Main advantages of electromagnetic flowmeters

(i) Unlike other methods, there is nothing directly to impede the fluid flow.
(ii) There is a linear relationship between the fluid flow and the induced e.m.f.
(iii) Flow can be metered in either direction by using a centre-zero measuring instrument.

Applications of electromagnetic flowmeters are found in the measurement of speeds of slurries, pastes and viscous liquids, and they are also widely used in the water production, supply and treatment industry.

38.12   Hot-wire anemometer

A simple hot-wire anemometer consists of a small piece of wire which is heated by an electric current and positioned in the air or gas stream whose velocity is to be measured. The stream passing the wire cools it, the rate of cooling being dependent on the flow velocity. In practice there are various ways in which this is achieved:

(i)   If a constant current is passed through the wire, variation in flow results in a change of temperature of the wire and hence a change in resistance which may be measured by a Wheatstone bridge arrangement. The change in resistance may be related to fluid flow.

(ii) If the wire's resistance, and hence temperature, is kept constant, a change in fluid flow results in a corresponding change in current which can be calibrated as an indication of the flow rate.

(iii) A thermocouple may be incorporated in the assembly, monitoring the hot wire and recording the temperature which is an indication of the air or gas velocity.

Advantages of the hot-wire anemometer

(a)   Its size is small.
(b)   It has great sensitivity.

38.13   Choice of flowmeter

Problem 1.   Choose the most appropriate fluid flow measuring device for the following circumstances:

(a)   The most accurate, permanent installation for measuring liquid flow rate.
(b)   To determine the velocity of low-speed aircraft and ships.
(c)   Accurate continuous volumetric measurement of crude petroleum products in a duct of 500 mm bore.
(d)   To give a reasonable indication of the mean flow velocity, while maintaining a steady pressure difference on a hydraulic test rig.
(e)   For an essentially constant flow rate with reasonable accuracy in a large pipe bore, with a cheap and simple installation.

(a)   Venturimeter
(b)   Pitot-static tube
(c)   Turbine flowmeter
(d)   Float and tapered-tube flowmeter
(e)   Orifice plate

38.14   Multi-choice questions on the measurement of fluid flow

(Answers on page 357.)

1. The term 'flow rate' usually refers to:
   (a) mass flow rate   (b) velocity of flow
   (c) volumetric flow rate
2. The most suitable method for measuring the velocity of high-speed gas flow in a duct is:
   (a) venturimeter   (b) orifice plate
   (c) Pitot-static tube
   (d) float and tapered-tube meter
3. Which of the following statements is false? When a fluid moves through a restriction in a pipe
   (a) the fluid accelerates and the pressure increases
   (b) the fluid decelerates and the pressure decreases
   (c) the fluid decelerates and the pressure increases
   (d) the fluid accelerates and the pressure decreases
4. With an orifice plate in a pipeline the vena contracta is situated:
   (a) downstream at the position of minimum cross-sectional area of flow
   (b) upstream at the position of minimum cross-sectional area of flow
   (c) downstream at the position of maximum cross-sectional area of flow
   (d) upstream at the position of maximum cross-sectional area of flow

In questions 5 to 14, select the most appropriate device for the particular requirements from the following list:
(a) orifice plate
(b) turbine flowmeter
(c) flow nozzle
(d) pitometer
(e) venturimeter
(f) cup anemometer
(g) electromagnetic flowmeter
(h) pitot-static tube
(i) float and tapered-tube meter
(j) hot-wire anemometer
(k) deflecting vane flowmeter

5. Easy to install, reasonably inexpensive, for high-velocity flows.
6. To measure the flow rate of gas, incorporating a Wheatstone bridge circuit.
7. Very low flow rate of corrosive liquid in a chemical process.
8. To detect leakages from water mains.
9. To determine the flow rate of liquid metals without impeding its flow.
10. To measure the velocity of wind.
11. Constant flow rate, large bore pipe, in the general process industry.
12. To make a preliminary test of flow rate in order to specify permanent flow measuring equipment.
13. To determine the flow rate of fluid very accurately with low pressure loss.
14. To measure the flow rate of air in a ventilating duct.

1. Orifice plate
2. Venturimeter
3. Float and tapered-tube meter
4. Electromagnetic flowmeter
5. Pitot-static tube
6. Hot-wire anemometer
7. Turbine flowmeter
8. Deflecting vane flowmeter
9. Flow nozzles
10. Rotary vane positive displacement meter

38.15 Short answer questions on the measurement of fluid flow

In the flowmeters listed 1 to 10, state typical practical applications of each.

38.16 Further questions on the measurement of fluid flow

For the flow measurement devices listed 1 to 5, (a) describe briefly their construction, (b) state their principle of operation, (c) state their characteristics and limitations, (d) state typical practical applications, (e) discuss their advantages and disadvantages.

1. Orifice plate
2. Venturimeter
3 Pitot-static tube
4. Float and tapered-tube meter
5. Turbine flowmeter

39

The measurement of pressure

At the end of this chapter you should be able to:

- define pressure and state its unit
- define the 'bar'
- describe the construction and principle of operation of the following pressure indicating devices:

  (a) barometers (including Fortin and aneroid types)

  (b) manometers (U-tube and inclined types)

  (c) Bourdon pressure gauge

  (d) McLeod gauge

  (e) Pirani gauge

- distinguish between atmospheric, absolute and gauge pressures
- understand how a pressure gauge can be calibrated

39.1   Introduction

Pressure is the force exerted by a fluid per unit area. A fluid (i.e. liquid, vapour or gas) has a negligible resistance to a shear force, so that the force it exerts always acts at right angles to its containing surface.

The SI unit of pressure is the **pascal**, **Pa**, which is unit force per unit area, i.e. $1\ \mathbf{Pa} = 1\ \mathbf{N/m^2}$. The pascal is a very small unit and a commonly used larger unit is the *bar*, where

$$\boxed{1\ \mathbf{bar} = 10^5\ \mathbf{Pa}}$$

Atmospheric pressure is due to the mass of the air above the earth's surface. Atmospheric pressure changes continuously. A standard value of atmospheric pressure, called 'standard atmospheric pressure', is often used, having a value of 101 325 Pa or 1.013 25 bars or 1013.25 millibars. This latter unit, the millibar, is usually used in the measurement of meteorological pressures. (Note that when atmospheric pressure varies from 101 325 Pa it is no longer standard.)

Pressure indicating instruments are made in a wide variety of forms because of their many different applications. Apart from the obvious criteria such as pressure range, accuracy and response, many measurements also require special attention to material, sealing and temperature effects. The fluid whose pressure is being measured may be corrosive or may be at high temperatures. Pressure indicating devices used in science and industry include:

(i)   barometers (see section 39.2),
(ii)   manometers (see section 39.4),
(iii) Bourdon pressure gauge (see section 39.5), and
(iv) McLeod and Pirani gauges (see section 39.6).

39.2   Barometers

Introduction

A barometer is an instrument for measuring atmospheric pressure. It is affected by seasonal changes of temperature. Barometers are therefore also used for the measurement of altitude and also as one of the aids in weather forecasting. The value of atmospheric pressure will thus vary with climatic conditions, although not usually by more than about 10% of standard atmospheric pressure.

Construction and principle of operation

A simple barometer consists of a glass tube, just under 1 m in length, sealed at one end, filled with mercury and then inverted into a trough containing more mercury. Care must be taken to ensure that no air enters the tube during this latter process. Such a barometer is shown in Fig. 39.1(a) and it is seen that the level of the mercury column falls, leaving an empty space, called a vacuum. Atmospheric pressure acts on the surface of the mercury in the trough as shown and this pressure is equal to the pressure at the base of the column of mercury in the inverted tube, i.e. the pressure of the atmosphere is supporting the column of mercury. If the atmospheric pressure falls the barometer height h decreases. Similarly, if the atmospheric pressure rises then h increases. Thus atmospheric pressure can be measured in terms of the height of the mercury column. It may be shown that for mercury the height h is 760 mm at standard atmospheric pressure, i.e. a vertical column of mercury 760 mm high exerts a pressure equal to the standard value of atmospheric pressure.

There are thus several ways in which atmospheric pressure can be expressed:

Standard atmospheric pressure

= 101 325 Pa or 101.325 kPa

= 101 325 N/m^2 or 101.325 kN/m^2

= 1.013 25 bars or 1013.25 mbars

= 760 mm of mercury

Another arrangement of a typical barometer is shown in Fig. 39.1(b) where a U-tube is used instead of an inverted tube and trough, the principle being similar.

If, instead of mercury, water was used as the liquid in a barometer, then the barometric height h at standard atmospheric pressure would be 13.6 times more than for mercury, i.e. about 10.4 m high, which is not very practicable. This is because the relative density of mercury is 13.6.

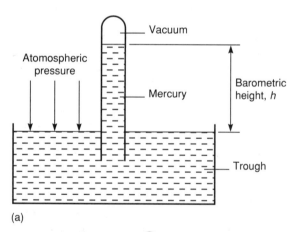

(a)

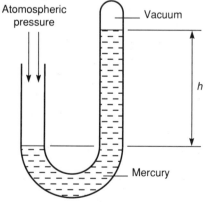

(b)

Figure 39.1

Types of barometer

The **Fortin barometer** is an example of a mercury barometer which enables barometric heights to be measured to a high degree of accuracy (in the order of one-tenth of a millimetre or less). Its construction is merely a more sophisticated arrangement of the inverted tube and trough shown in Fig. 39.1(a), with the addition of a vernier scale to measure the barometric height with great accuracy. A disadvantage of this type of barometer is that it is not portable.

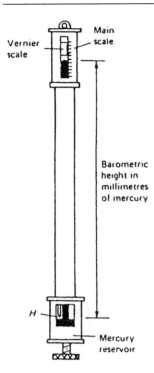

Figure 39.2

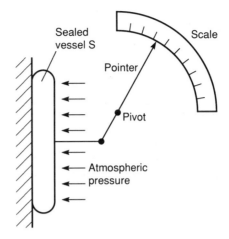

Figure 39.3

A **Fortin barometer** is shown in Fig. 39.2. Mercury is contained in a leather bag at the base of the mercury reservoir, and height, H, of the mercury in the reservoir can be adjusted using the screw at the base of the barometer to depress or release the leather bag. To measure the atmospheric pressure the screw is adjusted until the pointer at H is just touching the surface of the mercury and the height of the mercury column is then read using the main and vernier scales. The measurement of atmospheric pressure using a Fortin barometer is achieved much more accurately than by using a simple barometer.

A portable type often used is the **aneroid barometer**. Such a barometer consists basically of a circular, hollow, sealed vessel, S, usually made from thin flexible metal. The air pressure in the vessel is reduced to nearly zero before sealing, so that a change in atmospheric pressure will cause the shape of the vessel to expand or contract. These small changes can be magnified by means of a lever and be made to move a pointer over a calibrated scale. Fig. 39.3 shows a typical arrangement of an aneroid barometer. The scale is usually circular and calibrated in millimetres of mercury. These instruments require frequent calibration.

39.3   Absolute and gauge pressure

A barometer measures the true or absolute pressure of the atmosphere. The term **absolute pressure** means the pressure above that of an absolute vacuum (which is zero pressure). In Fig. 39.4 a pressure scale is shown with the line AB representing absolute zero pressure (i.e. a vacuum) and line CD representing atmospheric pressure. With most practical pressure-measuring instruments the part of the instrument which is subjected to the pressure being measured is also subjected to atmospheric pressure. Thus practical instruments actually determine the difference between the pressure being measured and atmospheric pressure. The pressure that the instrument is measuring is then termed the **gauge pressure**. In Fig. 39.4, the line EF represents an

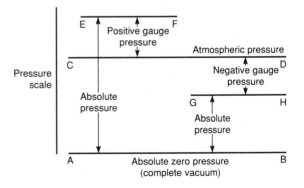

Figure 39.4

absolute pressure which has a value greater than atmospheric pressure, i.e. the 'gauge' pressure is positive.

Thus,

> **absolute pressure = gauge pressure + atmospheric pressure.**

Hence a gauge pressure of, say, 50 kPa recorded on an indicating instrument when the atmospheric pressure is 100 kPa is equivalent to an absolute pressure of 50 kPa + 100 kPa, or 150 kPa.

Pressure-measuring indicating instruments are referred to generally as **pressure gauges** (which acts as a reminder that they measure 'gauge' pressure).

It is possible, of course, for the pressure indicated on a pressure gauge to be below atmospheric pressure, i.e. the gauge pressure is negative. Such a gauge pressure is often referred to as a vacuum, even though it does not necessarily represent a complete vacuum at absolute zero pressure. Such a pressure is shown by the line GH in Fig. 39.4. An indicating instrument used for measuring such pressures is called a **vacuum gauge**.

A vacuum gauge indication of, say, 0.4 bar means that the pressure is 0.4 bar less than atmospheric pressure. If atmospheric pressure is 1 bar, then the absolute pressure is 1 − 0.4 or 0.6 bar.

39.4   The manometer

A manometer is a device for measuring or comparing fluid pressures, and is the simplest method of indicating such pressures.

U-tube manometer

A U-tube manometer consists of a glass tube bent into a U shape and containing a liquid such as mercury. A U-tube manometer is shown in Fig. 39.5(a). If limb A is connected to a container of gas whose pressure is above atmospheric, then the pressure of the gas will cause the levels of mercury to move as shown in Fig. 39.5(b), such that the difference in height is h_1. The measuring scale can be calibrated to give the gauge pressure of the gas as h_1 mm of mercury.

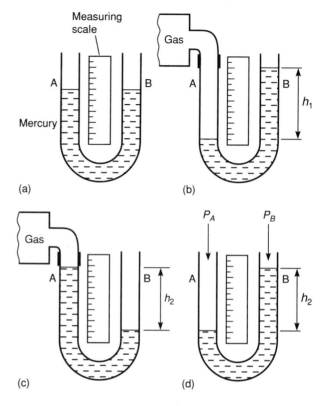

Figure 39.5

If limb A is connected to a container of gas whose pressure is below atmospheric then the levels of mercury will move as shown in Fig. 39.5(c), such that their pressure difference is h_2 mm of mercury.

It is also possible merely to compare two pressures, say, P_A and P_B, using a U-tube manometer. Fig. 39.5(d) shows such an arrangement with $(P_B - P_A)$ equivalent to h mm of mercury. One application of this differential pressure-measuring device is in determining the velocity of fluid flow in pipes (see Chapter 38).

For the measurement of lower pressures, water or paraffin may be used instead of mercury in the U-tube to give larger values of h and thus greater sensitivity.

Inclined manometers

For the measurement of very low pressures, greater sensitivity is achieved by using an inclined manometer, a typical arrangement of which is

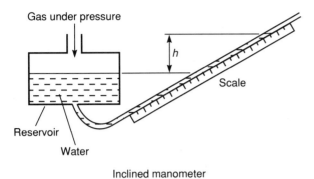

Inclined manometer

Figure 39.6

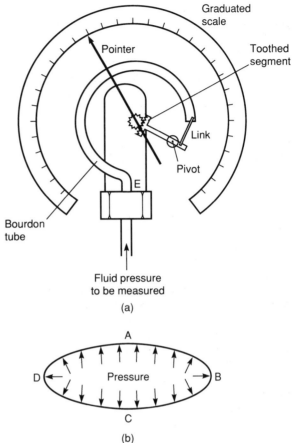

Fluid pressure
to be measured

(a)

(b)

Figure 39.7

shown in Fig. 39.6. With the inclined manometer the liquid used is water and the scale attached to the inclined tube is calibrated in terms of the vertical height h. Thus when a vessel containing gas under pressure is connected to the reservoir, movement of the liquid levels of the manometer occurs. Since small bore tubing is used the movement of the liquid in the reservoir is very small compared with the movement in the inclined tube and is thus neglected. Hence the scale on the manometer is usually used in the range 0.2 mbar to 2 mbar.

The pressure of a gas that a manometer is capable of measuring is naturally limited by the length of tube used. Most manometer tubes are less than 2 m in length and this restricts measurement to a maximum pressure of about 2.5 bar (or 250 kPa) when mercury is used.

39.5 The Bourdon pressure gauge

Pressures many times greater than atmospheric can be measured by the Bourdon pressure gauge, which is the most extensively used of all pressure-indicating instruments. It is a robust instrument. Its main component is a piece of metal tube (called the Bourdon tube), usually made of phosphor bronze or alloy steel, of oval or elliptical cross-section, sealed at one end and bent into an arc. In some forms the tube is bent into a spiral for greater sensitivity. A typical arrangement is shown in Fig. 39.7(a). One end, E, of the Bourdon tube is fixed and the fluid whose pressure is to be measured is connected to this end. The pressure acts at right angles to the metal tube wall as shown in the cross-section of the tube in Fig. 39.7(b). Because of its elliptical shape

it is clear that the sum of the pressure components, i.e. the total force acting on the sides A and C, exceeds the sum of the pressure components acting on ends B and D. The result is that sides A and C tend to move outwards and B and D inwards tending to form a circular cross-section. As the pressure in the tube is increased the tube tends to uncurl, or if the pressure is reduced the tube curls up further. The movement of the free end of the tube is, for practical purposes, proportional to the pressure applied to the tube, this pressure, of course, being the gauge pressure (i.e. the difference between atmospheric pressure acting on the outside of the tube and the applied pressure acting on the inside of the tube). By using a link, a pivot and a toothed segment as shown in Fig. 39.7(a), the movement can be converted into the rotation of a pointer over a graduated calibrated scale.

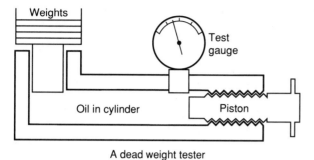

A dead weight tester

Figure 39.8

The Bourdon tube pressure gauge is capable of measuring high pressures up to 10^4 bar (i.e. 7600 m of mercury) with the addition of special safety features.

A pressure gauge must be calibrated, and this is done either by a manometer, for low pressures, or by a piece of equipment called a '**dead weight tester**'. This tester consists of a piston operating in an oil-filled cylinder of known bore, and carrying accurately known weights as shown in Fig. 39.8. The gauge under test is attached to the tester and the required pressure is applied by a screwed piston or ram, until the weights are just lifted. While the gauge is being read, the weights are turned to reduce friction effects.

39.6   Vacuum gauges

Vacuum gauges are instruments for giving a visual indication, by means of a pointer, of the amount by which the pressure of a fluid applied to the gauge is less than the pressure of the surrounding atmosphere. Two examples of vacuum gauges are the McLeod gauge and the Pirani gauge.

McLeod gauge

The McLeod gauge is normally regarded as a standard and is used to calibrate other forms of vacuum gauges. The basic principle of this gauge is that it takes a known volume of gas at a pressure so low that it cannot be measured, then compresses the gas in a known ratio until the pressure becomes large enough to be measured by an ordinary manometer. This device is used to measure low pressures, often in the range 10^{-6} to

1.0 mm of mercury. A disadvantage of the McLeod gauge is that it does not give a continuous reading of pressure and is not suitable for registering rapid variations in pressure.

Pirani gauge

The Pirani gauge measures the resistance and thus the temperature of a wire through which current is flowing. The thermal conductivity decreases with the pressure in the range 10^{-1} to 10^{-4} mm of mercury so that the increase in resistance can be used to measure pressure in this region. The Pirani gauge is calibrated by comparison with a McLeod gauge.

39.7   Multi-choice questions on the measurement of pressure

(Answers on page 357.)

1. Which of the following devices does not measure pressure?
   (a) barometer   (b) McLeod gauge
   (c) thermocouple   (d) manometer
2. A pressure of 10 kPa is equivalent to:
   (a) 10 millibars   (b) 1 bar
   (c) 0.1 bar   (d) 0.1 millibars
3. A pressure of 1000 mbars is equivalent to:
   (a) 0.1 kN/m^2 (b) 10 kPa
   (c) 1000 Pa   (d) 100 kN/m^2
4. Which of the following statements is false?
   (a) Barometers may be used for the measurement of altitude.
   (b) Standard atmospheric pressure is the pressure due to the mass of the air above the ground.
   (c) The maximum pressure that a mercury manometer, using a 1 m length of glass tubing, is capable of measuring is in the order of 130 kPa.
   (d) An inclined manometer is designed to measure higher values of pressure than the U-tube manometer.

In questions 5 and 6 assume that atmospheric pressure is 1 bar.

5. A Bourdon pressure gauge indicates a pressure of 3 bars. The absolute pressure of the system being measured is:
   (a) 1 bar   (b) 2 bars   (c) 3 bars   (d) 4 bars
6. In question 5, the gauge pressure is:
   (a) 1 bar   (b) 2 bars   (c) 3 bars   (d) 4 bars

In questions 7 to 11 select the most suitable pressure-indicating device from the following list.
(a) Mercury filled U-tube manometer
(b) Bourdon gauge
(c) McLeod gauge
(d) aneroid barometer
(e) Pirani gauge
(f) Fortin barometer
(g) water-filled inclined barometer

7. A robust device to measure high pressures in the range 0–30 MPa.
8. Calibration of a Pirani gauge.
9. Measurement of gas pressures comparable with atmospheric pressure.
10. To measure pressures of the order of 1 MPa.
11. Measurement of atmospheric pressure to a high degree of accuracy.
12. Figure 39.5(b) shows a U-tube manometer connected to a gas under pressure. If atmospheric pressure is 76 cm of mercury and h_1 is measured in centimetres then the gauge pressure (in cm of mercury) of the gas is:
    (a) h_1 (b) $h_1 + 76$ (c) $h_1 - 76$ (d) $76 - h_1$
13. In question 12 the absolute pressure of the gas (in cm of mercury) is:
    (a) h_1 (b) $h_1 + 76$ (c) $h_1 - 76$ (d) $76 - h_1$
14. Which of the following statements is true?
    (a) Atmospheric pressure of 101.325 kN/m^2 is equivalent to 101.325 millibars.
    (b) An aneroid barometer is used as a standard for calibration purposes.
    (c) In engineering, 'pressure' is the force per unit area exerted by fluids.
    (d) Water is normally used in a barometer to measure atmospheric pressure.

39.8 Short answer questions on the measurement of pressure

1. Define 'pressure' and state its unit.
2. Standard atmospheric pressure is 101 325 Pa. State this pressure in millibars.

3. Briefly describe how a barometer operates.
4. State the advantage of a Fortin barometer over a simple barometer.
5. What is the main disadvantage of a Fortin barometer?
6. Briefly describe an aneroid barometer.
7. What is meant by 'absolute pressure'?
8. What is meant by 'gauge pressure'?
9. What is a vacuum gauge?
10. Briefly describe the principle of operation of a U-tube manometer.
11. When would an inclined manometer be used in preference to a U-tube manometer?
12. Briefly describe the principle of operation of a Bourdon pressure gauge.
13. What is a 'dead weight tester'?
14. What is a 'Pirani gauge'?
15. What is a 'McLeod gauge' used for?

39.9 Further questions on the measurement of pressure

1. (a) What does the term pressure mean in engineering?
   (b) Why is atmospheric pressure measured as a height of mercury?
   (c) Distinguish between pressure and vacuum gauges.
2. Explain the difference between gauge pressure and absolute pressure. A pressure-indicating device records a pressure of 100 kPa. Assuming that atmospheric pressure is 1 bar determine the absolute pressure in units of:
   (a) kPa (b) bars.
                              [(a) 200 kPa (b) 2 bars]
3. Sketch a typical Bourdon pressure gauge and explain its principle of operation. Explain also how such a device may be calibrated.
4. Explain the principle of operation of: (a) a simple mercury barometer, and (b) a U-tube manometer. State for each whether the pressure measured is gauge or absolute.

40

The measurement of strain

At the end of this chapter you should be able to:

- define strain, stress and Young's modulus, stating their units

- understand the need for strain measurement

- understand the construction and principle of operation of the following extensometers:

  (a) Lindley

  (b) Huggenburger

  (c) Hounsfield

- describe a strain gauge, including advantages of foil strain gauge

- understand how a strain gauge measures strain via a Wheatstone bridge

- appreciate typical practical situations where strain gauges are used

40.1. Introduction

An essential requirement of engineering design is the accurate determination of stresses and strains in components under working conditions. 'Strength of materials' is a subject relating to the physical nature of substances which are acted upon by external forces. No solid body is perfectly rigid, and when forces are applied to it changes in dimensions occur. Such changes are not always perceptible to the human eye since they are so small. For example, a spanner will bend slightly when tightening a nut, and the span of a bridge will sag under the weight of a car.

Strain

The change in the value of a linear dimension of a body, say x, divided by the original value of the dimension, say l, gives a great deal of information about what is happening to the material itself.

This ratio is called **strain**, ϵ, and is dimensionless, i.e.

$$\epsilon = \frac{x}{l}$$

Stress

The force (F) acting on an area (A) of a body is called the **stress**, σ, and is measured in pascals (Pa) or newtons per square metre (N/m²), i.e.

$$\sigma = \frac{F}{A}$$

Young's modulus of elasticity

If a solid body is subjected to a gradually increasing stress, and if both the stress and the resulting strain are measured, a graph of stress against strain may be drawn. Up to a certain value of stress the graph is a straight line. That particular value is known as the **limit of proportionality** and its value varies for different materials. The gradient of the straight line is a constant known as **Young's modulus of elasticity**, E:

$$E = \frac{\text{stress}}{\text{strain}}$$

up to the limit of proportionality

$$E = \frac{F/A}{x/l} = \frac{Fl}{xA} \text{ Pa} \qquad (40.1)$$

Young's modulus of elasticity is a constant for a given material. As an example, mild steel has a value of E of about 200×10^9 Pa (i.e. 200 GPa).

Elastic limit

If on removal of external forces a body recovers its original shape and size, the material is said to be **elastic**. If it does not return to its original shape, it is said to be **plastic**. Copper, steel and rubber are examples of elastic materials while lead and plasticine are plastic materials. However, even for elastic materials there is a limit to the amount of strain from which it can recover its original dimensions. This limit is called the **elastic limit** of the material. The elastic limit and the limit of proportionality for all engineering materials are virtually the same. If a body is strained beyond the elastic limit permanent deformation will occur.

40.2   The need for strain measurement

In designing a structure, such as an electricity transmission tower carrying overhead power lines or support pillars and spans of new designs of bridges, the engineer is greatly concerned about the mechanical properties of the materials he is going to use. Many laboratory tests have been designed to provide important information about materials. Such tests include tensile, compression, torsion, impact, creep and fatigue tests and each attempts to provide information about the behaviour of materials under working conditions. (A typical tensile test is described in Chapter 18.)

It is possible to design a structure which is strong enough to withstand the forces encountered in service, but is, nonetheless, useless because of the amount of elastic deformation. Hence, tests made on materials up to the elastic limit are of great importance. A material which has a relatively high value of Young's modulus is said to have a high value of **stiffness**, stiffness being the ratio of force to extension (i.e. F/x) From equation (40.1)

$$E = \frac{\text{stress}}{\text{strain}} = \frac{F}{x} \cdot \frac{l}{A} = \text{stiffness} \times \frac{l}{A}$$

Thus, when the determination of Young's modulus of elasticity, E, of a material is required, an accurate stress/strain or load/extension graph must be obtained. The actual strain is very small and this means that very small extensions must be measured with a high degree of accuracy.

The measurement of extension, and thus strain, is achieved in the laboratory with an instrument called an **extensometer**. Although some extensometers can be used in such practical situations as a crane under load, it is more usual to use in these situations an electrical device called a **strain gauge**.

A knowledge of stress and strain is the foundation of economy and safety in design.

40.3   Extensometers

An extensometer is an instrument used in engineering and metallurgical design to measure accurately the minute elastic extensions of materials, in order to forecast their behaviour during use. There are several different designs of extensometer including the Lindley, the Huggenburger and the Hounsfield.

40.4   The Lindley extensometer

This is probably the most common type of extensometer used for measuring tensile strains. This instrument consists of two arms, A and B, connected by a strip of spring steel which acts as a hinge. The unstressed specimen of the material is clamped at points C and D by pointed screws, the distance between C and D usually being 50 mm. Thus 50 mm is termed the 'gauge length'. A dial test indicator is placed between the arms A and B as shown in the typical arrangement of the Lindley extensometer in Fig. 40.1.

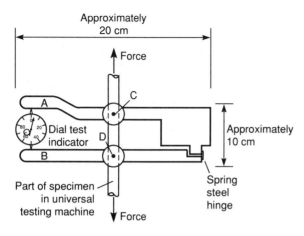

Figure 40.1

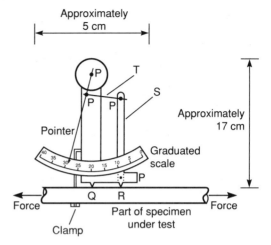

Figure 40.2

The point D is halfway between the hinge and the indicator, hence the movement of the pointer on the test indicator will record twice the extension of the specimen. However, the indicator is normally calibrated so that it indicates extension directly, each graduation representing an extension of 1 micron (i.e. 10^{-6} m or 0.001 mm). Extensions may be measured to an accuracy of 0.0001 mm using the Lindley extensometer.

40.5 The Huggenburger extensometer

This is a simple, rugged and accurate instrument which may be used to measure tensile or compressive strains. Its construction is based on a lever multiplying system capable of obtaining magnifications in the order of 2000. Figure 40.2 shows a simplified schematic arrangement of a front view of the Huggenburger extensometer clamped to a specimen, where Q and R are two knife edges, usually either 10 mm or 20 mm apart. Any strain encountered by the specimen under test will alter the gauge length QR. In Fig. 40.2, the specimen is shown in tension, thus QR will increase in length. This change is transmitted by pivots (labelled P) and levers S and T to the pointer, and is indicated on the scale according to the multiplication factor. This factor, of approximately 2000, is supplied to the instrument user by the supplier who calibrates each device after manufacture. This type of extensometer enables extensions to be recorded to an accuracy comparable with the Lindley extensometer and may be used in the laboratory or in the field.

40.6 The Hounsfield extensometer

This may be used in conjunction with a Hounsfield Tensometer (which is a universal portable testing machine capable of applying tensile or compressive forces to metals, plastics, textiles, timber, paper and so on), or with any other testing machine. The extensometer is a precision instrument for measuring the extension of a test specimen over a 50 mm gauge length, while the test specimen is loaded in the testing machine. The instrument can be attached to round specimens of material of up to 25 mm in diameter or rectangular sections of material of up to 25 mm square at precisely 50 mm gauge length without prior marking of the specimen. Figure 40.3 shows a typical Hounsfield extensometer viewed from two different elevations.

The gauge length rod is screwed into position, making the fixed centres exactly 50 mm apart. The extensometer is then clamped to the test piece before the gauge length rod is unscrewed. With the test piece still unloaded the micrometer is wound in until the platinum contacts meet, thus completing the circuit (shown by the lamp lighting up). The micrometer reading is then taken and the micrometer head unwound. After the load is placed on the specimen the micrometer head is again wound in and a new reading taken when the lamp lights. The difference between the two micrometer readings is an indication of the extension of the test piece for the particular load applied. Each division of the micrometer wheel is

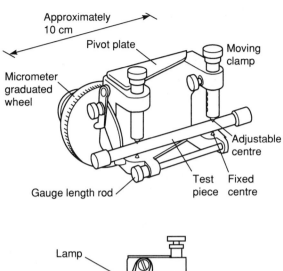

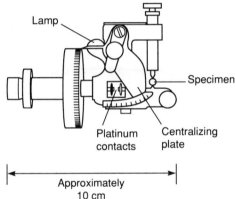

Figure 40.3

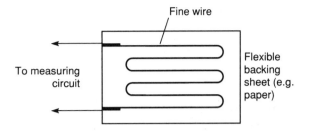

Figure 40.4

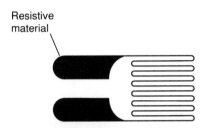

Figure 40.5

equal to 0.002 mm. The accuracy of the Hounsfield extensometer compares favourably with other extensometers, and an advantage, in certain circumstances, of this instrument is its small overall size.

40.7 Strain gauges

A strain gauge is an electrical device used for measuring mechanical strain, i.e. the change in length accompanying the application of a stress. The strain gauge consists essentially of a very fine piece of wire which is cemented, or glued strongly, to the part where the strain is to be measured. When the length of a piece of wire is changed, a change in its electrical resistance occurs, this change in resistance being proportional to the change in length of the wire. Thus, when the wire is securely cemented to the part which is being strained, a change of electrical

resistance of the wire occurs due to the change in length. By measuring this change of resistance the strain can be determined. The strain gauge was first introduced in the USA in 1939 and since that time it has come into widespread use, particularly in the aircraft industry, and is now the basis of one of the most useful of stress analysis techniques. A typical simple strain gauge is shown in Fig. 40.4.

A modern strain gauge is formed by rolling out a thin foil of the resistive material and then cutting away parts of the foil by a photo-etching process to create the required grid pattern. Such a device is called a **foil-strain gauge** and a typical arrangement is shown in Fig. 40.5. A foil-strain gauge has many advantages over the earlier method and these include:

(i) better adhesion between conductor and backing material,
(ii) better heat dissipation,
(iii) a more robust construction,
(iv) easier to attach leads to,
(v) accurate reproducibility of readings, and
(vi) smaller sizes are possible.

In order to obtain a deflection on a galvanometer, G, proportional to the strain occurring in the gauge, it must be connected into one arm of a Wheatstone bridge, as shown in Fig. 40.6. A

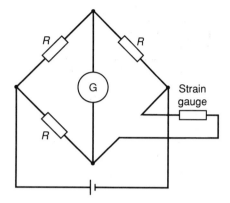

Figure 40.6

Wheatstone bridge circuit having four equal resistances in the arms has zero deflection on the galvanometer, but when the resistance of one or more of the arms changes, then the bridge galvanometer deflects from zero, the amount of deflection being a measure of the change in resistance. If the resistance change occurs in a strain gauge as a result of applied strain, then the bridge galvanometer deflection is a measure of the amount of strain. For very accurate measurements of strain there are a number of possible sophistications. These are not described in detail in this chapter but include:

(i)   the use of a temperature-compensating dummy gauge to make the bridge output independent of temperature, since the resistance of a gauge varies with temperature and such a resistance change may be misread as strain in the material,

(ii) an additional bridge balancing circuit to obtain zero galvanometer deflection for zero strain, and

(iii) the addition of an amplifier to amplify the signal output from the bridge in applications where the level of strain is such that the bridge deflection is too small to readily detect on a galvanometer.

A typical selection of **practical situations** where strain gauges are used include:

(i)   the airframe and skin of an aircraft in flight,

(ii) electricity pylons, cranes and support pillars and spans of new designs of bridges, where strain must be tested to validate the design, and

(iii) applications in harsh environments and remote positions, such as inside nuclear boilers, on turbine blades, in vehicle engines, on helicopter blades in flight and under water on oil rig platforms, where a knowledge of strain is required.

40.8 Multi-choice questions on the measurement of strain

(Answers on page 357.)

In questions 1 to 4 select from the following list, (a)–(d), the correct unit for the given physical quantities:
(a) newtons (b) pascals
(c) newtons per metre (d) dimensionless

1. Stress
2. Strain
3. Young's modulus of elasticity
4. Stiffness
5. A stress of 70 MPa is applied to a bar of aluminium. If Young's modulus for aluminium is 70 GPa the strain experienced by the bar is:
    (a) 0.1 (b) 0.01 (c) 0.001 (d) 0.0001
6. Which of the following statements is true?
    (a) Plasticine is an elastic material.
    (b) The symbol used for strain is σ.
    (c) Stiffness is the ratio of force to extension.
    (d) The elastic modulus of a material varies with length.
7. A Lindley extensometer is used to measure the strain of a 50 mm length of metal rod. Each of the 100 graduations on the dial indicator is calibrated to read an extension of one micron. The strain measured when the pointer indicates 50 is:
    (a) 0.001 (b) 0.01 (c) 50×10^{-6} (d) 0.05
8. Which of the following statements is false?
    (a) The Lindley extensometer is a mechanical device which uses a dial indicator.
    (b) A Wheatstone bridge is normally used as the measuring circuit of an extensometer.
    (c) To determine the strain in the airframe of an aircraft in flight, a foil strain gauge is likely to be used.
    (d) Steel is an example of an elastic material.

40.9   Short answer questions on the measurement of strain

1. What is meant by 'strain' in engineering design?
2. Define 'stress' and state its unit.
3. How can a value of Young's modulus of elasticity be determined for a particular metal?
4. Give two examples of (a) an elastic material (b) a plastic material.
5. Define 'stiffness' and state its unit.
6. What is an extensometer?
7. Name three extensometers.
8. What is a strain gauge?
9. State three practical examples where a strain gauge would be used.
10. Briefly explain, with a diagram, how a strain gauge is used in a Wheatstone bridge circuit to measure strain.

40.10   Further questions on the measurement of strain

1. Explain the need for strain measurement in engineering design.
2. Describe the principle of operation of a Lindley extensometer. Sketch the instrument showing relevant parts.
3. Compare the constructions of the Lindley and Huggenburger extensometers.
4. Explain how the Hounsfield extensometer is used to measure strain.
5. Explain with the aid of a sketch the principle of operation of a foil-strain gauge together with an appropriate measuring circuit. Give five typical practical situations where such a strain gauge may be used. What advantage does a foil-strain gauge possess over earlier methods?

Solutions

Answers to section 1, Chapters 1 to 6

Chapter 1 The solution of simple equations (page 8)

1. 1	2. 2	3. 6	4. –4
5. $1\frac{2}{3}$	6. $\frac{1}{2}$	7. 0	8. 3
9. –10	10. 6	11. –2	12. 2
13. –3	14. –2	15. $-4\frac{1}{2}$	16. 12
17. 15	18. $5\frac{1}{3}$	19. 13	20. 3
21. –6	22. 110	23. 9	24. $6\frac{1}{4}$
25. 3	26. 10	27. ±12	28. ±3
29. ±4	30. 10^{-7}	31. 8 m/s^2	32. 3.472

33. (a) 1.8 Ω (b) 30 Ω 34. 800 Ω 35. 0.004
36. 30 37. 45°C 38. 50

Chapter 2 The solution of simultaneous equations (page 13)

1. $a = 5, b = 2$ 2. $x = 1, y = 1$ 3. $s = 2, t = 3$
4. $x = 3, y = -2$ 5. $m = 2\frac{1}{2}, n = \frac{1}{2}$ 6. $a = 6, b = -1$
7. $x = 2, y = 5$ 8. $c = 2, d = -3$ 9. $p = -1, q = -2$
10. $a = \frac{1}{5}, b = 4$ 11. $I_1 = 6.47, I_2 = 4.62$
12. $u = 12, a = 4; v = 26$ 13. $m = -\frac{1}{2}, c = 3$

Chapter 3 Evaluation and transposition of formulae (page 19)

1. $A = 66.59$ cm^2 2. $C = 52.75$ mm 3. $R = 37.5$
4. $V = 2.61$ V 5. $F = 854.5$ 6. $I = 3.81$ A
7. $t = 14.79$ s 8. $E = 3.96$ J 9. $S = 17.25$ m
10. $d = c - e - a - b$ 11. $y = \frac{1}{3}(t - x)$
12. $r = c/2\pi$ 13. $x = (y - c)/m$ 14. $T = I/PR$
15. $R = E/I$ 16. $r = \dfrac{S - a}{S}$ 17. $C = \dfrac{5}{9}(F - 32)$
18. $x = \dfrac{d}{\lambda}(y + \lambda)$ 19. $f = \dfrac{3F - AL}{3}$ 20. $E = \dfrac{Ml^2}{8yI}$
21. $t = \dfrac{R - R_0}{R_0\alpha}$ 22. $R_2 = \dfrac{RR_1}{R_1 - R}$ 23. $R = \dfrac{E - e - Ir}{I}$
24. $l = \dfrac{t^2g}{r\pi^2}$ 25. $u = \sqrt{(v^2 - 2as)}$ 26. $R = \sqrt{\left(\dfrac{360A}{\pi\theta}\right)}$

27. $A = \sqrt{\left(\dfrac{xy}{m - n}\right)}$ 28. $L = \dfrac{mrCR}{\mu - m}$ 29. $v = \dfrac{u.f}{u - f}$; 30
30. $t_2 = t_1 + \dfrac{Q}{mC}$; 55 31. $v = \sqrt{\left(\dfrac{2\,dgh}{0.03L}\right)}$; 0.965
32. $l = \dfrac{8S^2}{3d} + d$; 2.725

Chapter 4 Straight line graphs (page 29)

1. 14.5 2. $\frac{1}{2}$ 3. (a) 4, –2 (b) –1, 0 (c) –3, –4 (d) 0, 4
4. (a) 6, –3 (b) –2, 4 (c) 3, 0 (d) 0, 7
5. (a) $\frac{3}{5}$ (b) –4 (c) $-1\frac{2}{3}$ 6. (a) –1.1 (b) –1.4 7. (2, 1)
8. (a) 40°C (b) 128Ω 9. (a) 0.25 (b) 12 (c) $F = 0.25L + 12$ (d) 89.5 N (e) 592 N (f) 212 N
10. –0.003, 8.73 11. (a) 22.5 m/s (b) 6.43 s (c) $v = 0.7t + 15.5$ 12. $m = 26.9l - 0.63$ 13. (a) 96 × 10^9 Pa (b) 0.000 22 (c) 28.8 × 10^6 Pa 14. (a) $\frac{1}{5}$ (b) 6 (c) $E = \frac{1}{5}L + 6$ (d) 12 N (e) 65 N
15. $a = 0.85, b = 12$, 254.3 kPa, 275.5 kPa, 280 K

Chapter 5 Trigonometry (page 46)

1. $\sin A = \frac{3}{5}$, $\cos A = \frac{4}{5}$, $\tan A = \frac{3}{4}$, $\sin B = \frac{4}{5}$, $\cos B = \frac{3}{5}$, $\tan B = \frac{4}{3}$
2. $\sin \theta = \frac{7}{25}$, $\cos \theta = \frac{24}{25}$
3. (a) 0.4540 (b) 0.1321 (c) –0.8399
4. (a) –0.5592 (b) 0.9307 (c) 0.2447
5. (a) –0.7002 (b) –1.1671 (c) 1.1612
6. (a) 0.8660 (b) –0.1010 (c) 0.5865
7. (a) 13.54°, 13°32', 0.236 rad
   (b) 34.20°, 34°12', 0.597 rad
   (c) 39.03°, 39°2', 0.681 rad
8. 227°4' or 312°56'
9. (a) 122°7' or 237°53' (b) 39°44' or 219°44'
10. (i) BC = 3.50 cm, AB = 6.10 cm, $B = 55°$
    (ii) FE = 5 cm, $\angle E = 53°8'$, $\angle F = 36°52'$
    (iii)GH = 9.841 mm, GI = 11.32 mm, $\angle H = 49°$
11. (a) $C = 83°$, $a = 14.1$ mm, $c = 28.9$ mm, area = 189 mm^2
    (b) $A = 52°2'$, $c = 7.568$ cm, $a = 7.152$ cm, area = 25.65 cm^2
    (c) $B = 38°30'$, $b = 10.62$ mm, $c = 7.074$ mm, area = 33.47 mm^2

12. (a) p = 13.2 cm, Q = 47°21′, R = 78°39′, area = 77.7 cm²
    (b) r = 52.1 mm, Q = 40°58′, P = 75°2′, area = 956 mm²
    (c) p = 6.127 m, Q = 30°49′, R = 44°11′, area = 6.938 m²
13. BF = 3.9 m, EB = 4.0 m
14. 6.35 m, 5.37 m
15. 80°25′, 59°23′, 40°12′
16. 40.25 cm, 126°3′

Chapter 6 Areas of irregular shapes (page 44)

1. 63 m   2. 4.70 ha   3. 143 m²   4. 0.093 A s

Answers to multi-choice questions

Chapter 7 SI units and density (page 50)

1. (c)   2. (d)   3. (b)   4. (c)   5. (b)

Chapter 8 Forces acting at a point (page 61)

1. (d)   2. (b)   3. (c)   4. (b)   5. (b)
6. (c)   7. (d)   8. (c)   9. (d)

Chapter 9 Speed and velocity (page 69)

1. (c)   2. (g)   3. (d)   4. (c)   5. (e)
6. (b)   7. (a)   8. (a)   9. (i)   10. (c)

Chapter 10 Acceleration (page 75)

1. (c)   2. (e)   3. (i)   4. (g)   5. (d)
6. (c)   7. (a)

Chapter 11 Force, mass and acceleration (page 81)

1. (b)   2. (b)   3. (a)   4. (a)   5. (b)
6. (c)   7. (d)   8. (d)   9. (c)   10. (b)

Chapter 12 Simply supported beams (page 88)

1. (a)   2. (c)   3. (a)   4. (d)   5. (a)
6. (d)   7. (c)   8. (a)   9. (d)   10. (c)

Chapter 13 Linear and angular motion (page 97)

1. (c)   2. (a)   3. (d)   4. (c)   5. (b)
6. (d)   7. (g)   8. (i)   9. (c)   10. (k)

Chapter 14 Friction (page 101)

1. (f)   2. (e)   3. (i)   4. (c)   5. (h)
6. (b)   7. (d)   8. (a)

Chapter 15 Work, energy and power (page 113)

1. (b)   2. (c)   3. (c)   4. (a)   5. (d)
6. (c)   7. (a)   8. (d)   9. (c)   10. (b)

Chapter 16 Simple machines (page 123)

1. (b)   2. (f)   3. (c)   4. (d)   5. (b)
6. (a)   7. (b)   8. (d)   9. (c)   10. (d)
11. (d)   12. (b)

Chapter 17 The effects of forces on materials (page 134)

1. (b)   2. (c)   3. (c)   4. (b)   5. (d)
6. (b)   7. (c)   8. (f)   9. (h)   10. (d)

Chapter 18 Tensile testing (page 141)

1. (f)   2. (d)   3. (g)   4. (b)

Chapter 19 Linear momentum and impulse (page 148)

1. (d)   2. (b)   3. (f)   4. (c)   5. (a)
6. (c)   7. (a)   8. (g)   9. (f)   10. (f)
11. (b)   12. (e)

Chapter 20 Torque (page 158)

1. (c)   2. (d)   3. (a)   4. (b)   5. (c)
6. (d)   7. (a)   8. (c)

Chapter 21 Pressure in fluids (page 165)

1. (b)   2. (d)   3. (a)   4. (a)   5. (c)
6. (d)   7. (b)

Chapter 22 Heat energy (page 173)

1. (d)   2. (b)   3. (a)   4. (c)   5. (b)
6. (b)

Chapter 23 Thermal expansion (page 180)

1. (b)   2. (c)   3. (a)   4. (d)   5. (b)
6. (c)   7. (c)   8. (a)   9. (c)   10. (b)

Chapter 24 Ideal gas laws (page 188)

1. (a)   2. (d)   3. (b)   4. (b)   5. (c)
6. (d)   7. (b)   8. (c)   9. (c)   10. (b)

Chapter 25 Properties of water and steam (page 198)

1. (b)   2. (a)   3. (d)   4. (g)   5. (f)
6. (b)   7. (a)   8. (j)   9. (f)   10. (c)

Chapter 26 An introduction to electric circuits (page 210)

1. (b)   2. (b)   3. (c)   4. (b)   5. (d)
6. (d)   7. (b)   8. (c)

Chapter 27 Resistance variation (page 217)

1. (c)   2. (d)   3. (b)   4. (d)   5. (d)
6. (c)   7. (b)

Chapter 28 Chemical effects of electricity (page 225)

1. (a)   2. (b)   3. (c)   4. (b)   5. (d)
6. (d)   7. (b)   8. (c)

Chapter 29 Series and parallel networks (page 237)

1. (a)   2. (c)   3. (c)   4. (c)   5. (a)
6. (b)   7. (d)   8. (b)

Chapter 30 Kirchhoff's laws (page 244)

1. (d)   2. (c)   3. (b)   4. (c)

Chapter 31 Electromagnetism (page 255)

1. (d)   2. (b)   3. (d)   4. (c)   5. (d)
6. (a)   7. (b)   8. (c)

Chapter 32 Electromagnetic induction (page 265)

1. (c)   2. (b)   3. (c)   4. (b)   5. (c)
6. (a)   7. (c)   8. (b)   9. (a)
10. (b) and (c)

Chapter 33 Alternating voltages and currents (page 274)

1. (c)   2. (d)   3. (d)   4. (a)   5. (d)
6. (c)   7. (b)   8. (c)   9. (b)

Chapter 34 Capacitors and their effects in electric circuits (page 289)

1. (b)   2. (a)   3. (b)   4. (c)   5. (a)
6. (b)   7. (b)   8. (c)   9. (c)   10. (c)
11. (b)   12. (c)   13. (b)   14. (c)

Chapter 35 Inductors and their effects in electric circuits (page 302)

1. (c)   2. (d)   3. (b)   4. (g)   5. (b)
6. (c)   7. (j)   8. (k)   9. (a)   10. (c)
11. (a)   12. (d)

Chapter 36 Electrical measuring instruments and measurements (page 318)

1. (d)   2. (a) or (c)   3. (b)   4. (b)
5. (c)   6. (f)   7. (c)   8. (a)   9. (i)
10. (j)   11. (g)   12. (c)   13. (b)   14. (p)
15. (d)   16. (o)   17. (n)

Chapter 37 The measurement of temperature (page 330)

1. (c)   2. (b)   3. (d)   4. (b)   5. (i)
6. (a)   7. (e)   8. (d)   9. (e) or (f)
10. (k)   11. (b)   12. (g)

Chapter 38 The measurement of fluid flow (page 339)

1. (c)	2. (c)	3. (d)	4. (a)	5. (c)
6. (j)	7. (i)	8. (d)	9. (g)	10. (f)
11. (a)	12. (h)	13. (e)	14. (k)	

Chapter 39 The measurement of pressure (page 346)

1. (c)	2. (c)	3. (d)	4. (d)	5. (d)
6. (c)	7. (b)	8. (c)	9. (a)	10. (c)
11. (f)	12. (a)	13. (b)	14. (c)	

Chapter 40 The measurement of strain (page 352)

1. (b)	2. (d)	3. (b)	4. (c)	5. (c)
6. (c)	7. (a)	8. (b)		

Index